D0152851

INTRODUCTION TO MATHEMATICAL PHYSICS

INTRODUCTION TO MATHEMATICAL PHYSICS

Methods and Concepts

CHUN WA WONG

Physics Department
University of California
Los Angeles, California

New York *Oxford*
OXFORD UNIVERSITY PRESS
1991

Oxford University Press

Oxford New York Toronto
Delhi Bombay Calcutta Madras Karachi
Petaling Jaya Singapore Hong Kong Tokyo
Nairobi Dar es Salaam Cape Town
Melbourne Auckland

and associated companies in
Berlin Ibadan

Published by Oxford University Press, Inc.,
200 Madison Avenue, New York, New York 10016

Library of Congress Cataloging-in-Publication Data

Wong, Chun Wa.
Introduction to mathematical Physics : methods and concepts / Chun Wa Wong.
p. cm. Includes bibliographical references and index.
ISBN 0-19-504473-8
1. Mathematical physics. I. Title.
QC20.W66 1991 530.1′5—dc20 90-45235

1 3 5 7 9 8 6 4 2

Printed in the United States of America
on acid-free paper

PREFACE

This book is based on lecture notes for two undergraduate courses on mathematical methods of physics that I have given at UCLA during the past twenty years. Most of the introductory topics in each chapter have been used at one time or another in these courses. The first of these courses was intended to be a beginning junior course to be taken before the physics core courses, while the second was an elective course for seniors. These courses have evolved over the years in response to the perceived needs of a changing student population. Our junior course is now a prerequisite for junior courses on electricity and magnetism and on quantum mechanics, but not for those on analytic mechanics. Our elective senior course is now concerned solely with functions of a complex variable. In this format, I am able to cover in one quarter most of the easier sections of the first three chapters in our junior course, and the entire Chapter 6 in the senior course. Most of our students are physics majors, but many are from engineering and chemistry, especially in the senior course. The latter course is also recommended to our own first-year graduate students who have not been exposed to complex analysis before.

The idea of teaching mathematical physics as a subject separate from the physics core courses is to help the students to appreciate the mathematical basis of physical theories and to acquire the expected level of competence in mathematical manipulations. I believe that our courses, like similar courses in other universities, have been quite successful. This is not to say that these courses are easy to teach. My experience has been that our junior course on mathematical methods of physics is one of the most difficult undergraduate courses to teach. Several factors combine to make it so challenging, including the diversity in the background and abilities of our students and the large number of topics that one would like to cover. Not the least of these factors is an adequate list of available textbooks at the right level and written in the right style. It is hoped that this book will address this need.

Each chapter in this book deals with a single subject. Chapter 1 on vectors and fields in space is concerned with the vector calculus needed for a study of electricity and magnetism. Chapter 2 on transformations, matrices, and operators contains a number of topics in algebra of special importance to the study of both classical and quantum mechanics. It is also concerned with the general mathematical structure of laws of physics. Chapter 3 on Fourier series and Fourier transforms prepares the students for their study of quantum mechanics. The treatment of differential equations in physics in Chapter 4 covers most of the basic mathematical concepts and analytic techniques needed to solve many equations of motion or equations of state in physics. Special functions are covered in Chapter 5 with emphasis on special techniques with which the properties of these functions can be extracted. Finally, Chapter 6 on functions of a complex variable gives a detailed introduction to complex analysis, which is so basic to the understanding of functions and their manipulations. This chapter provides a firmer mathematical foundation to students who intend to go on to graduate studies in the physical sciences. Many topics have been left out in order to have a book of manageable size. These include infinite series, tensor analysis, probability theory, the calculus of variations, numerical analysis, and computer mathematics.

The style of writing and the level of difficulty differ in different chapters, and even in different sections of the same chapter. As a rule, the pace is more leisurely and the derivations are more detailed in the more basic sections that have been used more heavily. My experience has been that even the most detailed derivation is not adequate for some of the students. There is really no substitute for a patient and perceptive instructor. On the other hand, the more advanced sections have been written in a rather concise style on the assumption that the reader who might want to read them can be expected to be more experienced and fearless; they might be able to bridge or tolerate most of the missing steps without too much anguish. Conciseness might also be a virtue if the book is to be used as a reference. It is hoped that, after the book has been used as a textbook, it will remain on the bookshelf as a reference book.

I am grateful to the many students who have taken my courses and used previous versions of my lecture notes. Their criticism and suggestions have helped me improve the text in numerous places. I want to thank the many teaching assistants who have been associated with these courses for their help with problems. I want to thank editors, reviewers, and colleagues who have read or used this book for their advice and suggestions. Two persons in particular must not remain unnamed. Mrs. Beatrice Blonsky typed successive versions of this book over a period of many years; but for her enthusiasm, this book would not have taken shape. Mr. Ron Bohm worked wonders at the daunting task of entering the manuscript into the computer. To each of these persons, named as well as unnamed, I want to express my deep and sincere appreciation.

Los Angeles
July 1990

C. W. W.

CONTENTS

INTRODUCTION TO MATHEMATICAL PHYSICS

1

VECTORS AND FIELDS IN SPACE

1.1. Concepts of space

Physics is concerned with the objective description of recurrent physical phenomena in space. Our intuitive appreciation of space arose in primordial times from the territorial imperative (i.e., the need to secure our immediate environment for our own safety and benefit). In time, it found a more objective realization in response to the search for vigor in reasoning in science and philosophy, and to the practical needs of mensuration in irrigation, construction, and navigation. Most of the great mathematicians of antiquity were geometers; we still remember the *Elements* of Euclid, the geometrical inventiveness of Archimedes, and perhaps also the conic sections of Apollonius.

The Greek mathematicians were followed by the Arabs, who made contributions in arithmetic and algebra. The introduction of Arabic textbooks on mathematics into Europe during the Renaissance stimulated the study of algebra. This in turn led to two important developments in the seventeenth century that laid the foundation of modern mathematics—the invention of analytic geometry and calculus. Both are crucial to the description of physical events in space.

In his book *Géométrie* (1637), Descartes brought the power of algebra to bear on geometrical problems. The book had a significant impact on the gradual evolution of calculus, which was established later in the century by Newton and by Leibnitz.

Newton is the pre-eminent genius of the ages. He made important discoveries in geometry, algebra, and calculus. We remember him even more for his contributions in physics.

Physics before Newton's time was composed of many empirical facts concerning physical phenomena both on earth and in the sky. Two major developments occurred in the century before Newton: Galileo had studied inertia, acceleration, and falling bodies, and Kepler had abstracted three empirical laws of planetary motion from the astronomical observations of Tycho Brahe. In his famous book *Principia Mathematica Philosophiae Naturalis,* published in 1687, Newton presented a unified theory of these diverse phenomena in the form of three laws of motion. The theory stated that, given the initial state of motion of a system in space and the forces acting on it, the state of motion at all other times could be computed exactly. He demonstrated the validity of his theory by showing that Kepler's empirical laws of planetary motion were consequences of a law of forces between massive objects—the law of universal gravitation. This deterministic or mechanical view of dynamics was highly successful; it was to provide the theoretical basis for the understanding of physical phenomena up to the end of the nineteenth century.

In its simplest form, Newtonian mechanics describes the motion of a point mass, say at position $\mathbf{r}(t)$. The mass may undergo acceleration as the result of the imposition of an external, action-at-a-distance force, provided, for example, by a second mass at a different point $\mathbf{r}'$ in space. In other words, there are two gravitating masses. The rest may be considered empty space.

The Newtonian concept of space proved quite adequate until the nineteenth century. Then a self-educated genius materialized in the person of Faraday, to make fundamental experimental discoveries on electromagnetic phenomena. Lacking the mathematical tools to handle the complicated interactions among magnets, charges, and currents, he made progress by visualizing them instead as lines of force emanating from these objects and forming webs or fields all over space, including the location at which the interaction took place.

Faraday's intuitive but successful picture of lines of force was given respectability by the mathematical physicist Maxwell. Stimulated by the work of Faraday and others, he proposed in 1864 that electromagnetic phenomena could be described by four mathematical equations, the Maxwell equations, satisfied by an electric field and a magnetic field. These fields were supposed to reside in a medium called the *ether,* which pervaded all space, including regions of vacuum. It was supposed to be capable of receiving and storing energy and of being set into vibrations. The speed of the resulting electromagnetic disturbance could be computed and was found to be close to that of light. Maxwell suggested that light itself was an electromagnetic wave, because its polarization was known to be affected by the electromagnetic properties of a medium. It was not long before nonoptical electromagnetic waves themselves were generated in the laboratory. This was achieved by Hertz, who found that they propagated with the expected speed. In this way, the reality of the electromagnetic fields in all space was convincingly demonstrated. Henceforth, physical space assumes a character rather different from the old mechanistic one, in which the only relevant points were those at which forces acted on masses.

A fundamental question was soon raised (by Maxwell in 1879): What was the motion of the earth relative to the ether? A definitive answer was given in 1887 by Michelson and Morley, who could find no relative motion.

The question of relative or absolute motion is an ancient one. The sun, the moon, the planets, and the stars march regularly across the sky once a day. The rational men of ancient Greece realized that it could be the earth rather than the heavens that rotated. The sun, the moon, and the planets also move regularly with respect to the stars, the sun's motion being an annual one. This was interpreted by some to mean that the sun moved around the earth, a view that found a quantitative expression in Ptolemy's model of planetary motion in the second century. By the sixteenth century, the accumulated inaccuracies of the model forced Copernicus to quantify the alternative Greek view that the earth and the planets moved around the sun instead.

After Newton, the dynamics of planetary motion was finally understood. It was found to be one of many manifestations of a universal dynamical theory of physical phenomena: Forces caused "absolute" motion, which was the same in all inertial frames moving relative to one another with constant velocities. At the same time, Newton still used the concept of "immovable" space, even though it had no dynamical significance. The concept of time to him was of course entirely independent of that of space.

In 1905, a bright star burst upon physics. An unknown patent examiner named Einstein published three fundamental papers on physics. Of these, the work on the electrodynamics of moving bodies was perhaps the most revolutionary. In this paper, he pointed out that Maxwell's highly successful theory of electrodynamics, like Newton's dynamics, depended only on relative motion and not on the idea of absolute rest. He then generalized this observation to the *principle of relativity* by stating that the laws of physics are the same in all inertial frames in which the equations of mechanics hold good and that the idea of absolute rest is superfluous. He further postulated that the speed of light in vacuum is independent of the motion of the source or of the observer. These postulates led him to a startling conclusion: Not only is there no absolute space, there is also no absolute time. That is, on going from a stationary frame to a moving frame of reference, one should find both space coordinates and time changed. In this theory, which Einstein called the *Special Theory of Relativity,* physics deals with physical events not in space, but in space-time.

The consequences of special relativity, many of which were worked out by Einstein himself, have all been verified experimentally to great precision. One of these, the equivalence between inertia and energy (the famous $E = mc^2$ relation), was to cast its influence far beyond the problems of physics by ushering in the age of nuclear weapons.

Remarkable as the union of space and time was, Einstein immediately objected to the preference given to inertial frames. The trouble is that the concept of an inertial frame is ill defined: In an inertial frame, a mass moves uniformly if not acted on by an external force, but we know that there is no external force only when the mass is moving uniformly. This is a classic example of circular reasoning.

A solution of the problem was obtained by Einstein in his *General Theory of Relativity* (1916). He started by noting that an observer falling freely in a gravitational field while inside a closed freely falling elevator would believe that his world was in an inertial frame. Indeed, Einstein saw no need to abandon this local inertial frame (or free-falling frame) in favor of an external (i.e., nonaccelerating), inertial frame in the formulation of physical laws. He was particularly impressed by the effect of gravity, because it had been known since Galileo that the acceleration of free fall was the same for all massive objects. (This comes about because the gravitational mass is always proportional to the inertial mass and may be taken to be equal to it.) This universal effect of gravity made it possible for Einstein to represent the curvature of the path of a gravitating mass as due to the inherent curvature of space itself. Thus space (or more precisely, space-time) itself took on an important dynamical attribute that was totally unsuspected previously.

Einstein spent many years of his life trying to show that electromagnetism was also a structure of space. In this he was unsuccessful. The modern successes in unifying electromagnetism with the weak interaction between subatomic particles are based instead on the space-time properties of the phase angles of fields describing physical properties in space. In other words, these interactions are the internal properties of fields residing in space. It is not yet clear whether gravity itself admits a similar description.

In this chapter, we make a modest beginning in studying the mathematics used to describe our increasingly subtle appreciation of the concept of physical space. The topics include vectors and fields in space and the use of curvilinear coordinates.

1.2. Vectors in space

Physics deals with physical events in space-time. In Newtownian mechanics, time is completely independent of space. It is characterized by a single number such as 5 in the statement, "It is now 5 minutes past the hour." Such single numbers are called *scalars.*

In contrast, a set of three numbers, called a *vector,* is needed to characterize a point in three-dimensional space. We use the equivalent notation

$$\begin{aligned}\mathbf{r} &= x\mathbf{i} + y\mathbf{j} + z\mathbf{k} = (x,y,z) \\ &= x\mathbf{e}_x + y\mathbf{e}_y + z\mathbf{e}_z \\ &= x_1\mathbf{e}_1 + x_2\mathbf{e}_2 + x_3\mathbf{e}_3 = (x_1,x_2,x_3) \qquad (1.1)\end{aligned}$$

to characterize a point whose rectangular (or Cartesian) coordinates as measured from an arbitrarily chosen origin are x,y,z or equivalent x_1,x_2,x_3. We call $\mathbf{r}$ the position vector of a point in space. The numbers (x_1,x_2,x_3) making up a vector $\mathbf{r}$ are called its *components.*

Unlike a scalar, a vector $\mathbf{r}$ has both a length and a direction. We know from geometry that the length in a position vector $\mathbf{r}$ is

$$r = |\mathbf{r}| = (x^2 + y^2 + z^2)^{1/2}. \qquad (1.2)$$

Its direction is then given by the *unit vector*

$$\mathbf{e}_r = \mathbf{e}(\mathbf{r}) = \mathbf{r}/r = (x\mathbf{i} + y\mathbf{j} + z\mathbf{k})/r \tag{1.3}$$

along the direction of $\mathbf{r}$. It is obviously a vector of unit length parallel to $\mathbf{r}$. Thus

$$\mathbf{r} = r\mathbf{e}(\mathbf{r}). \tag{1.4}$$

In particular, the vectors

$$\mathbf{i} = \mathbf{e}_x = \mathbf{e}_1, \qquad \mathbf{j} = \mathbf{e}_y = \mathbf{e}_2, \qquad \mathbf{k} = \mathbf{e}_z = \mathbf{e}_3, \tag{1.5}$$

in Eq. (1.1) are unit vectors along the arbitrarily chosen x, y, and z axes, respectively. These axes are chosen to be perpendicular to each other so that

$$\mathbf{e}_i \cdot \mathbf{e}_j = \delta_{ij} = \begin{cases} 0, & i \neq j \\ 1, & i = j. \end{cases} \tag{1.6}$$

The symbol δ_{ij} is called a *Kronecker delta symbol.* (They also form a right-handed coordinate system; that is, rotation from $\mathbf{e}_1$ to $\mathbf{e}_2$ causes a right-handed screw to advance along the $\mathbf{e}_3$ direction.) An arbitrary vector $\mathbf{A}$ can thus be written in terms of either its rectangular components (A_x, A_y, A_z) or its length A and direction $\mathbf{e}(\mathbf{A})$

$$\begin{aligned} \mathbf{A} &= A_x\mathbf{i} + A_y\mathbf{j} + A_z\mathbf{k}, \qquad \text{or} \qquad \sum_{i=1}^{3} A_i\mathbf{e}_i \\ &= A\mathbf{e}(\mathbf{A}). \end{aligned} \tag{1.7}$$

1.2.1. *Algebra of vectors*

In the above discussion, we have used the following two basic algebraic operations that define an algebra of vectors: vector addition:

$$\mathbf{C} = \mathbf{A} + \mathbf{B} = (A_x + B_x, A_y + B_y, A_z + B_z), \tag{1.8}$$

and scalar multiplication:

$$\mathbf{C} = \lambda\mathbf{A} = (\lambda A_x, \lambda A_y, \lambda A_z). \tag{1.9}$$

By an *algebra* we mean that the results of these operations are objects (i.e., vectors) similar to the original objects, which are vectors.

A zero or null vector $\mathbf{0}$ is defined:

$$\mathbf{0} = 0\mathbf{i} + 0\mathbf{j} + 0\mathbf{k} = (0,0,0). \tag{1.10}$$

As a consequence, the negative $-\mathbf{A}$ of a vector $\mathbf{A}$ is also defined:

$$(-\mathbf{A}) + \mathbf{A} = \mathbf{0}, \tag{1.11}$$

that is,

$$-\mathbf{A} = -A_x\mathbf{i} - A_y\mathbf{j} - A_z\mathbf{k} = -A\mathbf{e}(\mathbf{A}). \tag{1.12}$$

Since Eq. (1.2) shows that $-\mathbf{A}$ has the same length A as $\mathbf{A}$, we find

$$\mathbf{e}(-\mathbf{A}) = -\mathbf{e}(\mathbf{A}), \tag{1.13}$$

that is, $-\mathbf{A}$ points in a direction opposite that of $\mathbf{A}$.

Example 1.2.1. The sum of the vectors

$$\mathbf{A} = \mathbf{i} + \mathbf{j}$$
$$\mathbf{B} = \mathbf{j} + 3\mathbf{k}$$

is

$$\mathbf{C} = \mathbf{A} + \mathbf{B} = \mathbf{i} + 2\mathbf{j} + 3\mathbf{k}.$$

This vector sum has a length

$$C = (C_x^2 + C_y^2 + C_z^2)^{1/2} = (1 + 4 + 9)^{1/2} = (14)^{1/2},$$

and a direction

$$\mathbf{e}(\mathbf{C}) = \frac{\mathbf{C}}{C} = \frac{1}{\sqrt{14}}(\mathbf{i} + 2\mathbf{j} + 3\mathbf{k}).$$

1.2.2. Geometry of space

The description of a vector $\mathbf{A}$ by its length A and its direction $\mathbf{e}(\mathbf{A})$ is basically geometrical, since geometry deals with sizes and shapes, that is, with properties in space. In particular, the concept of *length* is a special case of the concept of *scalar product* between two vectors

$$\mathbf{A} \cdot \mathbf{B} = A_x B_x + A_y B_y + A_z B_z. \tag{1.14}$$

It involves the scalar product

$$\mathbf{A} \cdot \mathbf{A} = A_x^2 + A_y^2 + A_z^2 = A^2. \tag{1.15}$$

The concept of direction is most easily understood in two-dimensional space. We decompose $\mathbf{e}(\mathbf{A})$ into its x,y components:

$$\begin{aligned} \mathbf{e}(\mathbf{A}) &= \cos\theta_{Ax}\,\mathbf{e}_x + \sin\theta_{Ax}\,\mathbf{e}_y \\ &= \cos\theta_{Ax}\,\mathbf{e}_x + \cos\theta_{Ay}\,\mathbf{e}_y, \end{aligned} \tag{1.16}$$

where θ_{Ai} is the direction angle between $\mathbf{e}(\mathbf{A})$ and the ith axis. These components of $\mathbf{e}(\mathbf{A})$ are called its *direction cosines* (Fig. 1.1). Each of these direction cosines can be isolated by using the scalar product operation. For example,

$$\mathbf{e}(\mathbf{A}) \cdot \mathbf{e}_x = (\cos\theta_{Ax}\,\mathbf{e}_x + \cos\theta_{Ay}\,\mathbf{e}_y) \cdot \mathbf{e}_x = \cos\theta_{Ax}. \tag{1.17}$$

It is now easy to generalize the direction-cosine decomposition of a unit vector $\mathbf{e}(\mathbf{A})$ to three-dimensional space by induction. The result is

$$\mathbf{e}(\mathbf{A}) = \cos\theta_{Ax}\,\mathbf{e}_x + \cos\theta_{Ay}\,\mathbf{e}_y + \cos\theta_{Az}\,\mathbf{e}_z. \tag{1.18}$$

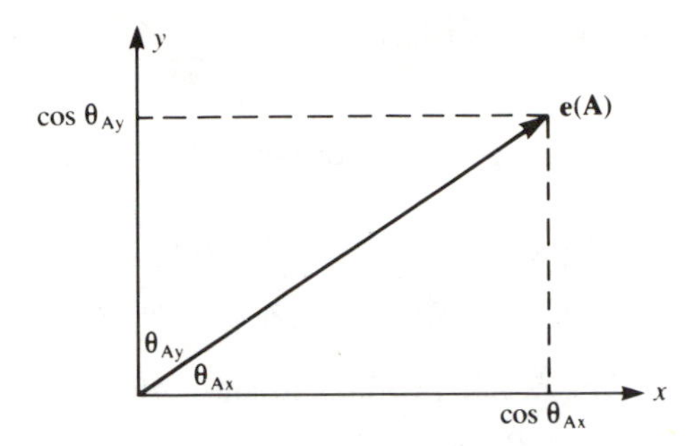

Fig. 1.1. Direction cosines.

Substitution of Eq. (1.17) into Eq. (1.18) gives the result

$$\begin{aligned}\mathbf{e}(\mathbf{A}) &= [\mathbf{e}(\mathbf{A})\cdot\mathbf{e}_x]\mathbf{e}_x + [\mathbf{e}(\mathbf{A})\cdot\mathbf{e}_y]\mathbf{e}_y + [\mathbf{e}(\mathbf{A})\cdot\mathbf{e}_z]\mathbf{e}_z \\ &= \mathbf{e}(\mathbf{A})(\cdot\,\mathbf{e}_x\mathbf{e}_x + \cdot\,\mathbf{e}_y\mathbf{e}_y + \cdot\,\mathbf{e}_z\mathbf{e}_z).\end{aligned} \tag{1.19}$$

Since this equation holds for any vector $\mathbf{e}(\mathbf{A})$, or any vector $\mathbf{A}$, we have obtained the formal identity

$$1 = \cdot\,\mathbf{e}_x\mathbf{e}_x + \cdot\,\mathbf{e}_y\mathbf{e}_y + \cdot\,\mathbf{e}_z\mathbf{e}_z, \tag{1.20}$$

called a *completeness relation*. It expresses symbolically the well-known result that a three-dimensional vector has three components. It further states that its x component, for example, can be obtained by the *projection* of $\mathbf{A}$ on $\mathbf{e}_x$, that is, by the scalar product

$$\mathbf{A}\cdot\mathbf{e}_x = A_x = A\mathbf{e}(\mathbf{A})\cdot\mathbf{e}_x = A\cos\theta_{Ax}. \tag{1.21}$$

Finally, we note that the scalar product, Eq. (1.14), between two vectors can be written in the familiar cosine form

$$\mathbf{A}\cdot\mathbf{B} = AB\mathbf{e}(\mathbf{A})\cdot\mathbf{e}(\mathbf{B}) = AB\cos\theta_{\mathbf{AB}}, \tag{1.22}$$

where $\theta_{\mathbf{AB}}$ is the angle between the vectors $\mathbf{A}$ and $\mathbf{B}$.

Example 1.2.2. Find the projection of $\mathbf{A} = \mathbf{i} + 2\mathbf{j} + 3\mathbf{k}$ on $\mathbf{B} = \mathbf{j} + 2\mathbf{k}$.

The projection of $\mathbf{A}$ on $\mathbf{B}$ is the same as the projection of $\mathbf{A}$ on $\mathbf{e}_\mathbf{B} = \mathbf{B}/B$, where $B = \sqrt{5}$. It is

$$\begin{aligned}\mathbf{A}\cdot\mathbf{e}_\mathbf{B} &= \mathbf{A}\cdot\mathbf{B}/B = (\mathbf{i} + 2\mathbf{j} + 3\mathbf{k})\cdot(\mathbf{j} + 2\mathbf{k})/\sqrt{5} \\ &= 8/\sqrt{5}.\end{aligned}$$

Example 1.2.3. Obtain the unit vector $\mathbf{e}$ in space that makes an equal angle with each of the three coordinate axes $\mathbf{e}_1$, $\mathbf{e}_2$, and $\mathbf{e}_3$. What is this angle?

We are given the information that $\theta_1 = \theta_2 = \theta_3 = \theta$, where θ_i is the angle between $\mathbf{e}$ and $\mathbf{e}_i$. Hence

$$\cos\theta_1 = \cos\theta_2 = \cos\theta_3 = \cos\theta = a,$$

where a is some constant. That is,

$$\mathbf{e} = a\mathbf{e}_1 + a\mathbf{e}_2 + a\mathbf{e}_3$$

has equal components along all three axes. Since $\mathbf{e}$ has unit length, we must have

$$1 = (a^2 + a^2 + a^2)^{1/2} = \sqrt{3}\, a,$$
$$a = 1/\sqrt{3} = \cos\theta, \qquad \text{or} \qquad \theta = 55°.$$

Example 1.2.4. Use the scalar product to show that

$$\cos(\alpha + \beta) = \cos\alpha \cos\beta - \sin\alpha \sin\beta.$$

For the two-dimensional vectors of Fig. 1.2, we see that

$$\theta_{Ax} = \alpha, \qquad \theta_{Bx} = \beta, \qquad \theta_{AB} = \alpha + \beta,$$
$$\theta_{Ay} = \frac{\pi}{2} + \alpha, \qquad \theta_{By} = \frac{\pi}{2} - \beta.$$

Since we are interested in $\cos(\alpha + \beta)$, we should examine the scalar product

$$\mathbf{A} \cdot \mathbf{B} = AB\cos(\alpha + \beta) = A_x B_x + A_y B_y.$$

Hence

$$\begin{aligned}\cos(\alpha + \beta) &= \left(\frac{A_x}{A}\right)\left(\frac{B_x}{B}\right) + \left(\frac{A_y}{A}\right)\left(\frac{B_y}{B}\right)\\ &= \cos\theta_{Ax}\cos\theta_{Bx} + \cos\theta_{Ay}\cos\theta_{By}\\ &= \cos\alpha\cos\beta + (-\sin\alpha)(\sin\beta).\end{aligned}$$

1.2.3. Vector product

One additional vector operation is important for vectors in three-dimensional space: the *vector product*

$$\begin{aligned}\mathbf{B} \times \mathbf{C} &= \mathbf{i}(B_y C_z - B_z C_y) + \mathbf{j}(B_z C_x - B_x C_z) + \mathbf{k}(B_x C_y - B_y C_x)\\ &= \begin{vmatrix} \mathbf{i} & \mathbf{j} & \mathbf{k} \\ B_x & B_y & B_z \\ C_x & C_y & C_z \end{vmatrix} = \begin{vmatrix} \mathbf{e}_1 & \mathbf{e}_2 & \mathbf{e}_3 \\ B_1 & B_2 & B_3 \\ C_1 & C_2 & C_3 \end{vmatrix} = -(\mathbf{C} \times \mathbf{B}).\end{aligned} \qquad (1.23)$$

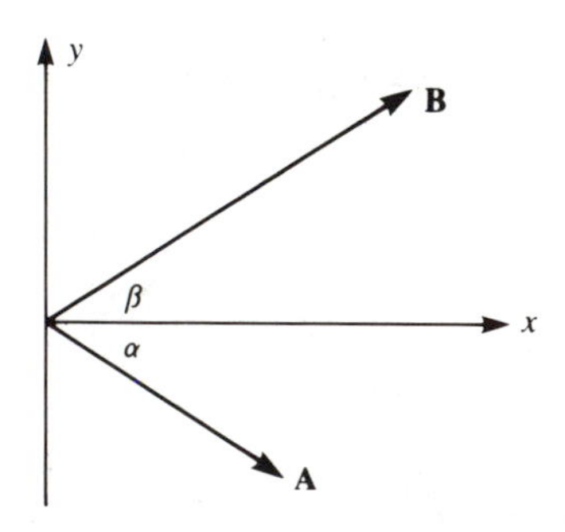

Fig. 1.2. Sum of angles.

In this equation, the six terms of the second expression have been represented by a 3×3 array called a *determinant.* A 3×3 determinant can also be written in terms of 2×2 determinants

$$\begin{vmatrix} \mathbf{i} & \mathbf{j} & \mathbf{k} \\ B_x & B_y & B_z \\ C_x & C_y & C_z \end{vmatrix} = \mathbf{i} \begin{vmatrix} B_y & B_z \\ C_y & C_z \end{vmatrix} + \mathbf{j} \begin{vmatrix} B_z & B_x \\ C_z & C_x \end{vmatrix} + \mathbf{k} \begin{vmatrix} B_x & B_y \\ C_x & C_y \end{vmatrix},$$

where

$$\begin{vmatrix} a & b \\ c & d \end{vmatrix} = ad - bc.$$

The reader should think of the six-term expression shown in Eq. (1.23) whenever a 3×3 determinant appears in this chapter. For example, the antisymmetric property shown in the last expression of this equation comes from a change of sign of the six-term expression when the symbols B and C are interchanged. A more detailed discussion of determinants can be found in Section 2.3.

Example 1.2.5.

$$\mathbf{e}_1 \times \mathbf{e}_1 = \begin{vmatrix} \mathbf{e}_1 & \mathbf{e}_2 & \mathbf{e}_3 \\ 1 & 0 & 0 \\ 1 & 0 & 0 \end{vmatrix} = 0,$$

because each of the six terms is zero, while

$$\mathbf{e}_1 \times \mathbf{e}_2 = \begin{vmatrix} \mathbf{e}_1 & \mathbf{e}_2 & \mathbf{e}_3 \\ 1 & 0 & 0 \\ 0 & 1 & 0 \end{vmatrix} = \mathbf{e}_3, \qquad \mathbf{e}_2 \times \mathbf{e}_3 = \mathbf{e}_1, \qquad \mathbf{e}_3 \times \mathbf{e}_1 = \mathbf{e}_2. \quad (1.24)$$

If in $\mathbf{B} \times \mathbf{C}$ we put

$$\mathbf{e}(\mathbf{B}) = \mathbf{e}_1, \qquad \mathbf{e}(\mathbf{C}) = \cos\alpha\, \mathbf{e}_1 + \sin\alpha\, \mathbf{e}_2, \quad (1.25)$$

where $\alpha = \theta_{\mathbf{BC}}$, then

$$\begin{aligned} \mathbf{B} \times \mathbf{C} &= BC\mathbf{e}_1 \times (\cos\alpha\, \mathbf{e}_1 + \sin\alpha\, \mathbf{e}_2) \\ &= BC \sin\alpha\, \mathbf{e}_3. \end{aligned} \quad (1.26)$$

Thus $\mathbf{B} \times \mathbf{C}$ has a length $BC \sin\alpha$ and a direction $\mathbf{e}_3$ perpendicular to the plane containing $\mathbf{B}$ and $\mathbf{C}$. Its length $BC \sin\alpha$ is equal to the area of the parallelogram with sides $\mathbf{B}$ and $\mathbf{C}$ shown by the shaded area in Fig. 1.3. The *triple scalar product*

$$\mathbf{A} \cdot (\mathbf{B} \times \mathbf{C}) = A\, |\mathbf{B} \times \mathbf{C}| \cos\beta = ABC \sin\alpha \cos\beta \quad (1.27)$$

may be interpreted as the volume of the parallelipiped shown in the figure with the vectors $\mathbf{A}$, $\mathbf{B}$, and $\mathbf{C}$ on three of its sides.

The vector

$$\mathbf{D} = \mathbf{e}_1 \mathrm{D}_1 + \mathbf{e}_2 \mathrm{D}_2 + \mathbf{e}_3 \mathrm{D}_3$$

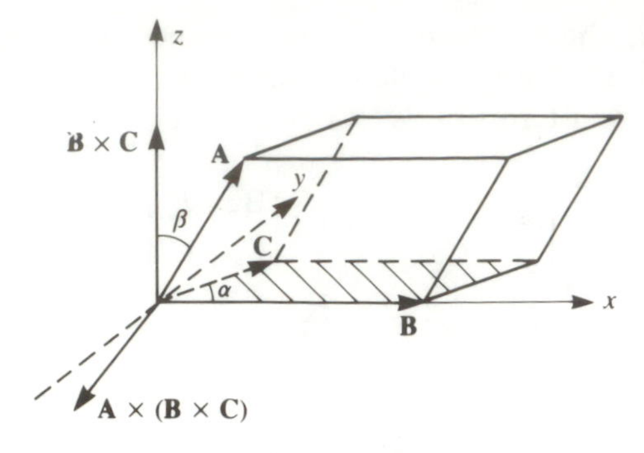

Fig. 1.3. The vector product, the triple scalar product, and the triple vector product.

and the scalar product

$$\mathbf{A} \cdot \mathbf{D} = A_1 D_1 + A_2 D_2 + A_3 D_3$$

look very similar in structure. They differ only by the substitutions $\mathbf{e}_i \leftrightarrow A_i$. The same substitution shows that $\mathbf{A} \cdot (\mathbf{B} \times \mathbf{C})$ can be written down directly from Eq. (1.23) as

$$\mathbf{A} \cdot (\mathbf{B} \times \mathbf{C}) = \begin{vmatrix} A_1 & A_2 & A_3 \\ B_1 & B_2 & B_3 \\ C_1 & C_2 & C_3 \end{vmatrix}.$$

Since a determinant changes sign when two of its rows are interchanged, we find

$$\begin{aligned} \mathbf{A} \cdot (\mathbf{B} \times \mathbf{C}) &= -\mathbf{A} \cdot (\mathbf{C} \times \mathbf{B}) = -\mathbf{B} \cdot (\mathbf{A} \times \mathbf{C}) = -\mathbf{C} \cdot (\mathbf{B} \times \mathbf{A}) \\ &= \mathbf{B} \cdot (\mathbf{C} \times \mathbf{A}) = \mathbf{C} \cdot (\mathbf{A} \times \mathbf{B}). \end{aligned}$$

Since the order of a scalar product is unimportant, we have

$$\mathbf{A} \cdot (\mathbf{B} \times \mathbf{C}) = (\mathbf{B} \times \mathbf{C}) \cdot \mathbf{A} = (\mathbf{A} \times \mathbf{B}) \cdot \mathbf{C} = (\mathbf{C} \times \mathbf{A}) \cdot \mathbf{B}.$$

Example 1.2.6. If **a**, **b**, and **c** are the position vectors of the points A, B, and C in space, what is the area of the triangle ABC?

From Fig. 1.4, we see that the sides of triangle ABC may be described by the vectors

$$\mathbf{s}_1 = \mathbf{a} - \mathbf{b}$$

$$\mathbf{s}_2 = \mathbf{b} - \mathbf{c}$$

$$\mathbf{s}_3 = \mathbf{c} - \mathbf{a}.$$

These vectors are coplanar because

$$\mathbf{s}_1 + \mathbf{s}_2 + \mathbf{s}_3 = (\mathbf{a} - \mathbf{b}) + (\mathbf{b} - \mathbf{c}) + (\mathbf{c} - \mathbf{a}) = 0,$$

or

$$\mathbf{s}_3 = -(\mathbf{s}_1 + \mathbf{s}_2).$$

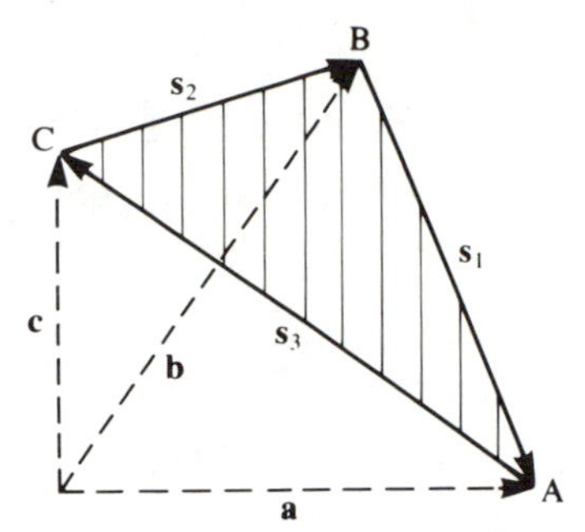

Fig. 1.4. Area of the triangle ABC.

The area of the triangle is one-half of the magnitude of the vector product

$$\mathbf{s}_1 \times \mathbf{s}_2 = (\mathbf{a} - \mathbf{b}) \times (\mathbf{b} - \mathbf{c}) = \mathbf{a} \times \mathbf{b} + \mathbf{b} \times \mathbf{c} + \mathbf{c} \times \mathbf{a}.$$

We also note that the right-hand side of this expression is unchanged under the *cyclic permutation* $\mathbf{abc} \to \mathbf{bca}$ (i.e., under the substitutions $\mathbf{a} \to \mathbf{b}$, $\mathbf{b} \to \mathbf{c}$, $\mathbf{c} \to \mathbf{a}$) or $\mathbf{abc} \to \mathbf{cab}$ (i.e., under the substitutions $\mathbf{a} \to \mathbf{c}$, $\mathbf{b} \to \mathbf{a}$, $\mathbf{c} \to \mathbf{b}$). As a result, the left-hand side is also unchanged under the cyclic permutation $123 \to 231$ or $123 \to 312$. That is,

$$\mathbf{s}_1 \times \mathbf{s}_2 = \mathbf{s}_2 \times \mathbf{s}_3 = \mathbf{s}_3 \times \mathbf{s}_1.$$

These equations can also be proved directly. For example,

$$\mathbf{s}_1 \times \mathbf{s}_2 = -\mathbf{s}_1 \times (\mathbf{s}_1 + \mathbf{s}_3) = -\mathbf{s}_1 \times \mathbf{s}_3 = \mathbf{s}_3 \times \mathbf{s}_1.$$

There is also a *triple vector product* $\mathbf{A} \times (\mathbf{B} \times \mathbf{C})$. Since $\mathbf{B} \times \mathbf{C}$ is perpendicular to the BC plane containing $\mathbf{B}$ and $\mathbf{C}$, $\mathbf{A} \times (\mathbf{B} \times \mathbf{C})$ must lie on the BC plane in a direction perpendicular to $\mathbf{A}$. This direction is shown in Fig. 1.3. A more precise statement of this result is given by the *BAC rule*

$$\mathbf{A} \times (\mathbf{B} \times \mathbf{C}) = \mathbf{B}(\mathbf{A} \cdot \mathbf{C}) - \mathbf{C}(\mathbf{A} \cdot \mathbf{B}). \qquad (1.28)$$

This rule can be proved easily with the help of Eq. (1.26)

$$\mathbf{A} \times (\mathbf{B} \times \mathbf{C}) = \begin{vmatrix} \mathbf{e}_1 & \mathbf{e}_2 & \mathbf{e}_3 \\ A_1 & A_2 & A_3 \\ 0 & 0 & BC \sin \alpha \end{vmatrix} = BC \sin \alpha \, (A_2 \mathbf{e}_1 - A_1 \mathbf{e}_2)$$

$$= BCA_2 \sin \alpha \, \mathbf{e}_1 - BCA_1 (\sin \alpha \, \mathbf{e}_2).$$

Since Eq. (1.25) gives

$$\mathbf{e}_1 = \mathbf{e}(\mathbf{B}), \qquad \sin \alpha \, \mathbf{e}_2 = \mathbf{e}(\mathbf{C}) - \cos \alpha \, \mathbf{e}_1,$$

we find that

$$\begin{aligned}\mathbf{A}\times(\mathbf{B}\times\mathbf{C}) &= \mathbf{B}(CA_2\sin\alpha + CA_1\cos\alpha) - \mathbf{C}BA_1\\ &= \mathbf{B}(A_2C_2 + A_1C_1) - \mathbf{C}(\mathbf{A}\cdot\mathbf{B})\\ &= \mathbf{B}(\mathbf{A}\cdot\mathbf{C}) - \mathbf{C}(\mathbf{A}\cdot\mathbf{B}).\end{aligned}$$

Other triple vector products can be deduced from the BAC rule. For example

$$(\mathbf{A}\times\mathbf{B})\times\mathbf{C} = -\mathbf{C}\times(\mathbf{A}\times\mathbf{B}) = \mathbf{C}\times(\mathbf{B}\times\mathbf{A}),$$

where the different forms are equal to one another by virtue of Eq. (1.23). The BAC rule can now be applied by simply interchanging the symbols **A** and **C** in every term of Eq. (1.28):

$$\mathbf{C}\times(\mathbf{B}\times\mathbf{A}) = \mathbf{B}(\mathbf{C}\cdot\mathbf{A}) - \mathbf{A}(\mathbf{C}\cdot\mathbf{B}). \tag{1.29}$$

These results show that in general

$$(\mathbf{A}\times\mathbf{B})\times\mathbf{C} \neq \mathbf{A}\times(\mathbf{B}\times\mathbf{C}). \tag{1.30}$$

These expressions are not the same in general because $\mathbf{A}\times(\mathbf{B}\times\mathbf{C})$ lies on the BC plane, according to Eq. (1.28), while $(\mathbf{A}\times\mathbf{B})\times\mathbf{C}$ lies on the AB plane. Only when **B** is perpendicular to both **A** and **C** will the two expressions be equal, for then the second term in both Eqs. (1.28) and (1.29) vanishes.

Example 1.2.7. If **A** is an arbitrary vector and **e** is an arbitrary unit vector, show that

$$\mathbf{A} = \mathbf{e}(\mathbf{A}\cdot\mathbf{e}) + \mathbf{e}\times(\mathbf{A}\times\mathbf{e}).$$

This follows directly from the BAC rule

$$\mathbf{e}\times(\mathbf{A}\times\mathbf{e}) = \mathbf{A}(\mathbf{e}\cdot\mathbf{e}) - \mathbf{e}(\mathbf{A}\cdot\mathbf{e}).$$

If we take $\mathbf{e} = \mathbf{e}_x$, we see by inspection that the $\mathbf{A}\cdot\mathbf{e}$ term is $A_x\mathbf{e}_x$. Hence the remaining term must be $A_y\mathbf{e}_y + A_z\mathbf{e}_z$ (i.e., that part of **A** that lies on the plane perpendicular to **e**). In particular, if **A** is on the xy plane, $A_z = 0$. Then

$$\mathbf{A}\times\mathbf{e} = (A_x\mathbf{e}_x + A_y\mathbf{e}_y)\times\mathbf{e}_x = -A_y\mathbf{e}_z,$$

and

$$\mathbf{e}\times(\mathbf{A}\times\mathbf{e}) = -A_y\mathbf{e}_x\times\mathbf{e}_z = A_y\mathbf{e}_y,$$

as expected.

Problems

1.2.1. If $\mathbf{A} = (1,2,3)$ and $\mathbf{B} = (3,1,1)$, calculate $\mathbf{A}+\mathbf{B}$, $\mathbf{A}-\mathbf{B}$, $\mathbf{A}\cdot\mathbf{B}$, the projections of **A** on **B** and of **B** on **A**, $\mathbf{A}\times\mathbf{B}$, $|\mathbf{A}\times\mathbf{B}|$, and $\mathbf{e}(\mathbf{A}\times\mathbf{B})$.

1.2.2. By using suitable vectors, prove the trigonometric identities
(a) $\cos(\alpha-\beta)=\cos\alpha\cos\beta+\sin\alpha\sin\beta$,
(b) $\sin(\alpha-\beta)=\sin\alpha\cos\beta-\cos\alpha\sin\beta$.

1.2.3. Prove that the diagonals of a parallelogram bisect each other.

1.2.4. Take one corner of the unit cube as the origin, and its three adjacent sides as the x,y,z axes. From the origin, four diagonals can be drawn across the cube: three on three faces and one across the body to the opposite corner. Calculate all the angles between pairs of these diagonals.

1.2.5. If $\mathbf{a}$, $\mathbf{b}$, and $\mathbf{c}$ are the position vectors of the points A, B, and C in space, the area of the triangle ABC is $\mathbf{a}\times\mathbf{b}+\mathbf{b}\times\mathbf{c}+\mathbf{c}\times\mathbf{a}$ (Example 1.2.6). Show that the perpendicular distance from the origin to the plane containing this triangle is $\mathbf{a}\cdot(\mathbf{b}\times\mathbf{c})/|\mathbf{a}\times\mathbf{b}+\mathbf{b}\times\mathbf{c}+\mathbf{c}\times\mathbf{a}|$.

1.2.6. Let $\mathbf{r}$ be the position vector of any point on a plane in space and let $\mathbf{a}$ be the position vector of that point on the plane nearest the origin. Show that $\mathbf{a}\cdot\mathbf{r}=a^2=\text{constant}$.

1.2.7. If $\mathbf{a}$ is the position vector of a fixed point A in space, what is the nature of the surface defined by the arbitrary position vector $\mathbf{r}$ satisfying the equation $|\mathbf{r}-\mathbf{a}|=\text{constant}=b$?

1.2.8. Use the BAC rule to prove the Jacobi identity $\mathbf{a}\times(\mathbf{b}\times\mathbf{c})+\mathbf{b}\times(\mathbf{c}\times\mathbf{a})+\mathbf{c}\times(\mathbf{a}\times\mathbf{b})=0$.

1.2.9. If an unknown vector $\mathbf{X}$ satisfies the relations $\mathbf{X}\cdot\mathbf{b}=\beta$, $\mathbf{X}\times\mathbf{b}=\mathbf{c}$, express $\mathbf{X}$ in terms of β, $\mathbf{b}$, and $\mathbf{c}$. (Hint: Decompose $\mathbf{X}$ along the three perpendicular directions along $\mathbf{b}$, $\mathbf{c}$, and $\mathbf{b}\times\mathbf{c}$, or use the *BAC rule*.)

1.2.10. If $\mathbf{D}$ is a linear combination of three arbitrary noncoplanar vectors $\mathbf{A}$, $\mathbf{B}$, $\mathbf{C}$: $\mathbf{D}=a\mathbf{A}+b\mathbf{B}+c\mathbf{C}$, show that $a=\mathbf{D}\cdot(\mathbf{B}\times\mathbf{C})/\mathbf{A}\cdot(\mathbf{B}\times\mathbf{C})$. Obtain corresponding expressions for b and c.

1.2.11. Describe and explain a test for coplanarity of three arbitrary vectors $\mathbf{A}$, $\mathbf{B}$, and $\mathbf{C}$.

1.2.12. Show that $(\mathbf{A}\times\mathbf{B})\times(\mathbf{C}\times\mathbf{D})$ lies on the line of intersection of the plane containing $\mathbf{A}$ and $\mathbf{B}$ and the plane containing $\mathbf{C}$ and $\mathbf{D}$.

1.2.13. If $\mathbf{r}$ and $\mathbf{v}=(d/dt)\mathbf{r}$ are both functions of time, show that

$$\frac{d}{dt}[\mathbf{r}\times(\mathbf{v}\times\mathbf{r})]=r^2\mathbf{a}+(\mathbf{r}\cdot\mathbf{v})\mathbf{v}-(v^2+\mathbf{r}\cdot\mathbf{a})\mathbf{r},$$

where $\mathbf{a}$ is the acceleration.

1.3. Permutation symbols

Our intuitive (i.e., geometrical) understanding of physical space is of great antiquity. Archimedes, perhaps the greatest mathematician of ancient Greece, was known for his geometrical work, including formulas for areas of figures and volumes of solid bodies. The well-known principle in hydrostatics bearing his name demonstrates his geometric insight. In contrast, the algebraic description of space, which we have used in the preceding section, is of relatively recent origin. It evolved from the algebra of Arabic mathematicians of the ninth through the eleventh centuries. (The word *algebra* itself arose from the Arabic word *al-jabr,* meaning the reunion of broken parts, which had appeared in the title of a well-known Arabic book of the ninth century on algebra.) The word *vector* was coined by Hamilton in the nineteenth century.

It is easy enough to see why the algebraic description of space, being so symbolic and abstract, has taken so long to develop. It is perhaps more difficult to appreciate its power. In this section, we further illustrate the power of algebraic recombinations by reconsidering the vector product of vectors in space.

Let us first recall that the determinantal form, Eq. (1.23), of the vector product can become very cumbersome when one deals with several successive vector products. (For example, try to write down a determinantal expression for $\mathbf{A}\times[\mathbf{B}\times(\mathbf{C}\times\mathbf{D})]$.) The structure of the vector product is very simple, however, It is a vector in space and must have three components. For example, one of the equations in Eq. (1.24) can be written in the form

$$\mathbf{e}_1 \times \mathbf{e}_2 = \varepsilon_{121}\mathbf{e}_1 + \varepsilon_{122}\mathbf{e}_2 + \varepsilon_{123}\mathbf{e}_3,$$

where the components of the vector product have been denoted ε_{12k}. The first two indices (12) refer to unit vectors on the left-hand side (LHS), while the last index k is that of the vector component on the right-hand side (RHS). Since $\mathbf{e}_1 \times \mathbf{e}_2 = \mathbf{e}_3$ (a result referred to as the right-hand rule for vector products), we must have

$$\varepsilon_{121} = 0, \qquad \varepsilon_{122} = 0, \qquad \varepsilon_{123} = 1.$$

There are $3 \times 3 = 9$ vector products $\mathbf{e}_i \times \mathbf{e}_j$, and a total of $9 \times 3 = 27$ ε_{ijk} components. Equation (1.24) shows that these components are zero except the six in which $i \neq j \neq k$, namely

$$\varepsilon_{123} = \varepsilon_{231} = \varepsilon_{312} = 1, \qquad \varepsilon_{213} = \varepsilon_{321} = \varepsilon_{132} = -1. \tag{1.31}$$

We shall see that the six sets of indices appearing in Eq. (1.31) are called the *permutations,* or rearrangements, of the three objects 1, 2, and 3. For this reason the ε_{ijk} are called *permutation symbols.*

1.3.1. Permutations

Let us examine the six sets of indices in Eq. (1.31) more carefully. The first set, 123, is an ordered sequence of three objects 1, 2, and 3. The remaining five sets are also ordered sequences of these objects, but they differ from 123 in the ordering. They are said to be *permutations* (or rearrangements) of the original ordering 123. The ordering 123 is itself a permutation of each of the other five sets. In other words, all six sets are permutations of one another.

We now show that there are only six permutations of three distinct objects such as 1, 2, and 3. In ordering these three objects, we can put any one of the three in the first position, and any one of the remaining two objects in the second position. The last remaining object must be put in the last position. Thus the number of permutations equals $3 \times 2 \times 1 = 6$.

Permutations from an original ordered sequence such as 123 can be achieved by successive *transpositions,* or interchanges, of neighboring objects. A permutation is said to be *even* (or odd) if an *even* (or odd) number of transpositions is required for the rearrangement. For example, 213 is an odd permutation of 123 because one transposition $12 \rightarrow 21$ will do the job. On the other hand, 231 is an even permutation and requires the transpositions $12 \rightarrow 21$ and then $13 \rightarrow 31$. Equation (1.31) shows that the permutation symbols of three indices have the value 1 for even permutations of 123 and the value -1 for odd permutations of 123.

We can also start with the ordering ijk instead. An even permutation of ijk will not change the value of the resulting permutation symbol, while an odd permutation gives a permutation symbol of the opposite sign. That is,

$$\varepsilon_{ijk} = \varepsilon_{jki} = \varepsilon_{kij} = -\varepsilon_{jik} = -\varepsilon_{kji} = -\varepsilon_{ikj}. \tag{1.32}$$

Permutation symbols such as ε_{112} and ε_{111} with two or more identical indices cannot be permutations of 123. They have the value 0, but they also satisfy Eq. (1.32).

1.3.2. Vector products with permutation symbols

It will be useful to write the nine equations implied by Eq. (1.24) in the abstract form

$$\mathbf{e}_i \times \mathbf{e}_j = \sum_{k=1}^{3} \varepsilon_{ijk} \mathbf{e}_k, \qquad i,j,k = 1,2, \text{ or } 3. \tag{1.33}$$

Using this notation, we can write the vector product $\mathbf{B} \times \mathbf{C}$ as

$$\begin{aligned} \mathbf{B} \times \mathbf{C} &= \left(\sum_{m=1}^{3} B_m \mathbf{e}_m \right) \times \left(\sum_{n=1}^{3} C_n \mathbf{e}_n \right) \\ &= \sum_{m,n} B_m C_n (\mathbf{e}_m \times \mathbf{e}_n) \\ &= \sum_{m,n,j} (B_m C_n \varepsilon_{mnj}) \mathbf{e}_j. \end{aligned}$$

This shows that $\mathbf{B} \times \mathbf{C}$ has components

$$(\mathbf{B} \times \mathbf{C})_j = \sum_{m,n} B_m C_n \varepsilon_{mnj} = \sum_{m,n} \varepsilon_{jmn} B_m C_n, \tag{1.34}$$

because jmn is an even permutation of mnj. For example,

$$(\mathbf{B}\times\mathbf{C})_1 = \sum_{m,n} \varepsilon_{1mn}B_mC_n = \varepsilon_{123}B_2C_3 + \varepsilon_{132}B_3C_2$$
$$= B_2C_3 - B_3C_2,$$

in agreement with the usual geometrical result.

There are many different ways of writing $\mathbf{B}\times\mathbf{C}$ in this notation, since the different permutations of mnj are in a sense equivalent. The following expressions can be obtained with the help of Eq. (1.32) or by relabeling.

$$\mathbf{B}\times\mathbf{C} = \sum_{m,n,j} B_mC_n\mathbf{e}_j \begin{cases} \varepsilon_{mnj} \\ \varepsilon_{njm} \\ \varepsilon_{jmn} \\ (-)\varepsilon_{nmj} \\ (-)\varepsilon_{jnm} \\ (-)\varepsilon_{mjn} \end{cases} = \sum_{m,n,j} \varepsilon_{mnj} \begin{cases} B_mC_n\mathbf{e}_j \\ B_nC_j\mathbf{e}_m \\ B_jC_m\mathbf{e}_n \\ (-)B_nC_m\mathbf{e}_j \\ (-)B_jC_n\mathbf{e}_m \\ (-)B_mC_j\mathbf{e}_n \end{cases}, \tag{1.35}$$

where the second set of relations involves the permutation of indices in the $BC\mathbf{e}$ factor rather than in ε.

Why should we use such an abstract and obscure notation to perform an operation for which we already have a nice intuitive geometrical understanding? The answer has to do with successive vector products such as $\mathbf{A}\times[\mathbf{B}\times(\mathbf{C}\times\mathbf{D})]$. Equation (1.35) shows that each vector product gives rise to an ε symbol, so that an expression with n vector products is a sum of terms each involving n of these symbols. Such sums of products can often be simplified with the help of the following three reduction formulas:

$$\sum_{k=1}^{3} \varepsilon_{mnk}\varepsilon_{ijk} = \varepsilon_{mn1}\varepsilon_{ij1} + \varepsilon_{mn2}\varepsilon_{ij2} + \varepsilon_{mn3}\varepsilon_{ij3}$$
$$= \delta_{mi}\delta_{nj} - \delta_{mj}\delta_{ni}, \tag{1.36a}$$

$$\sum_{j,k} \varepsilon_{mjk}\varepsilon_{njk} = 2\delta_{mn}, \tag{1.36b}$$

$$\sum_{i,j,k} \varepsilon_{ijk}^2 = 6. \tag{1.36c}$$

The δ symbols on the right are Kronecker δ symbols.

These reduction formulas can be proved in the following way. Equation (1.36c) is just a sum of six contributions of 1 from each of the six permutations of 123. Equation (1.36b) gives zero if $m \neq n$, because then each term in the sum contains a permutation symbol with repeated indices. If $m = n$ ($=1$ say), there are only two nonzero terms ($jk = 23$ or 32), each with a contribution of 1. Equation (1.36a) gives zero if ij is not a permutation of mn, because then each term in the sum contains a permutation symbol with repeated indices. If ij is a permutation of mn, the permutation is even if $ij = mn$, and odd if $ij = nm$. These account for the two terms on the RHS. These arguments are somewhat abstract. The reader who is not sure what they mean should write out some or all of these sums explicitly. This will be left as an exercise (Problem 1.3.4).

Thus, under appropriate conditions, the ε symbols can be made to disappear two by two from the expression. Every time this occurs, the reduced expression contains two vector products fewer than before. In this way, a highly complicated expression can be simplified algebraically (i.e., painlessly). This reduction is illustrated by the following example:

Example 1.3.1. Prove the *BAC* rule for $\mathbf{A} \times (\mathbf{B} \times \mathbf{C})$.

$$\begin{aligned}
\mathbf{A} \times (\mathbf{B} \times \mathbf{C}) &= \Big(\sum_i A_i \mathbf{e}_i\Big) \times \Big(\sum_j (\mathbf{B} \times \mathbf{C})_j \mathbf{e}_j\Big), \qquad \text{using Eq. (1.34)} \\
&= \sum_{i,j,k} A_i (\mathbf{B} \times \mathbf{C})_j \varepsilon_{ijk} \mathbf{e}_k, \qquad \text{using Eq. (1.33)} \\
&= \sum_{i,j,k} A_i \Big(\sum_{m,n} B_m C_n \varepsilon_{mnj}\Big) \varepsilon_{ijk} \mathbf{e}_k, \qquad \text{using Eq. (1.34)} \\
&= -\sum_{i,k} \sum_{m,n} A_i B_m C_n \mathbf{e}_k (\delta_{mi}\delta_{nk} - \delta_{mk}\delta_{ni}), \qquad \text{using Eq. (1.36a).}
\end{aligned}$$

The overall negative sign in the last expression comes from writing ε_{ijk} as $-\varepsilon_{ikj}$. The sum over i and k (or m and n) can next be performed readily with the help of the expressions

$$\begin{aligned}
\sum_i A_i \delta_{mi} &= A_1 \delta_{m1} + A_2 \delta_{m2} + A_3 \delta_{m3} \\
&= A_m, \\
\sum_k \mathbf{e}_k \delta_{nk} &= \mathbf{e}_n,
\end{aligned}$$

since δ_{mi} vanishes unless $m = i$. This means that the summation simply picks up the term A_m or $\mathbf{e}_n$. Hence

$$\begin{aligned}
\mathbf{A} \times (\mathbf{B} \times \mathbf{C}) &= -\sum_{i,m,n} A_i B_m C_n (\delta_{mi} \mathbf{e}_n - \mathbf{e}_m \delta_{ni}) \\
&= -\sum_{m,n} B_m C_n (A_m \mathbf{e}_n - \mathbf{e}_m A_n) \\
&= -(\mathbf{A} \cdot \mathbf{B}) \sum_n C_n \mathbf{e}_n + \mathbf{B} \sum_n C_n A_n \\
&= \mathbf{B}(\mathbf{A} \cdot \mathbf{C}) - \mathbf{C}(\mathbf{A} \cdot \mathbf{B}), \qquad (1.37)
\end{aligned}$$

where the last step is made by rearranging a sum of 9 terms into a product of sums of 3 terms each

$$\begin{aligned}
-\sum_{m,n} A_m B_m C_n \mathbf{e}_n &= -\Big(\sum_n C_n \mathbf{e}_n\Big)\Big(\sum_m A_m B_m\Big) \\
&= -\mathbf{C}(\mathbf{A} \cdot \mathbf{B}).
\end{aligned}$$

(If you are not sure what this means, you should write out all nine terms explicitly by letting both m and n go through the values of 1, 2, and 3.)

The purely formal manipulations used in this algebraic method are not without disadvantages. Problems can arise because the calculation is too abstract or mechanical. As a result, the geometrical significance of the intermediate steps or of the final results is easily lost. It is usually a good policy to use the elementary geometrical method of Section 1.1 whenever possible, and to call upon the abstract algebraic method only to handle the really complicated expressions. When the expression is very complicated, the algebraic method is usually the only decent method of calculation.

Problems

1.3.1. Write the nine numbers ε_{1jk} as a 3×3 table (or matrix), with j labeling the rows and k labeling the columns. Do this for ε_{2jk} and ε_{3jk}.

1.3.2. Show that there are $n!$ permutations for n distinct objects.

1.3.3. Obtain all permutations of the four distinct objects a, b, c, and d. Obtain all permutations of the four objects a, a, c, and d.

1.3.4. Show that

$$\varepsilon_{ijk}\varepsilon_{mnl} = \begin{vmatrix} \delta_{im} & \delta_{in} & \delta_{il} \\ \delta_{jm} & \delta_{jn} & \delta_{jl} \\ \delta_{km} & \delta_{kn} & \delta_{kl} \end{vmatrix}$$

by noting the following:

(a) Both sides vanish if either permutation symbol has two or more indices in common.

(b) If both permutation symbols involve distinct indices, mnl can only be one of the six permutations of ijk.

1.3.5. Show that

$$\sum_{j,k} \varepsilon_{mjk}\varepsilon_{njk} = 2\delta_{mn},$$

$$\sum_{l} \varepsilon_{mnl}\varepsilon_{ijl} = \delta_{mi}\delta_{nj} - \delta_{mj}\delta_{ni}.$$

by using

(a) The result of Problem 1.3.4 and

(b) The arguments given in the text but expressed in greater detail.

For method (a), you will need the identity $\sum_l \delta_{ln}\delta_{jl} = \delta_{nj}$.

1.3.6. Show that $(\mathbf{A} \times \mathbf{B}) \cdot (\mathbf{C} \times \mathbf{D}) = (\mathbf{A} \cdot \mathbf{C})(\mathbf{B} \cdot \mathbf{D}) - (\mathbf{A} \cdot \mathbf{D})(\mathbf{B} \cdot \mathbf{C})$.

1.3.7. Show that $(\mathbf{A} \times \mathbf{B}) \times (\mathbf{C} \times \mathbf{D}) = (\mathbf{A} \cdot \mathbf{B} \times \mathbf{D})\mathbf{C} - (\mathbf{A} \cdot \mathbf{B} \times \mathbf{C})\mathbf{D} = (\mathbf{A} \cdot \mathbf{C} \times \mathbf{D})\mathbf{B} - (\mathbf{B} \cdot \mathbf{C} \times \mathbf{D})\mathbf{A}$ by using

(a) The *BAC* rule; and

(b) Permutation symbols.

1.4. Vector differentiation of a scalar field

Newton was the first scientist to appreciate the relationships between infinitely small changes in related physical properties, as well as the cumulative effects of an infinitely large number of such small changes. Indeed, his Method of Fluxions or generalized velocities (discovered in 1665–66, but published only in 1736, 9 years after his death) represents the first invention of calculus. Using this method, he established the law of universal gravitation and obtained from it Kepler's empirical laws on planetary motion.

However, it was the mathematician Leibniz, the co-inventor of calculus and the master of mathematical notations, who first published in 1684–86 an account of this new mathematical theory. Leibniz introduced the symbols we now use in calculus and in many other branches of mathematics, as well as the names *differential* and *integral calculus*. The first text book on calculus, published in 1696 by l'Hospital, already contained much of what is now undergraduate calculus.

It is now easy for use to take a scalar function of a single variable, such as the distance of fall of an apple as a function of time, and differentiate it successively to get its velocity, acceleration, etc. (A *scalar function* is one whose functional value is specified by a scalar, i.e., a single number. A scalar function may still be multivalued; the important feature is that each of these multiple values is a scalar.) It is almost as easy to deal with a scalar function of several variables. For example, the total differential of a smooth (i.e., differentiable) scalar function $\phi(s,t,u)$ of three independent variables s,t,u

$$d\phi(s,t,u) = \frac{\partial \phi}{\partial s} ds + \frac{\partial \phi}{\partial t} dt + \frac{\partial \phi}{\partial u} du$$

is expressible in terms of the partial derivatives.

1.4.1. Scalar field

A scalar function $\phi(x,y,z)$ of the position coordinates (x,y,z) is called a *scalar field*. That is, it is a rule for associating a scalar $\phi(x,y,z)$ with each point $\mathbf{r} = (x,y,z)$ in space. For this reason, it is also called a *scalar point function*. Scalar fields are of obvious importance in physics, which deals with events in space.

By a smooth scalar field, we mean a scalar field that is differentiable with respect to the position coordinates (x,y,z). The total differential corresponding to an infinitesimal change $d\mathbf{r} = (dx,dy,dz)$ in position is then

$$d\phi(x,y,z) = \frac{\partial \phi}{\partial x} dx + \frac{\partial \phi}{\partial y} dy + \frac{\partial \phi}{\partial z} dz. \tag{1.38}$$

It is useful to express $d\phi$ as a scalar product of two vectors:

$$d\phi(x,y,z) = [\nabla\phi(x,y,z)] \cdot d\mathbf{r}, \tag{1.39}$$

where

$$\nabla\phi(x,y,z) = \frac{\partial\phi}{\partial x}\mathbf{i} + \frac{\partial\phi}{\partial y}\mathbf{j} + \frac{\partial\phi}{\partial z}\mathbf{k} \equiv \nabla\phi(\mathbf{r}) \tag{1.40}$$

is a *vector field*, or vector point function. By this we mean that to each point $\mathbf{r}$ in space we associated a vector $\nabla\phi(x,y,z)$ as specified by its three components $(\partial\phi/\partial x, \partial\phi/\partial y, \partial\phi/\partial z)$.

1.4.2. The gradient operator ∇

The operation that changes a scalar field to a vector field in Eq. (1.40) has been denoted by the symbol

$$\nabla = \mathbf{i}\frac{\partial}{\partial x} + \mathbf{j}\frac{\partial}{\partial y} + \mathbf{k}\frac{\partial}{\partial z}, \tag{1.41}$$

called a *gradient operator*. The vector field $\nabla\phi(\mathbf{r})$ itself is called the *gradient* of the scalar field $\phi(\mathbf{r})$. The operator ∇ contains both partial differential operators and a direction, and is known as a *vector differential operator*. Both features are important in the generated vector field $\nabla\phi(\mathbf{r})$, which like other vectors has a length $|\nabla\phi(\mathbf{r})|$ and a direction $\mathbf{e}(\nabla\phi)$:

$$|\nabla\phi(\mathbf{r})| = \left[\left(\frac{\partial\phi}{\partial x}\right)^2 + \left(\frac{\partial\phi}{\partial y}\right)^2 + \left(\frac{\partial\phi}{\partial z}\right)^2\right]^{1/2}, \qquad \mathbf{e}(\nabla\phi) = \frac{\nabla\phi(\mathbf{r})}{|\nabla\phi(\mathbf{r})|}. \tag{1.42}$$

All calculations in the following examples are performed in rectangular coordinates.

Example 1.4.1. $\phi(\mathbf{r}) = \phi(r) = r^2 = x^2 + y^2 + z^2$:

$$\frac{\partial}{\partial x}r^2 = 2x, \qquad \frac{\partial}{\partial y}r^2 = 2y, \qquad \frac{\partial}{\partial z}r^2 = 2z,$$

$$\therefore\ \nabla r^2 = 2x\mathbf{i} + 2y\mathbf{j} + 2z\mathbf{k} = 2\mathbf{r}.$$

Thus

$$|\nabla r^2| = 2r, \qquad \mathbf{e}(\nabla r^2) = \mathbf{e}_r. \tag{1.43}$$

Example 1.4.2. $\phi(\mathbf{r}) = r$:

$$\frac{\partial}{\partial x}r^2 = \frac{\partial r}{\partial x}\frac{d}{dr}r^2 = 2r\frac{\partial r}{\partial x}, \qquad \text{(chain rule)}$$

$$\therefore\ \nabla r = \frac{1}{2r}\nabla r^2 = \mathbf{e}_r.$$

Thus ∇r is a unit vector everywhere.

Example 1.4.3. $\phi(\mathbf{r}) = f(r)$:

$$\nabla f(r) = \frac{df(r)}{dr}(\nabla r)$$
$$= \frac{df}{dr}\mathbf{e}_r. \tag{1.44}$$

Example 1.4.4. $\phi(\mathbf{r}) = \exp(-ar^2)$:

$$\nabla[\exp(-ar^2)] = \frac{d}{dr}\exp(-ar^2)(\nabla r)$$
$$= \mathbf{e}_r(-2ar)\exp(-ar^2).$$

Example 1.4.5. $\phi(x,y) = xy$ in a two-dimensional space. $\nabla\phi(x,y) = \mathbf{i}y + \mathbf{j}x$. Hence

$$|\nabla\phi| = (y^2 + x^2)^{1/2} = r, \qquad \mathbf{e}(\nabla\phi) = \mathbf{i}\frac{y}{r} + \mathbf{j}\frac{x}{r}.$$

It is clear from Eq. (1.40) that $\nabla\phi(\mathbf{r})$ at the point $\mathbf{r}$ is a vector. So is $d\mathbf{r}$. Hence

$$d\phi(\mathbf{r}) = [\nabla\phi(\mathbf{r})]\cdot d\mathbf{r} = |\nabla\phi|\, dr[\mathbf{e}(\nabla\phi)\cdot\mathbf{e}(d\mathbf{r})]$$
$$= |\nabla\phi|\, dr\cos\theta, \tag{1.45}$$

where $dr = |d\mathbf{r}|$. If the displacement $d\mathbf{r}$ in three-dimensional space lies on a surface on which $\phi(\mathbf{r})$ is a constant (say C), then

$$d\phi(\mathbf{r}) = \phi(\mathbf{r} - d\mathbf{r}) - \phi(\mathbf{r}) = C - C = 0.$$

Since both $|d\mathbf{r}|$ and $|\nabla\phi|$ are not necessarily zero, we must conclude that in general

$$\cos\theta = 0, \quad \text{or} \quad \theta = \pi/2, \qquad \text{when} \quad d\phi = 0.$$

Thus, $\mathbf{e}(\nabla\phi)$ must be perpendicular to any $d\mathbf{r}$ on a surface of constant ϕ, that is, $\mathbf{e}(\nabla\phi)$ is normal to this surface.

Some of the most common scalar fields in physics are potential fields, which give potential energies of systems in space. As a result, the surfaces of constant ϕ in a three-dimensional space are often called *equipotential surfaces*. In two-dimensional spaces, the equipotentials are lines. A good example of the latter is a *contour line* on a topological map, giving the elevation (or gravitational potential) of a point on the earth's surface.

Example 1.4.6. If $\phi(x,y) = axy$ gives the elevation (in m) on a topological map, the "equipotential" condition is $\phi = axy$, or $y = \phi/ax$, where a is a constant (in units of m^{-1}). The contour lines are the hyperbolas shown in Fig. 1.5, where $a = 10^{-4}\,\mathrm{m}^{-1}$ is used. The arrows point to the valleys on both sides of the mountain pass.

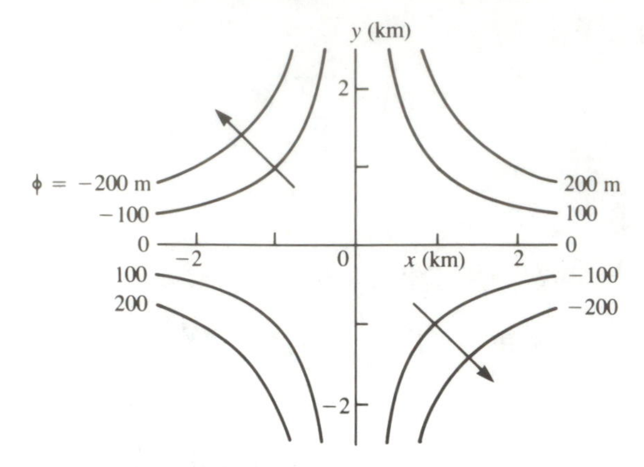

Fig. 1.5. The contour lines of a mountain pass or "saddle."

We have seen that $\nabla\phi(\mathbf{r})$ is everywhere perpendicular to the equipotential passing through $\mathbf{r}$. If $\phi(\mathbf{r})$ is a gravitational potential,

$$F(\mathbf{r}) = -m\nabla\phi(\mathbf{r})$$

gives the gravitational force on a point mass m. Hence forces are also perpendicular to equipotentials. A mass will be accelerated along $\mathbf{e}(F(\mathbf{r}))$ at the point $\mathbf{r}$. A slowly moving mass will therefore "flow" along a *flow curve,* or a *line of force,* whose tangent at $\mathbf{r}$ gives the direction $\mathbf{e}(F(\mathbf{r}))$ of the force acting on it. Flow curves are shown as rivers on a map.

The change $d\phi$ shown in Eq. (1.45) depends also on the direction $\mathbf{e}(d\mathbf{r})$ of the displacement. If $d\mathbf{r}$ is along a flow curve in the "uphill" direction $\mathbf{e}(\nabla\phi)$, we find $\mathbf{e}(d\mathbf{r}) = \mathbf{e}(\nabla\phi)$, and $\cos\theta = 1$. The change $d\phi$ then takes on its maximal value

$$(d\phi)_{\max} = |\nabla\phi|\, dr.$$

This means that

$$|\nabla\phi| = \frac{(d\phi)_{\max}}{dr}.$$

Thus at the point $\mathbf{r}$, the scalar field $\phi(\mathbf{r})$ changes most rapidly along $\mathbf{e}(\nabla\phi)$ with the maximal slope of $|\nabla\phi|$. This direction is perpendicular to the equipotential and is antiparallel to the flow curve, or line force, passing through $\mathbf{r}$.

Example 1.4.7. Calculate the directional derivative of the scalar field $\phi(\mathbf{r}) = x^2 + y^2 + z^2$ along the direction of the vector $\mathbf{A} = \mathbf{i} + 2\mathbf{j} + 3\mathbf{k}$ at the point $\mathbf{r} = (2,0,0)$.

By a directional derivative we mean

$$\frac{d\phi}{ds} = \nabla\phi \cdot \mathbf{e}(\mathbf{A}),$$

that is, the rate of change of ϕ along the direction $\mathbf{e}(\mathbf{A})$. Hence

$$\frac{d\phi}{ds} = (2x\mathbf{i} + 2y\mathbf{j} + 2z\mathbf{k}) \cdot (\mathbf{i} + 2\mathbf{j} + 3\mathbf{k})/\sqrt{14}$$
$$= (2x + 4y + 6z)/\sqrt{14}.$$

This gives a value of $4/\sqrt{14}$ at $\mathbf{r} = (2,0,0)$. By the way, the equipotentials of ϕ are spherical surfaces.

1.4.3. The operator ∇^2

The scalar product

$$\nabla^2 = \nabla \cdot \nabla = \frac{\partial^2}{\partial x^2} + \frac{\partial^2}{\partial y^2} + \frac{\partial^2}{\partial z^2} \tag{1.46}$$

is a scalar differential operator called the *Laplacian* named after Laplace, a French mathematician of the eighteenth century. Being a scalar, it will not change the vectorial character of the field on which it operates. Thus $\nabla^2\phi(\mathbf{r})$ is a scalar field if $\phi(\mathbf{r})$ is a scalar field, while $\nabla^2[\nabla\phi(\mathbf{r})]$ is a vector field because $\nabla\phi(\mathbf{r})$ is a vector field.

The Laplacian is also unchanged when the sign of one or more of the coordinates x,y,z are changed. A function $\phi(\mathbf{r})$ that is even (or odd) under one of these space reflection operations ($x \leftrightarrow -x$, etc.) will remain even (or odd) after the ∇^2 operation. The evenness or oddness of a function or operator under a sign change of its variable is called its *parity*. Hence the ∇^2 operator is said to have even parity.

∇^2 is the simplest differential operator that will not change the vectorial and parity properties of a field on which it operates. These are some of the reasons why ∇^2 appears so frequently in the equations of motion or the equations of state of physical systems.

Example 1.4.8.

$$\nabla^2 r^2 = \left(\frac{\partial^2}{\partial x^2} + \frac{\partial^2}{\partial y^2} + \frac{\partial^2}{\partial z^2}\right) r^2 = 6,$$

since $(\partial^2/\partial x^2)r^2 = 2$.

Problems

1.4.1. Show that the vector field $\mathbf{A}(\mathbf{r}) = \mathbf{r} \times \nabla\phi(\mathbf{r})$ is orthogonal to both $\mathbf{r}$ and $\nabla\phi(\mathbf{r})$:

$$\mathbf{A}(\mathbf{r}) \cdot \mathbf{r} = 0, \qquad \mathbf{A}(\mathbf{r}) \cdot \nabla\phi(\mathbf{r}) = 0.$$

1.4.2. If $r^2 = x^2 + y^2 + z^2$, calculate
(a) $\nabla(\ln r)$,
(b) $\nabla(3z^2 - r^2)$,
(c) $\nabla^2(\ln r)$, and
(d) $\nabla^2(3z^2 - r^2)$.

1.4.3. If $\phi(x,y,z) = [(x-1)^2 + y^2 + z^2]^{-1/2} + [(x+1)^2 + y^2 + z^2]^{-1/2}$, calculate $\nabla\phi$ at $\mathbf{r} = (x,y,z) = (1,1,1)$.

1.4.4. If $\phi(\mathbf{r})$ and $\psi(\mathbf{r})$ are two scalar fields such that $\nabla\phi(\mathbf{r}) \times \nabla\psi(\mathbf{r}) = 0$ over all space, how are their equipotential surfaces and lines of force related?

1.4.5. Show that the equation $x + y + z = 3$ is satisfied by any point $\mathbf{r} = (x,y,z)$ on a plane tangent to a sphere at the point (1,1,1) on the surface of the sphere, if the sphere has radius $\sqrt{3}$ and is centered at the origin. [Hint: consider the scalar field $\phi(x,y,z) = x + y + z$.]

1.4.6. Show that the point $(0,1,\sqrt{2})$ lies on the curve of intersection of the surfaces defined by the equations

$$x^2 + y^2 + z^2 = 3$$
$$2x + 2y + z^2 = 4.$$

Calculate the angle between the normals to these surfaces at this point.

1.4.7. Calculate the gravitational potential due to a homogeneous cylinder of density ρ, radius a, and height h, at exterior points along its axis. Calculate the corresponding gravitational force field.

1.5. Vector differentiation of a vector field

A vector function $\mathbf{F}(t)$ of a single variable t is made up of three components $F_x(t), F_y(t), F_z(t)$, each of which is a scalar function of t. Similarly, a vector function of several variables is made up of components, each of which is a scalar function of these variables.

Of particular interest in physics are vector functions of the position vector $\mathbf{r} = (x,y,z)$, or vector fields. A vector field

$$\mathbf{V}(\mathbf{r}) = (V_x(x,y,z), V_y(x,y,z), V_z(x,y,z))$$

is simply a rule for assigning the values $V_i(x,y,z)$, $i = x,y,z$, to the components of the vector at $\mathbf{r}$. A vector field is differentiable, if all the partial derivatives of its components exist.

Examples of vector fields are $\nabla\phi(\mathbf{r})$, where $\phi(\mathbf{r})$ is a scalar field, and a force field $\mathbf{F}(\mathbf{r})$, which defines a force $\mathbf{F}$ at every point $\mathbf{r}$ in space.

Vector differential operations on vector fields are more complicated because of the vectorial nature of both the operator and the field on which it operates. Since there are two types of products involving two vectors, namely the scalar product and the vector (i.e., cross) product, vector differential operations on vector fields can also be separated into two types called the *divergence* and the *curl.*

The *divergence* of a vector field $\mathbf{V}(\mathbf{r})$ is defined by the scalar product

$$\begin{aligned}\nabla \cdot \mathbf{V}(\mathbf{r}) &= \left(\mathbf{i}\frac{\partial}{\partial x} + \mathbf{j}\frac{\partial}{\partial y} + \mathbf{k}\frac{\partial}{\partial z}\right) \cdot (\mathbf{i}V_x + \mathbf{j}V_y + \mathbf{k}V_z) \\ &= \frac{\partial V_x}{\partial x} + \frac{\partial V_y}{\partial y} + \frac{\partial V_z}{\partial z}. \end{aligned} \tag{1.47}$$

The result is a scalar field. It differs from the scalar differential operator

$$\mathbf{V}(\mathbf{r}) \cdot \nabla = V_x \frac{\partial}{\partial x} + V_y \frac{\partial}{\partial y} + V_z \frac{\partial}{\partial z}, \tag{1.48}$$

as one can see in the following example.

Example 1.5.1.

$$\mathbf{r}(\nabla \cdot \mathbf{r}) = \mathbf{r}\left(\frac{\partial x}{\partial x} + \frac{\partial y}{\partial y} + \frac{\partial z}{\partial z}\right) = 3\mathbf{r},$$

$$\begin{aligned}(\mathbf{r} \cdot \nabla)\mathbf{r} &= \left(x\frac{\partial}{\partial x} + y\frac{\partial}{\partial y} + z\frac{\partial}{\partial z}\right)(\mathbf{i}x + \mathbf{j}y + \mathbf{k}z) = \mathbf{r} \\ &\neq \mathbf{r}(\nabla \cdot \mathbf{r}).\end{aligned}$$

The *curl* (or *rotation*) of a vector field $\mathbf{V}(\mathbf{r})$ is defined by the vector product

$$\begin{aligned}\nabla \times \mathbf{V}(\mathbf{r}) &= \begin{vmatrix} \mathbf{i} & \mathbf{j} & \mathbf{k} \\ \frac{\partial}{\partial x} & \frac{\partial}{\partial y} & \frac{\partial}{\partial z} \\ V_x & V_y & V_z \end{vmatrix} \\ &= \mathbf{i}\left(\frac{\partial V_z}{\partial y} - \frac{\partial V_y}{\partial z}\right) + \mathbf{j}\left(\frac{\partial V_x}{\partial z} - \frac{\partial V_z}{\partial x}\right) + \mathbf{k}\left(\frac{\partial V_y}{\partial x} - \frac{\partial V_x}{\partial y}\right) \\ &= \sum_{m,n,l} \varepsilon_{mnl}\mathbf{e}_m \frac{\partial}{\partial x_n} V_l. \end{aligned} \tag{1.49}$$

The result is a vector field, not the vector differential operator

$$\mathbf{V}(\mathbf{r}) \times \nabla = \sum_{m,n,l} \varepsilon_{mnl}\mathbf{e}_m V_n \frac{\partial}{\partial x_l}.$$

Example 1.5.2. $\mathbf{e}_1 \times (\nabla \times \mathbf{r}) = 0$, because

$$\nabla \times \mathbf{r} = \begin{vmatrix} \mathbf{i} & \mathbf{j} & \mathbf{k} \\ \frac{\partial}{\partial x} & \frac{\partial}{\partial y} & \frac{\partial}{\partial z} \\ x & y & z \end{vmatrix} = 0.$$

However,

$$(\mathbf{e}_1 \times \boldsymbol{\nabla}) \times \mathbf{r} = \begin{vmatrix} \mathbf{i} & \mathbf{j} & \mathbf{k} \\ 1 & 0 & 0 \\ \dfrac{\partial}{\partial x} & \dfrac{\partial}{\partial y} & \dfrac{\partial}{\partial z} \end{vmatrix} \times \mathbf{r} = \left(-\mathbf{j}\frac{\partial}{\partial z} + \mathbf{k}\frac{\partial}{\partial y}\right) \times \mathbf{r}$$

$$= -2\mathbf{i} \neq \mathbf{e}_1 \times (\boldsymbol{\nabla} \times \mathbf{r}).$$

The differences come from the fact that both differential operations and vector products depend on the positions of the objects in the expression.

Some of the most interesting vector fields in physics satisfy the property that the results of such vector differential operations vanish everywhere in space. Such special fields are given special names:

If $\boldsymbol{\nabla} \cdot \mathbf{V}(\mathbf{r}) = 0$, $\mathbf{V}$ is said to be *solenoidal* (or divergence-free).
If $\boldsymbol{\nabla} \times \mathbf{V}(\mathbf{r}) = 0$, $\mathbf{V}$ is said to be *irrotational.*

Example 1.5.3. Show that the gradient of any scalar field $\phi(\mathbf{r})$ is irrotational and that the curl of any vector field $\mathbf{V}(\mathbf{r})$ is solenoidal.

$$\boldsymbol{\nabla} \times (\boldsymbol{\nabla}\phi) = \begin{vmatrix} \mathbf{i} & \mathbf{j} & \mathbf{k} \\ \dfrac{\partial}{\partial x} & \dfrac{\partial}{\partial y} & \dfrac{\partial}{\partial z} \\ \dfrac{\partial}{\partial x} & \dfrac{\partial}{\partial y} & \dfrac{\partial}{\partial z} \end{vmatrix} \phi(x,y,z) = 0 \tag{1.50}$$

because there are two identical rows in the determinant. Alternatively,

$$\boldsymbol{\nabla} \times (\boldsymbol{\nabla}\phi) = \sum_{i,j,k} \varepsilon_{ijk}\mathbf{e}_i \frac{\partial}{\partial x_j}\frac{\partial}{\partial x_k}\phi(x,y,z) = 0,$$

because ε_{ijk} is antisymmetric in j,k, while $\partial^2/\partial x_j\,\partial x_k$ is symmetric. Thus each term in the sum is always cancelled by another term:

$$\varepsilon_{ijk}\frac{\partial}{\partial x_j}\frac{\partial}{\partial x_k} + \varepsilon_{ikj}\frac{\partial}{\partial x_k}\frac{\partial}{\partial x_j} = 0.$$

Similarly,

$$\boldsymbol{\nabla} \cdot (\boldsymbol{\nabla} \times \mathbf{V}) = \sum_i \frac{\partial}{\partial x_i}(\boldsymbol{\nabla} \times \mathbf{V})_i$$

$$= \sum_i \frac{\partial}{\partial x_i}\left(\sum_{j,k} \varepsilon_{ijk}\frac{\partial}{\partial x_j} v_k\right) = 0, \tag{1.51}$$

because ε_{ijk} is antisymmetric in i,j. The physical interpretation of these results will be given in the next two sections.

Finally, both the scalar operator ∇^2 and the vector operator $\boldsymbol{\nabla}$ can operate on a vector field $\mathbf{V}(\mathbf{r})$:

$$\nabla^2\mathbf{V}(\mathbf{r}) = \mathbf{i}(\nabla^2 V_x) + \mathbf{j}(\nabla^2 V_y) + \mathbf{k}(\nabla^2 V_z) \tag{1.52}$$

is obviously a vector field, while

$$\begin{aligned}\boldsymbol{\nabla}\mathbf{V}(\mathbf{R}) &= \mathbf{i}(\boldsymbol{\nabla}V_x) + \mathbf{j}(\boldsymbol{\nabla}V_y) + \mathbf{k}(\boldsymbol{\nabla}V_z)\\ &= \mathbf{i}\left(\mathbf{i}\frac{\partial V_x}{\partial x} + \mathbf{j}\frac{\partial V_x}{\partial y} + \mathbf{k}\frac{\partial V_x}{\partial z}\right)\\ &\quad + \mathbf{j}\left(\mathbf{i}\frac{\partial V_y}{\partial x} + \mathbf{j}\frac{\partial V_y}{\partial y} + \mathbf{k}\frac{\partial V_y}{\partial z}\right)\\ &\quad + \mathbf{k}\left(\mathbf{i}\frac{\partial V_z}{\partial x} + \mathbf{j}\frac{\partial V_z}{\partial y} + \mathbf{k}\frac{\partial V_z}{\partial z}\right)\end{aligned} \tag{1.53}$$

is a more complicated mathematical object known as a *dyadic field.* (A *dyad* or *dyadic* is a bilinear combination of vectors. In three-dimensional space it has nine components associated with the nine unit dyads $\mathbf{e}_i\mathbf{e}_j$.)

1.5.1. Using vector differential operators

In dealing with vector differential operators (VDO), it is important to remember that they are both differential operators and vectors. For example, the ordering of factors is important in differentiation:

$$f\frac{\partial}{\partial x}(gh) = fh\left(\frac{\partial g}{\partial x}\right) + fg\left(\frac{\partial h}{\partial x}\right) \neq \frac{\partial}{\partial x}(fgh). \tag{1.54}$$

The bookkeeping of vector components could be quite unpleasant if one used the cumbersome determinantal notation for vector products. In all but the simplest cases, the use of permutation symbols is far more convenient.

Example 1.5.4. If $\phi(\mathbf{r})$ is a scalar field and $\mathbf{V}(\mathbf{r})$ is a vector field, show that

$$\boldsymbol{\nabla}\times(\phi\mathbf{V}) = \phi(\boldsymbol{\nabla}\times\mathbf{V}) + (\boldsymbol{\nabla}\phi)\times\mathbf{V}.$$

We first note that both sides of the equation are vector fields.

Method 1.

$$\boldsymbol{\nabla}\times(\phi\mathbf{V}) = \begin{vmatrix}\mathbf{i} & \mathbf{j} & \mathbf{k}\\ \frac{\partial}{\partial x} & \frac{\partial}{\partial y} & \frac{\partial}{\partial z}\\ \phi V_x & \phi V_y & \phi V_z\end{vmatrix} = \phi\begin{vmatrix}\mathbf{i} & \mathbf{j} & \mathbf{k}\\ \frac{\partial}{\partial x} & \frac{\partial}{\partial y} & \frac{\partial}{\partial z}\\ V_x & V_y & V_z\end{vmatrix} + \begin{vmatrix}\mathbf{i} & \mathbf{j} & \mathbf{k}\\ \frac{\partial\phi}{\partial x} & \frac{\partial\phi}{\partial y} & \frac{\partial\phi}{\partial z}\\ V_x & V_y & V_z\end{vmatrix}$$

where the last step has been made with the help of identities of the form

$$\frac{\partial}{\partial x}(\phi V_y) = \phi \frac{\partial V_y}{\partial x} + \left(\frac{\partial \phi}{\partial x}\right) V_y.$$

Method 2.

$$\begin{aligned}\nabla \times (\phi \mathbf{V}) &= \sum_{i,j,k} \varepsilon_{ijk} \mathbf{e}_i \frac{\partial}{\partial x_j}(\phi V_k) \\ &= \phi \sum_{i,j,k} \varepsilon_{ijk} \mathbf{e}_i \frac{\partial}{\partial x_j} V_k + \sum_{i,j,k} \varepsilon_{ijk} \mathbf{e}_i \left(\frac{\partial \phi}{\partial x_j}\right) V_k \\ &= \phi(\nabla \times \mathbf{V}) + (\nabla \phi) \times \mathbf{V}. \end{aligned} \tag{1.55}$$

Other useful relations involving vector differential operations on products of fields are

$$\nabla \cdot (\phi \mathbf{V}) = \phi(\nabla \cdot \mathbf{V}) + (\nabla \phi) \cdot \mathbf{V}, \tag{1.56}$$

$$\nabla \cdot (\mathbf{A} \times \mathbf{B}) = \mathbf{B} \cdot (\nabla \times \mathbf{A}) - \mathbf{A} \cdot (\nabla \times \mathbf{B}), \tag{1.57}$$

$$\nabla \times (\nabla \times \mathbf{A}) = \nabla(\nabla \cdot \mathbf{A}) - \nabla^2 \mathbf{A}. \tag{1.58}$$

Example 1.5.5.

$$\begin{aligned}\nabla^2 f(r) &= \nabla \cdot \nabla f(r) = \nabla \cdot \left(\mathbf{e_r} \frac{df(r)}{dr}\right) \\ &= \mathbf{e_r} \cdot \left(\nabla \frac{df(r)}{dr}\right) + \frac{df(r)}{dr} \nabla \cdot \mathbf{e_r}.\end{aligned}$$

Now according to Eq. (1.44)

$$\nabla\left(\frac{df}{dr}\right) = \frac{d^2 f}{dr^2} \mathbf{e}_r.$$

Therefore the first term is $d^2 f/dr^2$. The second term requires the calculation of

$$\begin{aligned}\nabla \cdot \mathbf{e_r} &= \nabla \cdot \left(\frac{\mathbf{r}}{r}\right) = \frac{1}{r} \nabla \cdot \mathbf{r} + \mathbf{r} \cdot \left(\nabla \frac{1}{r}\right) \\ &= \frac{3}{r} + \mathbf{r} \cdot \mathbf{e_r}\left(-\frac{1}{r^2}\right) = \frac{2}{r}.\end{aligned}$$

Hence

$$\nabla^2 f(r) = \frac{d^2}{dr^2} f(r) + \frac{2}{r} \frac{d}{dr} f(r). \tag{1.59}$$

Example 1.5.6. Show that if $\mathbf{B}(\mathbf{r})$ and $\mathbf{C}(\mathbf{r})$ are vector fields, then

$$\nabla \times (\mathbf{B} \times \mathbf{C}) = \mathbf{B}(\nabla \cdot \mathbf{C}) + (\mathbf{C} \cdot \nabla)\mathbf{B} - [\mathbf{C}(\nabla \cdot \mathbf{B}) + (\mathbf{B} \cdot \nabla)\mathbf{C}]. \tag{1.60}$$

Method 1.

$$\nabla \times (\mathbf{B} \times \mathbf{C}) = \sum_{i,j,k} \varepsilon_{ijk} \mathbf{e}_i \frac{\partial}{\partial x_j} (\mathbf{B} \times \mathbf{C})_k, \qquad \text{using Eq. (1.34)}$$

$$= \sum_{i,j,k} \varepsilon_{ijk} \mathbf{e}_i \frac{\partial}{\partial x_j} \left(\sum_{m,n} \varepsilon_{kmn} B_m C_n \right)$$

$$= \sum_{i,j,m,n} \left(\sum_k \varepsilon_{ijk} \varepsilon_{kmn} \right) \mathbf{e}_i \frac{\partial}{\partial x_j} (B_m C_n)$$

$$= \sum_{i,j,m,n} (\delta_{im}\delta_{jn} - \delta_{in}\delta_{jm}) \mathbf{e}_i \frac{\partial}{\partial x_j} (B_m C_n), \qquad \text{using Eqs. (1.32) and (1.36a)}$$

$$= \sum_{i,j} \mathbf{e}_i \frac{\partial}{\partial x_j} (B_i C_j - B_j C_i)$$

$$= \sum_{i,j} \left(\mathbf{e}_i B_i \frac{\partial C_j}{\partial x_j} + C_j \frac{\partial B_i}{\partial x_j} \mathbf{e}_i - \mathbf{e}_i C_i \frac{\partial B_j}{\partial x_j} - B_j \frac{\partial C_i}{\partial x_j} \mathbf{e}_i \right). \tag{1.61}$$

These four terms correspond exactly to the four terms of Eq. (1.60). We note that, when permutation symbols are used, the proof proceeds in a very straightforward and orderly manner.

Method 2. Many of the steps used to obtain Eq. (1.61) are already contained in the BAC rule of Eq. (1.28). The only novel feature is that $\mathbf{A}$ is now the vector differential operator ∇ here placed to the left of the vector fields $\mathbf{B}(\mathbf{r})$ and $\mathbf{C}(\mathbf{r})$. The chain rule of differential calculus, Eq. (1.54), now requires the replacements

$$\mathbf{B}(\mathbf{A} \cdot \mathbf{C}) = (\mathbf{C} \cdot \mathbf{A})\mathbf{B} \rightarrow \mathbf{B}(\nabla \cdot \mathbf{C}) + (\mathbf{C} \cdot \nabla)\mathbf{B}$$

$$-\mathbf{C}(\mathbf{A} \cdot \mathbf{B}) = -(\mathbf{B} \cdot \mathbf{A})\mathbf{C} \rightarrow -\mathbf{C}(\nabla \cdot \mathbf{B}) - (\mathbf{B} \cdot \nabla)\mathbf{C},$$

with two terms appearing for each term of the BAC rule. Thus Eq. (1.60) obtains. We shall refer to this procedure as the *operator form* of the BAC rule.

Problems

1.5.1. Derive the following identities:
(a) $\nabla \cdot (\mathbf{A} \times \mathbf{B}) = \mathbf{B} \cdot (\nabla \times \mathbf{A}) - \mathbf{A} \cdot (\nabla \times \mathbf{B})$;
(b) $\nabla \times (\nabla \times \mathbf{A}) = \nabla(\nabla \cdot \mathbf{A}) - \nabla^2 \mathbf{A}$;
(c) $\nabla(\mathbf{A} \cdot \mathbf{B}) = (\mathbf{B} \cdot \nabla)\mathbf{A} + (\mathbf{A} \cdot \nabla)\mathbf{B} + \mathbf{B} \times (\nabla \times \mathbf{A}) + \mathbf{A} \times (\nabla \times \mathbf{B})$.

1.5.2. If $\mathbf{A}(\mathbf{r})$ is irrotational, show that $\mathbf{A} \times \mathbf{r}$ is solenoidal.

1.5.3. A rigid body is rotating with constant angular velocity $\boldsymbol{\omega}$ about an axis that passes through the origin. Show that the velocity $\mathbf{v}$ of any point $\mathbf{r}$ in the rigid body is given by the equation

$$\mathbf{v} = \boldsymbol{\omega} \times \mathbf{r}.$$

Show that

(a) $\mathbf{v}$ is solenoidal;

(b) $\nabla \times \mathbf{v} = 2\boldsymbol{\omega}$; and

(c) $\nabla(\boldsymbol{\omega} \cdot \mathbf{r}) = \boldsymbol{\omega}$.

1.5.4. Calculate the divergence of the inverse-power force $\mathbf{F} = k\mathbf{e}_r/r^n$ in rectangular coordinates. Show that for $n = 2$ the force is solenoidal, except at $\mathbf{r} = 0$. Show that all these forces are irrotational.

1.5.5. If $\mathbf{B}(\mathbf{r})$ is both irrotational and solenoidal, show that for a constant vector $\mathbf{m}$

$$\nabla \times (\mathbf{B} \times \mathbf{m}) = \nabla(\mathbf{B} \cdot \mathbf{m}).$$

1.5.6. Use rectangular coordinates to show that $\nabla \cdot \mathbf{e}_r = 2/r$ and $\nabla \times \mathbf{e}_r = 0$.

1.6. Path-dependent scalar and vector integrations

Integrals are easy to visualize when the number of integration variables equals the number of variables in the integrand function. For example, the integrals

$$\int_a^b f(x)\,dx, \qquad \int_{\text{volume }\Omega} \phi(x,y,z)\,dx\,dy\,dz, \qquad \int_{\text{volume }\Omega} \mathbf{V}(x,y,z)\,dx\,dy\,dz$$

give a scalar constant, a scalar constant, and a vector constant, respectively.

Integrals can also be defined when the number of integration variables is less than the number of variables in the integrand, but the situation is more complicated. For example, the integral

$$I = \int_{(x_1,y_1)}^{(x_2,y_2)} \phi(x,y)\,dx$$

is not yet completely defined because we do not know the value of y in $\phi(x,y)$. What is needed in addition is a statement such as

$$y = y_c(x) \tag{1.62}$$

that specifies y for each value of x. The integrand then reduces to

$$\phi(x, y_c(x)) = f_c(x),$$

so that the integral becomes well defined:

$$\int_{x_1}^{x_2} f_c(x)\,dx = \int_c \phi(x,y)\,dx, \tag{1.63a}$$

where c refers to the constraint in Eq. (1.62).

The constraint [Eq. (1.62)] specifies a path c on the xy plane connecting the starting point (x_1,y_1) to the ending point (x_2,y_2). The x integration in Eq. (1.63a) is carried out along this path. For this reason we call I_c a *path-dependent* integral, or simply path integral. It is clear from their definition that both $f_c(x)$ and I_c can be expected to change when the path is changed even though the endpoints on the xy plane remain the same.

Example 1.6.1. Integrate $\int_{(0,0)}^{(1,1)} (x^2 + y^2)\, dx$ along the straight line (path 1) and the circular arc of radius 1 (path 2) shown in Fig. 1.6.

Path 1: $y = x$:

$$I_1 = \int_0^1 2x^2\, dx = \tfrac{2}{3}.$$

Path 2: $(x-1)^2 + y^2 = 1$:

$$\begin{aligned} I_2 &= \int_0^1 \{x^2 + [1 - (x-1)^2]\}\, dx \\ &= \int_0^1 2x\, dx = 1 \neq I_1. \end{aligned}$$

In a similar way, the line integral

$$I_c = \int_{(x_1,y_1,z_1)}^{(x_2,y_2,z_2)} \phi(x,y\, z)\, dx \tag{1.63b}$$

of a scalar field in space is a one-dimensional integral over a path c in three-dimensional space. It gives a single number since, along the path c, $y = y_c(x)$, $z = z_c(x)$; hence $\phi(x, y_c(x), z_c(x)) = f_c(x)$.

Example 1.6.2. The line integral

$$I_c = \int_{(0,0,0)}^{(1,1,1)} (x^2 + y^2 + z^2)\, dx$$

gives different results along the following paths:

1. Along the straight line connecting (0,0,0) and (1,1,1), we find $y = x$ and $z = x$. Hence

$$I_c = \int_0^1 3x^2\, dx = 1.$$

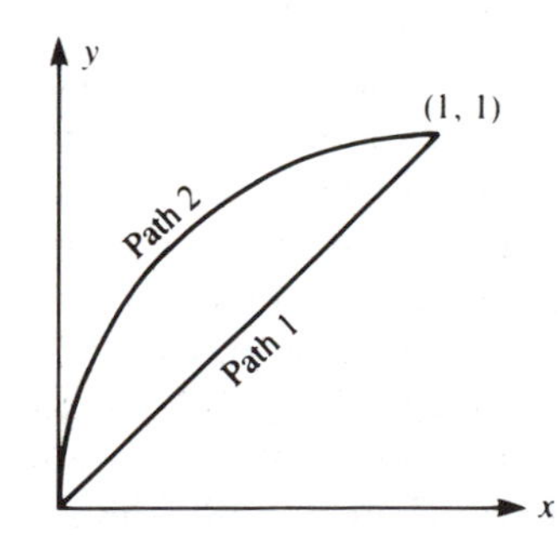

Fig. 1.6. Different paths for a path integral.

2. Along the following three edges of the cube (α) (0,0,0) to (1,0,0), (β) (1,0,0) to (1,1,0), and (γ) (1,1,0) to (1,1,1), we find $I_c = I_\alpha + I_\beta + I_\gamma = \frac{1}{3}$, because $I_\alpha = \int_0^1 x^2\,dx = \frac{1}{3}$,

$$I_\beta = \int_{(1,0,0)}^{(1,1,0)} (x^2 + y^2 + z^2)\,dx = 0,$$

$$I_\gamma = \int_{(1,1,0)}^{(1,1,1)} (x^2 + y^2 + z^2)\,dx = 0.$$

The above discussion also makes clear that the two-dimensional integral

$$I = \iint \phi(x,y,z)\,dx\,dy$$

of a scalar field in space cannot lead to a unique number unless a surface S has been defined on which $z = z_S(x,y)$ is known. In that case, the integrand becomes

$$\phi(x,y,z) = \phi(x,y,z_S(x,y)) = g_S(x,y).$$

The resulting integral then yields a single number, and is called a *surface integral.*

Example 1.6.3. the surface integral

$$I_S = \iint_S (x^2 + y^2 + z^2)\,dx\,dy$$

gives different results on the following surfaces:

1. Inside the circle $x^2 + y^2 = 1$ on the $z = 0$ plane:

$$I_S = \iint_{\text{circle}} (x^2 + y^2)\,dx\,dy = 2\pi \int_0^1 \rho^3\,d\rho = \frac{\pi}{2}.$$

2. On the surface of the hemisphere $x^2 + y^2 + z^2 = 1$ with $z \geqslant 0$:

$$I_S = \iint_{\text{hemisphere}} dx\,dy = \iint_{\text{circle}} dx\,dy = 2\pi \int_0^1 \rho\,d\rho = \pi.$$

We thus see that the specification of a path (in the case of a line integral) or a surface (in the case of a surface integral) gives us a rule for the elimination of undesirable variables (or variable) in favor of the integration variable (or variables). Such an elimination or substitution is conceptually very simple and elementary; there is really nothing interesting in the procedure itself. It is the scalar field involved that turns out to be interesting, because certain integrals of related scalar fields can be shown to be equal to each other. We call these formal mathematical relations *integral theorems.* They describe important mathematical properties of scalar and vector fields in space.

Before studying these integral theorems, we shall first show how more complicated path-dependent integrals can be constructed from the basic line and surface integrals discussed previously.

1.6.1. Vector line integrals

By a (vector) *line integral* is meant an integral in which the integration variable is the vector line element

$$d\mathbf{r} = \mathbf{i}\,dx + \mathbf{j}\,dy + \mathbf{k}\,dz$$

in space. Three types of line integrals can be distinguished

$$\int_c \phi(\mathbf{r})\,d\mathbf{r} = \mathbf{i}\int_c \phi\,dx + \mathbf{j}\int_c \phi\,dy + \mathbf{k}\int_c \phi\,dz = \mathbf{a}, \tag{1.64}$$

$$\int_c \mathbf{V}(\mathbf{r})\cdot d\mathbf{r} = \int_c V_x\,dx + \int_c V_y\,dy + \int_c V_z\,dz = w, \tag{1.65}$$

$$\int_c \mathbf{V}(\mathbf{r})\times d\mathbf{r} = \int_c \begin{vmatrix} \mathbf{i} & \mathbf{j} & \mathbf{k} \\ V_x & V_y & V_z \\ dx & dy & dz \end{vmatrix} = \mathbf{b}, \tag{1.66}$$

i.e., $b_x = \int_c (V_y\,dz - V_z\,dy)$, etc. All these integrals are path dependent in general and give for each chosen path a scalar or a vector constant.

1.6.2. Vector surface integrals

In describing the flow of fluids across a closed surface, it is important to distinguish a flow out of an enclosed volume from a flow into the volume. This distinction can be made conveniently by giving a surface element $d\sigma$ a direction $\mathbf{n}$ also, so that

$$d\boldsymbol{\sigma} = \mathbf{n}\,d\sigma \tag{1.67}$$

is a vector differential surface element. By convention $\mathbf{n}$ (or $\mathbf{e_n}$) is chosen to be the normal or perpendicular direction to the surface element, with the sign chosen so that $\mathbf{n}$ points outward away from the interior if $d\sigma$ is part of a closed surface. Since $d\boldsymbol{\sigma}$ is a vector, we may write in rectangular coordinates

$$d\boldsymbol{\sigma} = \mathbf{i}\,d\sigma_x + \mathbf{j}\,d\sigma_y + \mathbf{k}\,d\sigma_z, \tag{1.68}$$

where

$$d\sigma_x = \pm dy\,dz, \qquad d\sigma_y = \pm dz\,dx, \qquad d\sigma_z = \pm dx\,dy,$$

with the signs chosen to match the given $d\boldsymbol{\sigma}$ on the surface. We shall show how this is done in Eq. (1.74).

Three types of surface integrals can be defined

$$\int_S \phi(\mathbf{r})\, d\boldsymbol{\sigma} = \mathbf{i}\int_S \phi\, d\sigma_x + \mathbf{j}\int_S \phi\, d\sigma_y + \mathbf{k}\int_S \phi\, d\sigma_z = \mathbf{A}, \tag{1.69}$$

$$\int_S \mathbf{V}(\mathbf{r}) \cdot d\boldsymbol{\sigma} = \int_S V_x\, d\sigma_x + \int_S V_y\, d\sigma_y + \int_S V_z\, d\sigma_z = \Phi, \tag{1.70}$$

$$\int_S \mathbf{V}(\mathbf{r}) \times d\boldsymbol{\sigma} = \int_S \begin{vmatrix} \mathbf{i} & \mathbf{j} & \mathbf{k} \\ V_x & V_y & V_z \\ d\sigma_x & d\sigma_y & d\sigma_z \end{vmatrix} = \mathbf{B}, \tag{1.71}$$

with $B_x = \int_S (V_y\, d\sigma_z - V_z\, d\sigma_y)$, etc. Each of these integrals requires the specification of a surface S over which the integrand becomes a function of the two integration variables only. Also, we have simplified the notation by using only one integration symbol for a multivariable integral.

Example 1.6.4. Show that

$$\int_S \mathbf{r} \cdot d\boldsymbol{\sigma} = 2\pi a^3, \tag{1.72}$$

where S is the curved surface of a hemisphere of radius a.

On the hemispherical surface, the longitudinal displacement is $a\, d\theta$, while the latitudinal displacement is $a \sin\theta\, d\phi$. The resulting directed surface element is

$$d\boldsymbol{\sigma} = \mathbf{e}_r a^2\, d(\cos\theta)\, d\phi.$$

Hence

$$I = a^3 \int_0^1 d\cos\theta \int_0^{2\pi} d\phi = 2\pi a^3.$$

It is also instructive to do this problem in rectangular coordinates

$$\begin{aligned} I &= \int_S x\, d\sigma_x + \int_S y\, d\sigma_y + \int_S z\, d\sigma_z \\ &= I_x + I_y + I_z, \end{aligned} \tag{1.73}$$

where the surface S is defined by the equation

$$x^2 + y^2 + z^2 = a^2. \tag{1.74}$$

Suppose the hemisphere is the Northern Hemisphere, with the z axis passing through the North Pole, as shown in Fig. 1.7. Now $d\sigma_z = dx\, dy$ in the Northern Hemisphere because $\mathbf{r} \cdot \mathbf{k}$ is always positive. Hence

$$\begin{aligned} I_z &= \iint (a^2 - x^2 - y^2)^{1/2}\, dx\, dy \\ &= 2\pi \int_0^a (a^2 - \rho^2)^{1/2} \rho\, d\rho \\ &= 2\pi a^3/3 \end{aligned}$$

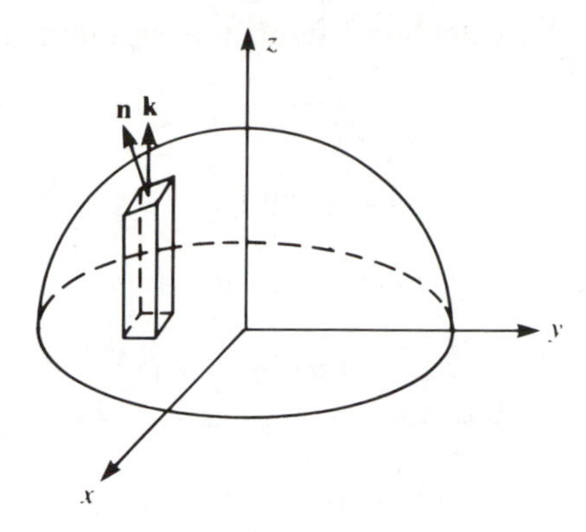

Fig. 1.7. A surface integral over a hemisphere.

after the integration over the equatorial circle of radius a in the xy plane.

The integral I_x is more complicated, because the $dy\,dz$ integration is performed over a simicircle in the yz plane. The hemispherical surface now separates into two equal pieces, one half lying in front of the yz plane in the positive x direction. Here $\mathbf{r}\cdot\mathbf{i}$ is positive, and hence $d\sigma_x = dy\,dz$. The remaining half lies behind the yz plane and has $d\sigma_x = -dy\,dz$. Thus

$$\begin{aligned} I_x &= I_{x^+} + I_{x^-} \\ &= \int_{\text{front}} x\,dy\,dz - \int_{\text{back}} x\,dy\,dz. \end{aligned} \tag{1.75}$$

Actually both halves contribute the same amount of $\pi a^3/3$, so that $I_x = 2\pi a^3/3$. Finally, I_y can be calculated in the same way as I_x also to yield $2\pi a^3/3$. Thus the total surface integral is

$$I = I_x + I_y + I_z = 2\pi a^3,$$

in agreement with Eq. (1.72).

A third method for doing this problem is perhaps the most general. It turns out to be possible to integrate over just the equatorial circle in the xy plane. This possibility arises because $d\boldsymbol{\sigma}$ is related to its projection on the xy plane through the equation

$$dx\,dy = |d\boldsymbol{\sigma}|\,\mathbf{n}\cdot\mathbf{k}, \tag{1.76}$$

where $\mathbf{n}$ is the normal direction to the spherical surface, as shown in Fig. 1.7. Hence

$$\begin{aligned} I &= \int \mathbf{r}\cdot\mathbf{n}\,|d\boldsymbol{\sigma}| \\ &= \int \mathbf{r}\cdot\mathbf{n}\,\frac{dx\,dy}{\mathbf{k}\cdot\mathbf{n}}. \end{aligned}$$

The normal $\mathbf{n}$ must be calculated from the gradient of the LHS of Eq. (1.74)

$$\nabla r^2 = 2\mathbf{r},$$

or $\mathbf{n} = \mathbf{e}_r$. Hence $\mathbf{r} \cdot \mathbf{n} = a\mathbf{k} \cdot \mathbf{n} = z/a$, and

$$I = a \int \frac{a}{z}\, dx\, dy.$$

When expressed in circular coordinates $\rho = (x^2 + y^2)^{1/2}$ and ϕ in the xy plane, this simplifies further to

$$\begin{aligned} I &= 2\pi a^2 \int_0^a (a^2 - \rho^2)^{-1/2} \rho\, d\rho \\ &= 2\pi a^3. \end{aligned}$$

Surface integrals tend to be difficult to calculate or to visualize. We shall see in the following two sections how it is sometimes possible to express certain special types of surface integrals in terms of simpler volume or line integrals.

In the rest of this section, we show a few common surface integrals that can be evaluated by using simple geometrical considerations alone.

Example 1.6.5. Show that

$$\oint_c \mathbf{r} \times d\mathbf{r} = 2 \int_S d\boldsymbol{\sigma}, \tag{1.77}$$

where c is an arbitrary closed curve (its closure being denoted by the small circle superposed on the integration sign), and S is any surface bounded by c.

A simple way to show this is to note that $\mathbf{r} \times d\mathbf{r}$ gives twice the vectorial area $d\boldsymbol{\sigma}$ of the triangle with $\mathbf{r}$ and $d\mathbf{r}$ as two of its sides. As the boundary c is traversed, we just add up the vectorial areas swept by the radius $\mathbf{r}$. The result holds for any choice of the origin from which $\mathbf{r}$ is measured.

It is instructive to demonstrate the result in a more convoluted way in order to illustrate a way of thinking that will be useful in more complicated situations. Consider first a small closed rectangular loop c on the plane $z = z_0 =$ constant, as shown by the arrows $A \to B \to C \to D \to A$ in Fig. 1.8. For this closed path

$$\begin{aligned} \oint_{\Delta c} \mathbf{r} \times d\mathbf{r} &= \oint_{\Delta c} \begin{vmatrix} \mathbf{i} & \mathbf{j} & \mathbf{k} \\ x & y & z_0 \\ dx & dy & 0 \end{vmatrix} \\ &= \mathbf{i}(-z_0) \oint_{\Delta c} dy + \mathbf{j} z_0 \oint_{\Delta c} dx + \mathbf{k} \oint_{\Delta c} (x\, dy - y\, dx). \end{aligned}$$

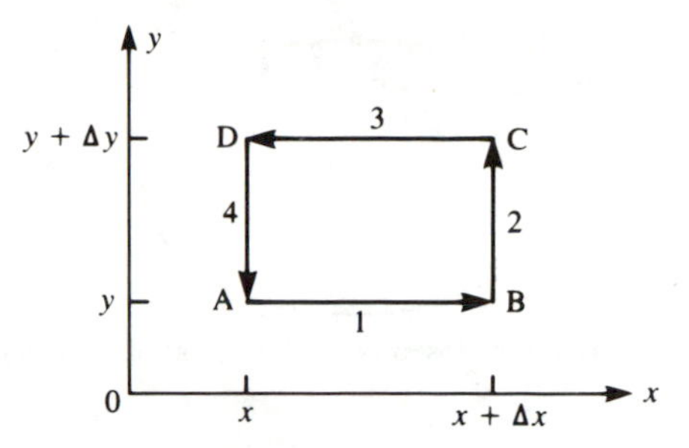

Fig. 1.8. A small rectangular closed loop.

The first two closed-path integrals vanish identically because

$$\oint dy = y(\text{final}) - y(\text{initial}) = 0;$$

similarly,

$$\oint dx = 0.$$

We are left with

$$\begin{aligned}\oint_{\Delta c} \mathbf{r} \times d\mathbf{r} &= \mathbf{k} \oint (x\, dy - y\, dx) \\ &= \mathbf{k}\{(-y\, \Delta x)_1 + [(x + \Delta x)\, \Delta y]_2 \\ &\quad + [-(y + \Delta y)(-\Delta x)]_3 + [x(-\Delta y)]_4\},\end{aligned}$$

where a subscript denotes the contribution of the side marked in Fig. 1.8. In this way we find

$$\oint_{\Delta c} \mathbf{r} \times d\mathbf{r} = \mathbf{k}(2\, \Delta \mathrm{x}\, \Delta y) = 2\mathbf{k}\, \Delta\sigma_z.$$

Since the planar curve can be tilted in space, we have actually proved the more general result

$$\oint_{\Delta c} \mathbf{r} \times d\mathbf{r} = 2\mathbf{n}\, \Delta\boldsymbol{\sigma} = 2\, \Delta\boldsymbol{\sigma},$$

where $\mathbf{n}$ is normal to the plane containing Δc.

We now go back to the original closed curve c, and break it up into small rectangular closed loops Δc_i, as shown in Fig. 1.9. Adjacent sides of neighboring loops always involve integrations in opposite directions, with zero net contribution. As a result, only the outer exposed sides of closed loops will contribute. These exposed sides add up to just the original closed curve c. Hence

$$\begin{aligned}\oint_c \mathbf{r} \times d\mathbf{r} &= \sum_i \lim_{\Delta c_i \to 0} \oint_{\Delta c_i} \mathbf{r} \times d\mathbf{r} \\ &= \sum_i \lim_{\Delta \sigma_i \to 0} 2\, \Delta\boldsymbol{\sigma}_i = 2 \int_S d\boldsymbol{\sigma}.\end{aligned}$$

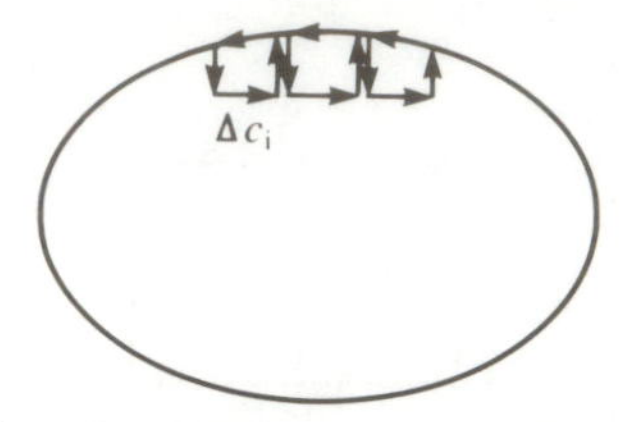

Fig. 1.9. Decomposition of a closed curve into small rectangular closed loops.

It is important to realize that each small closed loop Δc_i can be made planar by making it infinitesimally small. However, the finite closed curve c does not have to be planar. Furthermore, the enclosed surface S is not uniquely defined by the boundary curve c. For example, if c is the circle shown in Fig. 1.10, the surface S can be either the flat surface inside the circle shown in Fig. 1.10(a), or part of the spherical surface shown in Fig. 1.10(b). That the surface integral $\int d\boldsymbol{\sigma}$ gives the same result on either surface can be seen from the following examples.

Example 1.6.6. Show that

$$\oint_S d\boldsymbol{\sigma} = 0 \tag{1.78}$$

over a closed surface. (The closure of the surface is denoted by the small circle superposed on the integration sign.)

Let us first consider a small brick-like volume of sides $\Delta x, \Delta y, \Delta z$. Over the six sides of its closed surface ΔS, we find

$$\begin{aligned}\oint_{\Delta S} d\boldsymbol{\sigma} &= \mathbf{i}(\Delta y\ \Delta z - \Delta y\ \Delta z) + \mathbf{j}(\Delta z\ \Delta x - \Delta z\ \Delta x) + \mathbf{k}(\Delta x\ \Delta y - \Delta x\ \Delta y)\\ &= 0,\end{aligned}$$

where the contribution from a front surface of the brick is exactly cancelled by that from its back surface if the normal directions always point to the outside of the volume.

If the volume enclosed by the original closed surface is separated into

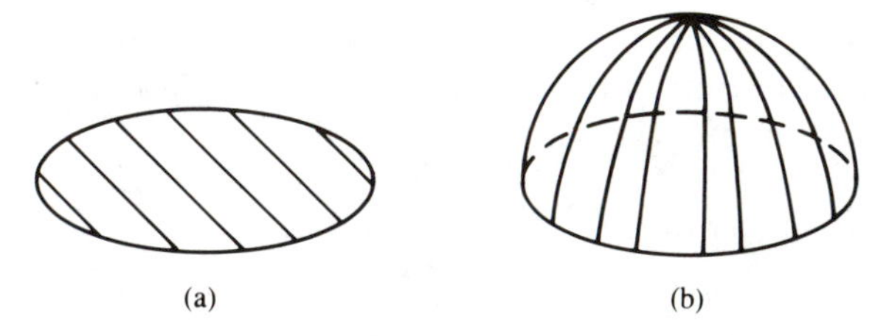

Fig. 1.10. Two different surfaces enclosed by the same boundary curve.

such small brick-like volumes with closed surfaces ΔS_i, we find that

$$\oint_S d\boldsymbol{\sigma} = \sum_i \oint_{\Delta S_i} d\boldsymbol{\sigma} = \sum_i 0 = 0.$$

Example 1.6.7. Show that the surface integral $\int d\boldsymbol{\sigma}$ over a surface bounded by a closed curve c is the same for any surface bounded by c.

Take two distinct surfaces S_1 and S_2 bounded by c, as shown, for example, in Fig. 1.10. These two surfaces together form a closed surface S, if the normal to one of the surfaces (say S_1) is reversed to make it always point away from the volume. We then find

$$\oint_S d\boldsymbol{\sigma} = 0 = -\int_{S_1} d\boldsymbol{\sigma} + \int_{S_2} d\boldsymbol{\sigma}.$$

Example 1.6.8. Show that

$$\oint_S \mathbf{r} \cdot d\boldsymbol{\sigma} = 3V, \tag{1.79}$$

where V is the volume enclosed by the closed surface S.

We start by noting that

$$\oint_{\Delta S} \mathbf{r} \cdot d\boldsymbol{\sigma} = \oint_{\Delta S} (x\, d\sigma_x + y\, d\sigma_y + z\, d\sigma_z) = 3\, \Delta x\, \Delta y\, \Delta z = 3\, \Delta V$$

for a brick-like volume $\Delta V = \Delta x\, \Delta y\, \Delta z$. The final result then follows from the usual rule of integral calculus.

Of these integrals of vector fields in space, the two involving scalar products, namely Eqs. (1.65) and (1.70), are so important as to deserve special names. The first

$$w = \int_c \mathbf{V}(\mathbf{r}) \cdot d\mathbf{r} \tag{1.80}$$

is called the *generalized work* done against the vector field $\mathbf{V}(\mathbf{r})$ along the path c. The reason for this name is that w is the mechanical work done when $\mathbf{V}$ is the mechanical force $\mathbf{F}(\mathbf{r})$. It is well known that, if the mechanical system is a conservative one, that is, if

$$\mathbf{F}(\mathbf{r}) = -\boldsymbol{\nabla}\phi(\mathbf{r})$$

is derivable from a potential field $\phi(\mathbf{r})$, then

$$\begin{aligned} w &= \int_c \mathbf{F} \cdot d\mathbf{r} = -\int_c \boldsymbol{\nabla}\phi(\mathbf{r}) \cdot d\mathbf{r} = -\int_c d\phi(\mathbf{r}) \\ &= -[\phi(\mathbf{r}_2) - \phi(\mathbf{r}_1)], \end{aligned}$$

where use has been made of Eq. (1.39). Under these circumstances, w depends only on the endpoints $\mathbf{r}_1$ and $\mathbf{r}_2$, but not on the path of integration. This is an example of the kind of mathematical results of interest in the study of integral properties of fields in space.

The closed-path integral

$$C = \oint_c \mathbf{V}(\mathbf{r}) \cdot d\mathbf{r} \tag{1.81}$$

is called the *circulation* of the vector field $\mathbf{V}(\mathbf{r})$ along the closed path c. For a conservative mechanical system, all circulations of the force field vanish. Conversely, if all circulations of a vector field vanish, the vector field must be proportional to the gradient of a scalar potential field. According to Eq. (1.50), such a vector field must be irrotational. In fluid mechanics, a fluid flow is said to be a potential flow or an irrotational flow in a region of space in which all circulations of its velocity field vanish.

The second important integral is the scalar Φ defined in Eq. (1.70). It is called the *flux* of the vector field $\mathbf{V}(\mathbf{r})$ across the surface S. Its physical significance is described in the next section.

Problems

1.6.1. Show the closed-path line integral $\oint \mathbf{r} \cdot d\mathbf{r} = 0$.

1.6.2. (a) Show that

$$\int_c \mathbf{V}(\mathbf{r}) \cdot d\mathbf{r} = x_0^2 y_0 - \tfrac{1}{3} y_0^3$$

if

$$V(\mathbf{r}) = 2xy\mathbf{i} + (x^2 - y^2)\mathbf{j},$$

and the path c is made up of two parts:

c_1: $(0,0) \to (x_0,0)$ on the x axis,

c_2: $(x_0,0) \to (x_0,y_0)$ parallel to the y axis.

(b) Show that the integral along the straight line connecting (0,0) to (x_0,y_0) gives the same result.

1.6.3. Evaluate the surface integral $\int_S x^2y^2z^2\, d\boldsymbol{\sigma}$, where $d\boldsymbol{\sigma}$ always points away from the volume and S is

(a) the curved-cylindrical surface of the cylinder $x^2 + y^2 = 1$ between $z = 0$ and 1;

(b) the curved spherical surface of the unit sphere for positive x,y,z, that is, in the first octant. Obtain the result in each of the remaining seven octants.

1.7. Flux, divergence, and Gauss's theorem

The *flux* (or *flow*) Φ_S of a vector field $\mathbf{j}(\mathbf{r})$ across a surface S is defined by the scalar integral

$$\Phi_S \equiv \int_S \mathbf{j}(\mathbf{r}) \cdot d\boldsymbol{\sigma}.$$

It can be visualized readily by considering a liquid of density $\rho(\mathbf{r})$ flowing in a pipe of rectangular cross section $\Delta y\ \Delta z$ with velocity

$$\mathbf{v}(\mathbf{r}) = \mathbf{i}v_x(\mathbf{r}),$$

as shown in Fig. 1.11. The vector field

$$\mathbf{j}(\mathbf{r}) = \rho(\mathbf{r})\mathbf{v}(\mathbf{r}) \tag{1.82}$$

is called a *mass flux density*. It has the dimension of a mass per second per unit area. The flux of $\mathbf{j}(\mathbf{r})$ across the shaded surface in Fig. 1.11 is the mass flux or flow (mass per second)

$$\Phi_2 = \int_{\Delta y\ \Delta z} \mathbf{j}(\mathbf{r}) \cdot d\boldsymbol{\sigma} \simeq \Delta y\ \Delta z\ \rho(\mathbf{r}_2)v_x(\mathbf{r}_2),$$

where $\mathbf{r}_2$ is a representative point on the surface. It gives the total mass crossing the surface per second along the direction of its normal ($\mathbf{n} = \mathbf{i}$). Similarly

$$\Phi_1 \simeq -\Delta y\ \Delta z\ \rho(\mathbf{r}_1)v_x(\mathbf{r}_1)$$

is the mass flux on the back surface, the negative sign appearing because $\mathbf{n} = -\mathbf{i}$ is so chosen that a flow *into* a closed volume is negative. Finally, we note that the total mass flux out of the volume is

$$\Phi_x = \Phi_2 + \Phi_1 \simeq \Delta y\ \Delta z\ \Delta j_x = \Delta\tau \frac{\Delta j_x}{\Delta x}, \tag{1.83}$$

where

$$\Delta\tau = \Delta x\ \Delta y\ \Delta z, \qquad \Delta j_x = \rho(\mathbf{r}_2)v_x(\mathbf{r}_2) - \rho(\mathbf{r}_1)v_x(\mathbf{r}_1).$$

1.7.1. Divergence

We can also consider a mass flux density for a general velocity field

$$\mathbf{v}(\mathbf{r}) = \mathbf{i}v_x(\mathbf{r}) + \mathbf{j}v_y(\mathbf{r}) + \mathbf{k}v_z(\mathbf{r})$$

in the body of the liquid itself, rather than in the rectangular pipe of Fig. 1.11. A rectangular volume of space in the fluid, of sides Δx, Δy, and Δz, can still be imagined. The only difference is that in general there will be mass fluxes across the remaining four sides. Hence the total mass flux out of the volume *per unit volume,* in the case of infinitesimal volumes, is

$$\lim_{\Delta\tau\to 0} \frac{1}{\Delta\tau} \oint_{\Delta S} \mathbf{j} \cdot d\boldsymbol{\sigma} = \lim_{\Delta x, \Delta y, \Delta z \to 0} \left(\frac{\Delta j_x}{\Delta x} + \frac{\Delta j_y}{\Delta y} + \frac{\Delta j_z}{\Delta z} \right) = \boldsymbol{\nabla} \cdot \mathbf{j}(\mathbf{r}), \tag{1.84}$$

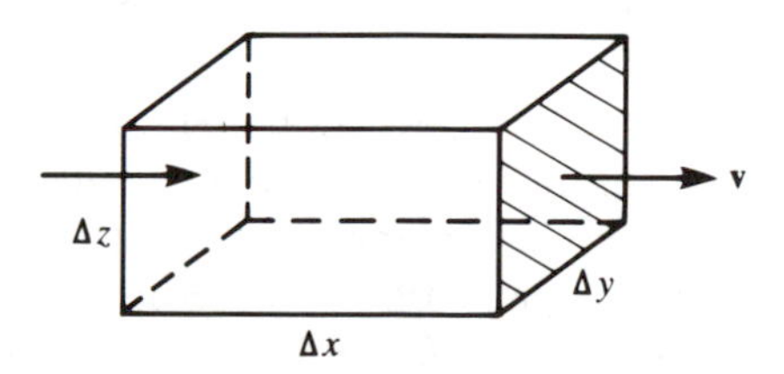

Fig. 1.11. Liquid flow in a rectangular pipe.

where the integration symbol means integration over a closed surface. The divergence of a vector field at a point $\mathbf{r}$ in space gives its total flux (or net *outflow*) per unit volume "coming out of the point $\mathbf{r}$," that is, passing through an infinitesimally small closed surface surrounding the point $\mathbf{r}$.

1.7.2. Gauss's theorem

A finite closed surface S can be subdivided into many smaller closed surfaces. This is done by dividing the enclosed volume into many smaller volumes. The closed surfaces of the small volumes are then the small closed surfaces in question. By applying Eq. (1.84) to each of these small closed surfaces, we find that

$$\oint_S d\boldsymbol{\sigma}\cdot\mathbf{j}(\mathbf{r}) = \sum_i \left(\lim_{\Delta\tau_i\to 0}\frac{1}{\Delta\tau_i}\oint_{\Delta S_i} d\boldsymbol{\sigma}\cdot\mathbf{j}(\mathbf{r})\right)$$
$$= \int_\Omega d\tau\,\nabla\cdot\mathbf{j}(\mathbf{r}). \tag{1.85}$$

This integral relation is known as *Gauss's theorem*. It states that the net outflow across a closed surface S is equal to the total divergence inside S. As a result, we may say that the enclosed divergence "causes" the outflow; that is, the enclosed divergence is a "source" of the outflow.

The most familiar example of the application of Gauss's theorem is in electrostatics, where the electric field intensity $\mathbf{E}(\mathbf{r})$ is known to be related to the charge density $\rho(\mathbf{r})$ as follows:

$$\nabla\cdot\mathbf{E}(\mathbf{r}) = \rho(\mathbf{r})/\varepsilon_0, \tag{1.86}$$

where ε_0 is the permittivity of free space. As a result,

$$\int_\Omega d\tau\frac{\rho}{\varepsilon_0} = \frac{Q}{\varepsilon_0} = \oint_S d\boldsymbol{\sigma}\cdot\mathbf{E}. \tag{1.87}$$

This *Gauss's law* of electrostatics states that the total flux of the electric field intensity coming out across a closed surface is proportional to the total charge Q enclosed by the surface. For this reason, we consider Q, or more specifically $\rho(\mathbf{r})$, to be the *scalar source* of $\mathbf{E}(\mathbf{r})$. In this language, a solenoidal (divergence-free) vector field is said to exist without the help of a scalar source, because by definition a solenoidal field is divergence-free. Its source density is not a scalar field, but a vector field, as we shall see in Section 1.9.

Example 1.7.1. Show that the electrostatic field intensity $\mathbf{E}(\mathbf{r})$ of a point charge at the origin has an inverse-square dependence on r.

We first note that the angular components of $\mathbf{E}$ must vanish, because with the point charge at the origin, there is no preferred angular direction. Hence $\mathbf{E}(\mathbf{r}) = E_r(\mathbf{r})\mathbf{e}_r$. Consider next a spherical surface of radius r surrounding Q. On this surface

$$d\boldsymbol{\sigma} = r^2 \, d^2\Omega \, \mathbf{e_r}, \tag{1.88}$$

where

$$d^2\Omega = d \cos\theta \, d\phi, \tag{1.89}$$

Ω being called a *solid angle*. Gauss's law then gives

$$\begin{aligned} \frac{Q}{\varepsilon_0} &= \int (r^2 \, d^2\Omega \, \mathbf{e_r}) \cdot [E_r(\mathbf{r})\mathbf{e_r}] \\ &= r^2 E_r(r) \int d^2\Omega = 4\pi r^2 E_r(r), \end{aligned}$$

since the full solid angle of the sphere is

$$\int d^2\Omega = \int_{-1}^{1} d\cos\theta \int_0^{2\pi} d\phi = (2)2\pi = 4\pi.$$

Hence

$$\mathbf{E}(\mathbf{r}) = \frac{Q}{4\pi\varepsilon_0} \frac{\mathbf{e_r}}{r^2}. \tag{1.90}$$

1.7.3. *Continuity equation*

If a fluid is in a region of space where there is neither a source nor a sink, the total mass flow out of a volume Ω expresses itself as a rate of *decrease* of the total mass in the volume. Mass conservation requires an exact balance between these effects:

$$\begin{aligned} -\frac{\partial}{\partial t} \int_\Omega \rho(\mathbf{r}) \, d\tau &= \int_S \rho(\mathbf{r})\mathbf{v}(\mathbf{r}) \cdot d\boldsymbol{\sigma} \\ &= \int_\Omega \boldsymbol{\nabla} \cdot (\rho\mathbf{v}) \, d\tau \end{aligned}$$

on applying Gauss's theorem. The first and the last expressions show that mass conservation requires that the equation

$$\frac{\partial \rho(\mathbf{r})}{\partial t} + \boldsymbol{\nabla} \cdot [\rho(\mathbf{r})\mathbf{v}(\mathbf{r})] = 0 \tag{1.91}$$

must be satisfied everywhere in the volume. This is called the *continuity equation* for a *conserved current*, the current being $\mathbf{j}(\mathbf{r}) = \rho(\mathbf{r})\mathbf{v}(\mathbf{r})$.

1.7.4. *Operator identity*

Gauss's theorem, Eq. (1.85), can be applied to the vector field $\mathbf{V}(\mathbf{r}) = \mathbf{e}_i \phi(\mathbf{r})$ to give

$$\oint_S d\sigma_i \phi(\mathbf{r}) = \oint_\Omega d\tau \frac{\partial}{\partial x_i} \phi(\mathbf{r}).$$

Since this is true for any i, it follows that

$$\oint_S d\boldsymbol{\sigma}\, \phi(\mathbf{r}) = \sum_i \mathbf{e}_i \oint_S d\sigma_i\, \phi(\mathbf{r})$$

$$= \int_\Omega d\tau\, \boldsymbol{\nabla}\phi(\mathbf{r}). \tag{1.92}$$

This result generalizes Gauss's theorem to the operator identity

$$\oint_S d\boldsymbol{\sigma} = \int_\Omega \boldsymbol{\nabla} \tag{1.93}$$

for operations on any field in space. It gives rise to other integral theorems such as the following:

$$\oint_S d\boldsymbol{\sigma} \times \mathbf{A}(\mathbf{r}) = \int_\Omega d\tau\, \boldsymbol{\nabla} \times \mathbf{A}(\mathbf{r}), \tag{1.94}$$

$$\oint_S d\boldsymbol{\sigma} \cdot (u\, \boldsymbol{\nabla} v - v\, \boldsymbol{\nabla} u) = \int_\Omega d\tau\, \boldsymbol{\nabla} \cdot (u\, \boldsymbol{\nabla} v - v\, \boldsymbol{\nabla} u)$$

$$= \int_\Omega d\tau\, (u\, \nabla^2 v - v\, \nabla^2 u). \tag{1.95}$$

The last identity is called *Green's theorem.*

The application of these formulas is illustrated in the following examples:

Example 1.7.2. Show that

$$\oint d\boldsymbol{\sigma} \cdot \frac{\mathbf{r}}{r^2} = \int d\tau \frac{1}{r^2}.$$

This is equivalent to showing that

$$\boldsymbol{\nabla} \cdot \left(\frac{\mathbf{r}}{r^2}\right) = \frac{1}{r^2},$$

a result that can be demonstrated readily in rectangular coordinates.

Example 1.7.3. Calculate $\oint_S d\boldsymbol{\sigma} \times \mathbf{r}$ over a closed surface.

$$\oint_s d\boldsymbol{\sigma} \times \mathbf{r} = \int d\tau\, \boldsymbol{\nabla} \times \boldsymbol{r} = 0$$

over any closed surface.

Example 1.7.4. Show that $\oint_S d\boldsymbol{\sigma} \cdot \boldsymbol{r}$ is three times the volume V enclosed by the closed surface S.

$$\oint_S d\boldsymbol{\sigma} \cdot \mathbf{r} = \int_R d\tau\, \boldsymbol{\nabla} \cdot \mathbf{r} = 3 \int d\tau = 3V.$$

Problems

1.7.1. Show that $\oint_S d\boldsymbol{\sigma} = 0$ over a closed surface S.

1.7.2. Calculate $\int d\boldsymbol{\sigma} \cdot \mathbf{V}(\mathbf{r})$ over the unit sphere for
(a) $\mathbf{V}(\mathbf{r}) = xy^2\mathbf{i} + yz^2\mathbf{j} + zx^2\mathbf{k}$,
(b) $\mathbf{V}(\mathbf{r}) = \mathbf{A}(\mathbf{r}) \times \mathbf{r}$, where $\mathbf{A}(\mathbf{r})$ is irrotational.

1.7.3. Calculate the total flux (or net outflow) over a spherical surface of radius R about the origin for the vector field

$$\mathbf{F}(\mathbf{r}) = \frac{\mathbf{r} - \mathbf{a}}{|\mathbf{r} - \mathbf{a}|^3}, \qquad \mathbf{a} = a\mathbf{i},$$

when $R < a$ and when $R > a$.

1.7.4. Verify Gauss's theorem by showing separately that over a sphere of radius R the surface integral $\int d\boldsymbol{\sigma} \cdot (\mathbf{r}/r^{n+1})$ and the volume integral $\int d\tau\, \boldsymbol{\nabla} \cdot (\mathbf{r}/r^{n+1})$ are both equal to $4\pi R^{2-n}$. Explain why this is true even for $n = 2$ when $\boldsymbol{\nabla} \cdot (\mathbf{r}/r^{n+1}) = 0$ for all finite r.

1.7.5. If $\mathbf{V}(x,y) = u(x,y)\mathbf{e}_x + v(x,y)\mathbf{e}_y$, use Gauss's theorem to show that

$$\oint_S \left(\frac{\partial u}{\partial x} + \frac{\partial v}{\partial y}\right) dx\, dy = \oint_c (u\, dy - v\, dx),$$

where c is the boundary curve of the surface. (This is known as Green's theorem in the plane.) [Hint: Show that, if the volume is a circular cylinder parallel to the z axis, the surface element in Gauss's theorem over the curved cylindrical surface is $d\boldsymbol{\sigma} = (dy\, \mathbf{e}_x - dx\, \mathbf{e}_y)\, dz$ because the normal $\mathbf{n}$ is perpendicular to $d\mathbf{r} = dx\, \mathbf{e}_x + dy\, \mathbf{e}_y$ on the circle where the curved cylinder surface intersects the xy plane.]

1.7.6. If $\mathbf{B} = \boldsymbol{\nabla} \times \mathbf{A}$ is the magnetic induction and $\mathbf{D} = \varepsilon_0 \mathbf{E}$ is the electric displacement satisfying the equation $\boldsymbol{\nabla} \cdot \mathbf{D} = \rho$, evaluate the net outflow integrals over a closed surface S

$$\oint_S \mathbf{B} \cdot d\boldsymbol{\sigma} \qquad \text{and} \qquad \oint_S \mathbf{D} \cdot d\boldsymbol{\sigma}.$$

Use these results to show that the normal component of $\mathbf{B}$ is continuous across the boundary between two media, while that of $\mathbf{D}$ has a discontinuity given by the surface charge density (per unit area) at the boundary.

1.7.7. Show that in electrostatics (where $\mathbf{E} = -\boldsymbol{\nabla}\phi$, $\mathbf{D} = \varepsilon_0 \mathbf{E}$, $\boldsymbol{\nabla} \cdot \mathbf{D} = \rho$)

$$\int \rho\phi\, d\tau = \int \mathbf{D} \cdot \mathbf{E}\, d\tau$$

if the electrostatic potential $\phi(r)$ vanishes sufficiently rapidly for large r. If $\phi(\mathbf{r}) \propto r^{-p}$, what inequality should p satisfy for the above identity to hold?

1.7.8. If $T(\mathbf{r})$ is the temperature field in space, the heat flux (or flow) at a point $\mathbf{r}$ is $\mathbf{q}(\mathbf{r}) = -K\,\nabla T(r)$, where K is the thermal conductivity (in calories per second per meter squared for a thickness of 1 m and a temperature difference of 1 K). When heat flows out of the region, its temperature decreases because its internal energy per unit mass e decreases by an amount $\Delta e = C\,\Delta T$, where C is the specific heat. From the energy conservation requirement

$$\int_\Omega \rho \frac{\partial e}{\partial t}\, d\tau = -\oint_S \mathbf{q} \cdot d\boldsymbol{\sigma},$$

obtain a continuity equation known as the *diffusion equation* for heat conduction.

1.8. Circulation, curl, and Stokes's theorem

The circulation Γ_c of a vector field $\mathbf{F}(\mathbf{r})$ along a closed curve c is defined by a line integral

$$\Gamma_c \equiv \oint_c (F_x\, dx + F_y\, dy + F_z\, dz). \tag{1.96}$$

The path c contains a direction of integration, the counter clockwise direction being taken to be positive. The integral changes sign when its direction is reversed:

$$\oint_c = -\oint_{-c}. \tag{1.97}$$

Example 1.8.1. A helical pipe of helical radius a carries water at a constant speed v. The velocity circulation after n turns of flow in a *counterclockwise* direction is

$$\Gamma_n = \oint_{n \text{ turns}} \mathbf{v} \cdot d\mathbf{r} = \oint_{n \text{ turns}} (v_x\, dx + v_y\, dy).$$

If

$$x = a \cos\theta, \qquad y = a \sin\theta,$$

$$v_x = -v \sin\theta, \qquad v_y = v \cos\theta,$$

$$\Gamma_n = av \oint_{n \text{ turns}} (\sin^2\theta + \cos^2\theta)\, d\theta = n(2\pi av).$$

Another way to visualize the circulation is to consider a small rectangular closed loop on the plane z = constant, such as the one shown in Fig. 1.8. We find

$$\Gamma = \oint_{\Delta c} (v_x\,dx + v_y\,dy)$$
$$\simeq \Delta x[v_x(x_1,y,z) - v_x(x_3,y+\Delta y,z)]$$
$$+ \Delta y[v_y(x+\Delta x,y_2,z) - v_y(x,y_4,z)],$$

where x_i,y_i are suitable points on side i. Hence

$$\Gamma = \Delta x\,\Delta y\left(-\frac{\Delta_y v_x}{\Delta y} + \frac{\Delta_x v_y}{\Delta x}\right) \simeq \Delta x\,\Delta y\left(-\frac{\partial v_x}{\partial y} + \frac{\partial v_y}{\partial x}\right), \qquad (1.98)$$

where the partials are calculated at a suitable point (x,y,z) on or inside the rectangular path.

1.8.1. *Curl*

The right-hand side of Eq. (1.98) involves $(\nabla \times \mathbf{v})_z$. Hence

$$\Gamma \simeq (\nabla \times \mathbf{v})_z(\Delta\sigma)_z, \qquad \text{where} \qquad (\Delta\sigma)_z = \Delta x\,\Delta y$$
$$= (\nabla \times \mathbf{v}) \cdot \Delta\boldsymbol{\sigma}$$
$$= (\nabla \times \mathbf{v}) \cdot \mathbf{n}\,\Delta\sigma,$$

where $\mathbf{n}$ ($=\mathbf{k}$ here) is the normal to the surface $\Delta\boldsymbol{\sigma}$ enclosed by the loop. Since a scalar product does not depend on the absolute orientations of the vectors, but only on their relative orientation (i.e., $\cos\theta$), it is clear that the resulting circulation

$$\Gamma = \oint_{\Delta c} \mathbf{v} \cdot d\mathbf{r} \simeq (\nabla \times \mathbf{v}) \cdot \mathbf{n}\,\Delta\sigma \qquad (1.99)$$

is actually independent of the choice of the coordinate system. As a consequence, the same result is obtained whether or not the small loop lies on the xy plane. If we now make the loop infinitesimally small, we obtain the identity

$$\lim_{\Delta\sigma \to \sigma} \frac{1}{\Delta\sigma} \oint \mathbf{v} \cdot d\mathbf{r} = [\nabla \times \mathbf{v}(\mathbf{r})] \cdot \mathbf{n}. \qquad (1.100)$$

Thus the curl of a vector field $\mathbf{v}(\mathbf{r})$ at a point $\mathbf{r}$ in space has the simple interpretation that its component along any direction $\mathbf{n}$ is given by the circulation per unit enclosed area of $\mathbf{v}$ around an infinitesimally small closed loop surrounding $\mathbf{r}$ on a plane perpendicular to $\mathbf{n}$. This circulation is to be calculated along the positive (i.e., right-handed) direction relative to $\mathbf{n}$.

1.8.2. Stokes's theorem

By "subdividing" a given closed loop into small closed loops c_i, we find

$$\oint_c \mathbf{v} \cdot d\mathbf{r} = \sum_i \Delta\sigma_i \left(\lim_{\Delta\sigma_i} \frac{1}{\Delta\sigma_i} \oint_{c_i} \mathbf{v} \cdot d\mathbf{r} \right)$$

$$= \int_S d\boldsymbol{\sigma} \cdot (\nabla \times \mathbf{v}). \qquad (1.101)$$

This integral relation is known as *Stokes's theorem*. It states that the circulation around a closed path is equal to the flow of the curl across the enclosed area. That is, the circulation is "caused" by a "flow of curl," and vice versa.

Example 1.8.2. Show that the gradient of a scalar field is irrotational:

$$\nabla \times \nabla\phi(\mathbf{r}) = 0.$$

This can be done by using Stokes's theorem to change a surface integral into a line integral:

$$\int_S d\boldsymbol{\sigma} \cdot [\nabla \times \nabla\phi(\mathbf{r})] = \oint_c d\mathbf{r} \cdot \nabla\phi(\mathbf{r}) = \oint_c d\phi(\mathbf{r})$$

$$= \phi(\mathbf{r}_0) - \phi(\mathbf{r}_0) = 0,$$

where $\mathbf{r}_0$ is *any* point on c at which the closed curve can both begin and end. Since this result holds for any surface S, $\nabla\phi(\mathbf{r})$ must be irrotational.

Thus, a conservative force, which is derivable from a potential $\phi(\mathbf{r})$, is irrotational. The converse of this is also true: Wherever $\nabla \times \mathbf{F} = 0$, $\mathbf{F}$ is conservative and is derivable from a potential.

In fluid dynamics, the curl of the velocity field $\mathbf{v}(\mathbf{r})$ (which gives the velocity of the fluid at the point $\mathbf{r}$) is called its *vorticity*, vortices being the whirls one creates in a cup of coffee on stirring it. If the velocity field is derivable from a potential

$$\mathbf{v}(\mathbf{r}) = -\nabla\phi(\mathbf{r}),$$

it must be irrotational, according to Eq. (1.50). For this reason, an irrotational flow is also called a *potential flow*. Its circulation vanishes. Being thus free of vortices and eddies, it describes a steady flow of the fluid.

One of Maxwell's equations in electromagnetism is Ampere's law:

$$\nabla \times \mathbf{B} = \mu_0 \mathbf{J}, \qquad (1.102)$$

where $\mathbf{B}$ is the magnetic induction, $\mathbf{J}$ is the current density (per unit area), and μ_0 is the permeability of free space. As a result, current densities may be visualized as "vortices" of $\mathbf{B}$. An application of Stokes's theorem to Eq. (1.102) gives *Ampere's circuital law*:

$$\oint_c \mathbf{B} \cdot d\mathbf{r} = \mu_0 \int \mathbf{J} \cdot d\boldsymbol{\sigma} = \mu_0 I. \tag{1.103}$$

This states that the circulation of the magnetic induction is proportional to the total current passing through the surface enclosed by c.

It is worth pointing out that Stokes's theorem in Eq. (1.101) is valid whether or not the closed curve c lies in a plane. This means that in general the surface S is not a planar surface. Indeed S does not have to be planar even when c is planar, because the small closed loops c_i can be taken out of the plane. Thus Stokes's theorem holds for any surface *bounded* by c.

Example 1.8.3. Calculate $\int_S \mathbf{v}(\mathbf{r}) \cdot d\boldsymbol{\sigma}$ for the vector field $\mathbf{v}(\mathbf{r}) = \boldsymbol{\omega} \times \mathbf{r}$, where $\boldsymbol{\omega}$ is a constant vector, over the surface S enclosed by a unit circle in the xy plane centered at the origin.

The result is unchanged if the surface bounded by c is deformed into a hemispherical surface. Then $d\boldsymbol{\sigma}$ is proportional to $\mathbf{r}$, and $\mathbf{v} \cdot d\boldsymbol{\sigma}$ vanishes everywhere on the hemisphere because $\boldsymbol{\omega} \times \mathbf{r} \cdot \mathbf{r} = 0$. Alternatively, $\boldsymbol{\omega} \times \mathbf{r}$ can be integrated over the original planar surface enclosed by the unit circle c. The integral vanishes because contributions from equal areas on opposite sides of the origin always cancel because $\boldsymbol{\omega} \times \mathbf{r}$ is odd in $\mathbf{r}$.

1.8.3. Operator identity

Stokes's theorem, Eq. (1.101), can be applied to the vector field $\mathbf{v}(\mathbf{r}) = \mathbf{e}_i \phi(\mathbf{r})$ to give

$$\oint_c dx_i\, \phi(\mathbf{r}) = \oint_S (d\boldsymbol{\sigma} \times \boldsymbol{\nabla})_i \phi(\mathbf{r}).$$

Since this is true for any i, it follows that

$$\oint_c d\mathbf{r}\, \phi(\mathbf{r}) = \sum_i \mathbf{e}_i \oint_c dx_i\, \phi(\mathbf{r})$$

$$= \int_S (d\boldsymbol{\sigma} \times \boldsymbol{\nabla}) \phi(\mathbf{r}). \tag{1.104}$$

This result generalizes Stokes's theorem to the operator identity

$$\oint_c d\mathbf{r} = \int_S (d\boldsymbol{\sigma} \times \boldsymbol{\nabla}) \tag{1.105}$$

for operations on any field in space. It gives rise to other integral theorems such as the following

$$\oint d\mathbf{r} \times \mathbf{V}(\mathbf{r}) = \int_S (d\boldsymbol{\sigma} \times \boldsymbol{\nabla}) \times \mathbf{V}(\mathbf{r}) \tag{1.106}$$

$$\begin{aligned}\oint d\mathbf{r} \cdot (u\, \boldsymbol{\nabla} v) &= \int_S (d\boldsymbol{\sigma} \times \boldsymbol{\nabla}) \cdot (u\, \boldsymbol{\nabla} v) \\ &= \int_S d\boldsymbol{\sigma} \cdot [\boldsymbol{\nabla} \times (u\, \boldsymbol{\nabla} v)] \\ &= \int_S d\boldsymbol{\sigma} \cdot [(\boldsymbol{\nabla} u) \times (\boldsymbol{\nabla} v)].\end{aligned} \tag{1.107}$$

Example 1.8.4. If $\mathbf{v}(\mathbf{r}) = \phi(\mathbf{r})\mathbf{e}_x - \psi(\mathbf{r})\mathbf{e}_y$, Stokes's theorem for this vector field reads

$$\begin{aligned}\oint_c \mathbf{v}(\mathbf{r}) \cdot d\mathbf{r} &= \oint_c [\phi(\mathbf{r})\, dx - \psi(\mathbf{r})\, dy] \\ &= \int_S [\boldsymbol{\nabla} \times \mathbf{v}(\mathbf{r})] \cdot d\boldsymbol{\sigma} \\ &= \int_S [(\boldsymbol{\nabla} \times \mathbf{v})_x\, dy\, dz + (\boldsymbol{\nabla} \times \mathbf{v})_y\, dz\, dx + (\boldsymbol{\nabla} \times \mathbf{v})_z\, dx\, dy] \\ &= \int_S \left[\frac{\partial \psi}{\partial z} dy\, dz + \frac{\partial \phi}{\partial z} dz\, dx - \left(\frac{\partial \psi}{\partial x} + \frac{\partial \phi}{\partial y}\right) dx\, dy\right].\end{aligned}$$

If $\phi(\mathbf{r})$ and $\psi(\mathbf{r})$ are both independent of z, the result simplifies to

$$\oint_c [\phi(x,y)\, dx - \psi(x,y)\, dy] = -\int_S \left(\frac{\partial \psi}{\partial x} + \frac{\partial \phi}{\partial y}\right) dx\, dy.$$

This is known as Green's theorem in the plane.

Problems

1.8.1. Use Stokes's theorem to calculate $\oint \mathbf{r} \cdot d\mathbf{r}$.

1.8.2. Use Stokes's theorem to show that $\int_{\Delta c} \mathbf{r} \times d\mathbf{r} = 2\, \Delta\boldsymbol{\sigma}$, where $\Delta\boldsymbol{\sigma}$ is the area enclosed by the small closed curve Δc.

1.8.3. Use Stokes's theorem to identify the conservative forces in the following:
- (a) $\mathbf{F}(\mathbf{r}) = x^n\mathbf{i} + y^n\mathbf{j} + z^n\mathbf{k}$,
- (b) $\mathbf{F}(\mathbf{r}) = z^n\mathbf{i} + x^n\mathbf{j} + y^n\mathbf{k}$,
- (c) $\mathbf{F}(\mathbf{r}) = e^x \sin y\, \mathbf{i} + e^x \cos y\, \mathbf{j}$,
- (d) $\mathbf{F}(\mathbf{r}) = r^n\mathbf{e}(\mathbf{r})$, $\quad n =$ any integer.

1.8.4. Use Stokes's theorem to calculate the circulation of the following vector fields along the closed unit circle in the yz plane centered at the origin.
- (a) $\mathbf{v}(\mathbf{r}) = yz\mathbf{i} + zx\mathbf{j} + xy\mathbf{k}$,
- (b) $\mathbf{v}(\mathbf{r}) = \boldsymbol{\omega} \times \mathbf{r}$, where $\boldsymbol{\omega}$ is a constant vector.

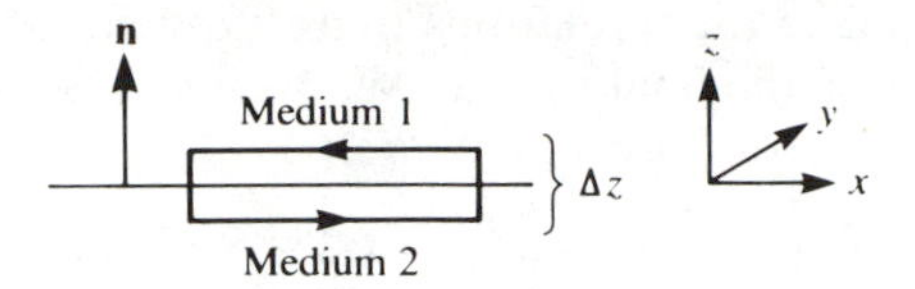

Fig. 1.12. Interface between two media.

1.8.5. A time-independent magnetic field $\mathbf{H}$ satisfies the equation $\nabla \times \mathbf{H} = \mathbf{J}$, where $\mathbf{J}$ is the current density (per unit area). By considering a narrow loop perpendicular to the interface between two media as shown in Fig. 1.12, show that $(H_1 - H_2)_x = J_y \, \Delta z$. Generalize this result to read $\mathbf{n} \times (\mathbf{H}_1 - \mathbf{H}_2) = \mathbf{K}$, where $\mathbf{K}$ is the surface current density (per unit length) at the interface.

1.9. Helmholtz's theorem

It is not accidental that the divergence and curl of a vector field play such important roles. Their significance is made clear by *Helmholtz's theorem*, which we now state without proof:

> A vector field is uniquely determined by its divergence and curl in a region of space, and its normal component over the boundary of the region. In particular, if both divergence and curl are specified everywhere and if they both disappear at infinity sufficiently rapidly, then the vector field can be written as a unique sum of an irrotational part and a solenoidal part.

In other words, we may write

$$\mathbf{V}(\mathbf{r}) = -\nabla\phi(\mathbf{r}) + \nabla \times \mathbf{A}(\mathbf{r}), \tag{1.108}$$

where $-\nabla\phi$ is the irrotational part and $\nabla \times \mathbf{A}$ is the solenoidal part. [This can be seen with the help of Eqs. (1.50) and (1.51), which show that both $\nabla \times (\nabla\phi)$ and $\nabla \cdot (\nabla \times \mathbf{A})$ vanish identically.] The fields $\phi(\mathbf{r})$ and $\mathbf{A}(\mathbf{r})$ are called the *scalar* and *vector potential*, respectively, of $\mathbf{V}(\mathbf{r})$.

It is useful to express the divergence and curl of $\mathbf{V}(\mathbf{r})$ in terms of these potentials. From Eq. (1.108), we obtain directly the results

$$\nabla \cdot \mathbf{V} = -\nabla^2\phi, \qquad \nabla \times \mathbf{V} = \nabla \times (\nabla \times \mathbf{A}) = \nabla(\nabla \cdot \mathbf{A}) - \nabla^2\mathbf{A}, \tag{1.109}$$

where the last step is made with the help of the operator form of the *BAC* rule discussed in Section 1.4. It turns out that the term $\nabla(\nabla \cdot \mathbf{A})$ vanishes under certain conditions (e.g., when $\nabla \times \mathbf{V}$ is either bounded in space or vanishes more rapidly than $1/r$ for large r). Then

$$\nabla \times \mathbf{V} = -\nabla^2\mathbf{A}. \tag{1.110}$$

In this way we can see that $\nabla \cdot \mathbf{V}$ is related to the scalar potential ϕ, while $\nabla \times \mathbf{V}$ is related to the vector potential $\mathbf{A}$.

It is possible to make these relations quite explicit, although we are not in a position to derive the final results. We begin by pointing out that Eq. (1.109) has the form of a *Poisson equation*

$$\nabla^2\phi(\mathbf{r}) = -s(\mathbf{r}), \qquad s(\mathbf{r}) = \nabla \cdot \mathbf{V}(\mathbf{r}), \tag{1.111}$$

where the charge density $\rho(\mathbf{r})$ of electrostatics is now denoted $s(\mathbf{r})$, the *scalar source density* of $\phi(\mathbf{r})$. From electrostatics, we know that the electrostatic potential due to a charge distribution is

$$\phi(\mathbf{r}) = \frac{1}{4\pi}\int \frac{s(\mathbf{r}_2)}{r_{12}}\, d\tau_2, \qquad r_{12} = |\mathbf{r} - \mathbf{r}_2|. \tag{1.112}$$

Hence the scalar potential in Eq. (1.108) can be written explicitly as

$$\phi(\mathbf{r}) = \frac{1}{4\pi}\int \frac{\nabla \cdot \mathbf{V}(\mathbf{r}_2)}{r_{12}}\, d\tau_2. \tag{1.113}$$

Similarly, Eq. (1.110) shows that each component $A_i(\mathbf{r})$ of the vector potential also satisfies a Poisson equation. Hence $\mathbf{A}(\mathbf{r})$ has the explicity solution

$$\mathbf{A}(\mathbf{r}) = \frac{1}{4\pi}\int \frac{\nabla \times \mathbf{V}(\mathbf{r}_2)}{r_{12}}\, d\tau_2. \tag{1.114}$$

Thus the divergence and curl (or vorticity) of a vector field can be interpreted as the source density of its scalar and vector potential, respectively.

Once the structure of a vector field is understood, it is easy to form a physical picture of a vector field used in physics. As an example, let us consider *Maxwell's equations* in vacuum:

$$\begin{aligned} \nabla \cdot \mathbf{E} &= \frac{\rho}{\varepsilon_0}, &\qquad \nabla \times \mathbf{E} &= -\frac{\partial \mathbf{B}}{\partial t}, \\ \nabla \cdot \mathbf{B} &= 0, &\qquad \nabla \times \mathbf{B} &= \mu_0\varepsilon_0\frac{\partial \mathbf{E}}{\partial t} + \mu_0\mathbf{J}. \end{aligned} \tag{1.115}$$

We see by inspection that these equations summarize the following *experimental* facts concerning the two vector fields **E** and **B** of electromagnetism:

1. The source of the electroscalar potential (i.e., the scalar potential of the electric field, or the electrostatic potential) is proportional to the charge density $\rho(\mathbf{r})$. The source of the electrovector potential is Faraday's induction term $(\partial/\partial t)\mathbf{B}(\mathbf{r})$.
2. There is no magnetoscalar potential, because the magnetic monopole density is found experimentally to vanish everywhere. The magnetovector potential (i.e., the vector potential **A**) originates from either an actual current density (Ampere's law) or Maxwell's displacement current density $\varepsilon_0\, \partial\mathbf{E}/\partial t$.

In other words, Maxwell's equations describe the physical nature of the four source densities of the electromagnetic fields.

Equations (1.115) are known as Maxwell's equations in honor of his discovery (in 1865) of the displacement current. In the absence of this displacement current term, the fourth equation (Ampere's law) gives a result

$$\nabla \cdot (\nabla \times \mathbf{B}) = 0 = \mu_0 \nabla \cdot \mathbf{J},$$

which violates charge conservation as expressed by the continuity equation, Eq. (1.91)

$$\nabla \cdot \mathbf{J} + \frac{\partial \rho}{\partial t} = 0$$

when the charge density ρ changes in time. The missing term $\partial \rho / \partial t$ when expressed in terms of $\varepsilon_0 \mathbf{E}$ with the help of Coulomb's law [the first of Eqs. (1.115)] gives rise to Maxwell's displacement current.

Problem

1.9.1. With the help of Helmholtz's theorem, show that a velocity field $\mathbf{v}(\mathbf{r})$ free of vortices is completely determined by a scalar potential.

1.10. Orthogonal curvilinear coordinate systems

In discussing vectors and fields in space, we have so far used only the rectangular or Cartesian coordinate system. This system has great intuitive appeal since it makes use of the straight lines and perpendicular directions of the flat space in which we live. The simplicity of vector differential and integral operators in rectangular coordinates reinforces our fondness for them.

However, many physical systems are not naturally rectangular. A sphere is a good example. The rectangular coordinates of a point on its surface are changing from point to point, but in spherical coordinates the spherical surface is specified simply as a surface of constant radius r. Thus the choice of coordinate systems can be important in the description of physical systems: A good choice may lead to greater simplification and insight in the description of physical properties.

There is of course a price to pay for this improvement. Coordinate systems other than the rectangular are less intuitive and harder to visualize. Integral and differential operators have more complicated forms, and these are harder to memorize. It is the purpose of this and the following section to show that the task is quite tractable, perhaps even enjoyable, when approached from a certain point of view. This turns out to be just a matter of changing directions and changing scales, as we shall see.

1.10.1. Generalized coordinates

Let us begin by noting that any three independent variables (u_1,u_2,u_3) can be used to form a coordinate system if they specify *uniquely* the position of a point in space. It is convenient to start with the familiar rectangular coordinates $(x,y,z)=(x_1,x_2,x_3)$ and specify the new, generalized coordinates in terms of these, that is,

$$u_i = u_i(x,y,z). \tag{1.116a}$$

For the transformations between these two coordinate systems to be well defined and unique, it is necessary that the inverse relations

$$x_i = x_i(u_1,u_2,u_3), \qquad i=1,2,3, \tag{1.116b}$$

also exist, and that all these relations are single-valued functions. For example, the spherical coordinates may be defined in terms of the rectangular coordinates by either of the following sets of equations (see Fig. 1.13):

$$\begin{aligned} r &= (x^2+y^2+z^2)^{1/2} \\ \theta &= \arctan[(x^2+y^2)/z^2]^{1/2} \\ \phi &= \arctan(y/x), \end{aligned} \tag{1.117a}$$

or

$$\begin{aligned} x &= r\sin\theta\cos\phi, \\ y &= r\sin\theta\sin\phi, \\ z &= r\cos\theta. \end{aligned} \tag{1.117b}$$

A coordinate axis now becomes a coordinate curve, along which only one of the coordinates is changing. The coordinate curves in spherical coordinates are radii (for variable r), longitudes (for variable θ), and latitudes (for variable ϕ).

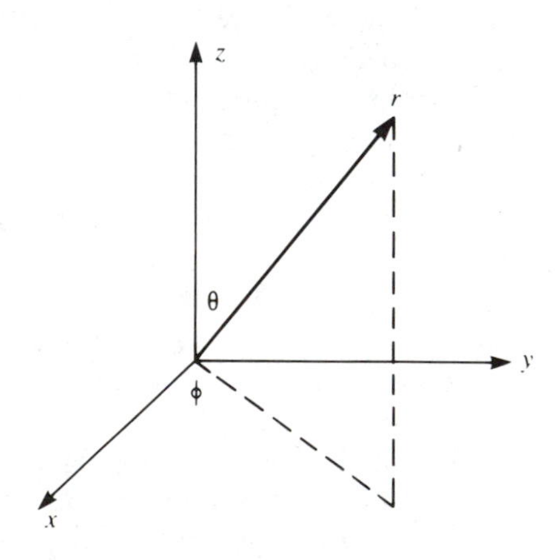

Fig. 1.13. Spherical coordinates as related to rectangular coordinates.

It is easy to find algebraic expressions describing these coordinate curves, because by definition only one of the coordinates changes along such a curve while all the others remain unchanged. For example, $\mathbf{r} = r\mathbf{e}_r$ in spherical coordinates. Then the partial derivatives

$$\frac{\partial}{\partial r}\mathbf{r} = \mathbf{e}_r, \qquad \frac{\partial}{\partial \theta}\mathbf{r} = r\frac{\partial \mathbf{e}_r}{\partial \theta}, \qquad \frac{\partial}{\partial \phi}\mathbf{r} = r\frac{\partial \mathbf{e}_r}{\partial \phi}$$

describe the vectorial changes along these coordinate curves. Each of these derivatives is a vector in space; it has a length h_i and a direction $\mathbf{e}_i$, $i = r$, θ, or ϕ. Their explicit forms can be obtained readily with the help of rectangular coordinates:

$$\begin{aligned}
\frac{\partial \mathbf{r}}{\partial r} &= \frac{\partial}{\partial r}(x\mathbf{e}_x + y\mathbf{e}_y + z\mathbf{e}_z) \\
&= \sin\theta\cos\phi\,\mathbf{e}_x + \sin\theta\sin\phi\,\mathbf{e}_y + \cos\theta\,\mathbf{e}_z = h_r\mathbf{e}_r, \\
\frac{\partial \mathbf{r}}{\partial \theta} &= r\cos\theta\cos\phi\,\mathbf{e}_x + r\cos\theta\sin\phi\,\mathbf{e}_y - r\sin\theta\,\mathbf{e}_z = h_\theta\mathbf{e}_\theta, \\
\frac{\partial \mathbf{r}}{\partial \phi} &= -r\sin\theta\sin\phi\,\mathbf{e}_x + r\sin\theta\cos\phi\,\mathbf{e}_y = h_\phi\mathbf{e}_\phi.
\end{aligned} \tag{1.118}$$

A simple calculation then yields

$$h_r = 1, \qquad h_\theta = r, \qquad h_\phi = r\sin\theta, \tag{1.119}$$

and therefore

$$\begin{aligned}
d\mathbf{r} &= \frac{\partial \mathbf{r}}{\partial r}dr + \frac{\partial \mathbf{r}}{\partial \theta}d\theta + \frac{\partial \mathbf{r}}{\partial \phi}d\phi \\
&= h_r\mathbf{e}_r\,dr + h_\theta\mathbf{e}_\theta\,d\theta + h_\phi\mathbf{e}_\phi\,d\phi \\
&= dr\,\mathbf{e}_r + r\,d\theta\,\mathbf{e}_\theta + r\sin\theta\,d\phi\,\mathbf{e}_\phi.
\end{aligned}$$

For arbitrary generalized coordinates (u_1, u_2, u_3), coordinate curves can be determined in a similar way by examining the displacement

$$d\mathbf{r} = \frac{\partial \mathbf{r}}{\partial u_1}du_1 + \frac{\partial \mathbf{r}}{\partial u_2}du_2 + \frac{\partial \mathbf{r}}{\partial u_3}du_3.$$

Each partial derivative

$$\frac{\partial \mathbf{r}}{\partial u_i} = \frac{\partial x}{\partial u_i}\mathbf{e}_x + \frac{\partial y}{\partial u_i}\mathbf{e}_y + \frac{\partial z}{\partial u_i}\mathbf{e}_z \tag{1.120a}$$

is a vector (the *tangent vector*) with a length and a direction:

$$\frac{\partial \mathbf{r}}{\partial u_i} = h_i(\mathbf{r})\mathbf{e}_i(\mathbf{r}), \tag{1.120b}$$

where

$$h_i(\mathbf{r}) = \left[\left(\frac{\partial x}{\partial u_i}\right)^2 + \left(\frac{\partial y}{\partial u_i}\right)^2 + \left(\frac{\partial z}{\partial u_i}\right)^2\right]^{1/2}.$$

In terms of these quantities, the original displacement may be written in the form

$$d\mathbf{r} = \sum_i h_i(\mathbf{r})\mathbf{e}_i(\mathbf{r})\, du_i = \sum_i \mathbf{e}_i(\mathbf{r})\, ds_i(\mathbf{r}). \tag{1.121}$$

Thus $\mathbf{e}_i(\mathbf{r})$ defines a coordinate curve, since it gives the unit vector tangent to the curve at $\mathbf{r}$. The infinitesimal scalar displacement

$$ds_i(\mathbf{r}) = h_i(\mathbf{r})\, du_i$$

gives the displacement along this coordinate curve. The function $h_i(\mathbf{r})$ is called a *scale factor,* since it ensures that the displacement has the dimension of a length irrespective of the nature and dimension of the generalized coordinate u_i itself.

Other geometrical quantities can be calculated readily in terms of these scale factors and unit tangent vectors. Thus the infinitesimal scalar displacement ds along a path in space is

$$(ds)^2 = d\mathbf{r} \cdot d\mathbf{r} = \sum_{i,j} g_{ij}\, du_i\, du_j, \tag{1.122}$$

where

$$g_{ij} = h_i h_j(\mathbf{e}_i \cdot \mathbf{e}_j)$$

are called the *metric coefficients* of the generalized coordinate system. The differential elements of surface and volume can be written down with the help of Eqs. (1.26) and (1.27) of Section 1.2:

$$\begin{aligned} d\boldsymbol{\sigma}_{ij} &= \left(\frac{\partial \mathbf{r}}{\partial u_i}\, du_i\right) \times \left(\frac{\partial \mathbf{r}}{\partial u_j}\, du_j\right) = h_i h_j(\mathbf{e}_i \times \mathbf{e}_j)\, du_i\, du_j \\ &= ds_i\, ds_j(\mathbf{e}_i \times \mathbf{e}_j), \end{aligned} \tag{1.123}$$

and

$$d\tau = ds_1\, ds_2\, ds_3\, \mathbf{e}_1 \cdot (\mathbf{e}_2 \times \mathbf{e}_3). \tag{1.124}$$

1.10.2. Orthogonal curvilinear coordinates

If at every point $\mathbf{r}$ the three unit tangets $\mathbf{e}_i(\mathbf{r})$ are orthogonal to one another, that is, if

$$\mathbf{e}_i(\mathbf{r}) \cdot \mathbf{e}_j(\mathbf{r}) = \delta_{ij}, \tag{1.125}$$

or equivalently if

$$\frac{\partial \mathbf{r}}{\partial u_i} \cdot \frac{\partial \mathbf{r}}{\partial u_j} \propto \delta_{ij},$$

then the generalized coordinates u_i are said to form an *orthogonal curvilinear coordinate* (OCC) *system.* In this system, the unit tangents $\mathbf{e}_1(\mathbf{r})$, $\mathbf{e}_2(\mathbf{r})$, and $\mathbf{e}_3(\mathbf{r})$ form a Cartesian coordinate system at every point $\mathbf{r}$. The only complication is that their orientations change from point to point in space.

In an OCC system, the metric coefficients

$$g_{ij} = h_i^2 \delta_{ij}, \qquad h_i^2 = \left(\frac{\partial x}{\partial u_i}\right)^2 + \left(\frac{\partial y}{\partial u_i}\right)^2 + \left(\frac{\partial z}{\partial u_i}\right)^2, \tag{1.126}$$

are diagonal, and the squared length

$$(ds)^2 = \sum_{i=1}^{3} h_i^2 \,(du_i)^2 = \sum_i (ds_i)^2 \tag{1.127}$$

does not contain cross terms. (The quantity ds_i always has the dimension of length.) The differential elements of surface and volume also simplify to

$$d\boldsymbol{\sigma}_{ij} = ds_i \, ds_j \sum_k \varepsilon_{ijk} \mathbf{e}_k, \tag{1.128}$$

and

$$d\tau = ds_1 \, ds_2 \, ds_3. \tag{1.129}$$

Thus the ds_i are very much like the rectangular coordinate dx_i. However, the tangent directions $\mathbf{e}_i(\mathbf{r})$ do change from point to point, except in the special case of rectangular coordinates for which they are constant unit vectors.

For spherical coordinates, direct calculations with the tangent vectors of Eq. (1.118) show that

$$\mathbf{e}_r \cdot \mathbf{e}_\theta = 0 = \mathbf{e}_\theta \cdot \mathbf{e}_\phi = \mathbf{e}_\phi \cdot \mathbf{e}_r,$$

and

$$\mathbf{e}_r \times \mathbf{e}_\theta = \mathbf{e}_\phi, \qquad \mathbf{e}_\theta \times \mathbf{e}_\phi = \mathbf{e}_r, \qquad \mathbf{e}_\phi \times \mathbf{e}_r = \mathbf{e}_\theta.$$

These orthogonality properties are easily visualized geometrically because $\mathbf{e}_r$ points radially outward from the origin, $\mathbf{e}_\theta$ points south along a longitude, while $\mathbf{e}_\phi$ points east along a latitude. Also with

$$ds_r = dr, \qquad ds_\theta = r\, d\theta, \qquad ds_\phi = r \sin\theta \, d\phi,$$

we find

$$\begin{aligned}
(ds)^2 &= (dr)^2 + (r\, d\theta)^2 + (r \sin\theta\, d\phi)^2, \\
d\boldsymbol{\sigma}_{r\theta} &= r\, dr\, d\theta\, \mathbf{e}_\phi = -d\boldsymbol{\sigma}_{\theta r}, \\
d\boldsymbol{\sigma}_{\theta\phi} &= r^2 \sin\theta\, d\theta\, d\phi\, \mathbf{e}_r = -d\boldsymbol{\sigma}_{\phi\theta}, \\
d\boldsymbol{\sigma}_{\phi r} &= r \sin\theta\, d\phi\, dr\, \mathbf{e}_\theta = -d\boldsymbol{\sigma}_{r\phi}, \\
d\tau &= r^2 \sin\theta\, dr\, d\theta\, d\phi.
\end{aligned}$$

We can also define coordinate surfaces on which one of the coordinates is constant, while the remaining two coordinates change. The coordinate surfaces for spherical coordinates are spherical surfaces of constant r, conical surfaces of constant θ, and half-planes of constant ϕ. For OCC systems, coordinate surfaces are perpendicular to coordinate curves along which the third coordinate changes value. For example, spherical surfaces of constant r are perpendicular to the coordinate curve $\mathbf{e}_r = \mathbf{e}_\mathbf{r}$.

Table 1.1. Two common orthogonal curvilinear coordinate systems

Coordinates	Cylindrical	Spherical
	ρ, ϕ, z	r, θ, ϕ
Definitions	$x = \rho \cos \phi$ $y = \rho \sin \phi$ $z = z$	$x = r \sin \theta \cos \phi$ $y = r \sin \theta \sin \phi$ $z = r \cos \theta$
Scale factors	$h_\rho = 1$ $h_\phi = \rho$ $h_z = 1$	$h_r = 1$ $h_\theta = r$ $h_\phi = r \sin \theta$

We list in Table 1.1 the definitions of u_i, and their scale factors, in two common OCC systems. The coordinate surfaces for cylindrical coordinates are the cylindrical surfaces of constant ρ, the planes of constant z, and the half-planes of constant ϕ.

Finally we should note that the coordinate $\mathbf{r(t)}$ of a physical event is a function of time. This can be differentiated with respect to time to give successively its velocity, acceleration, etc. If curvilinear coordinates are used, there will be contributions from the time variations of both the tangent vectors and the curvilinear coordinates themselves. More specifically, we find by dividing Eq. (1.121) by dt that

$$\mathbf{v} = \frac{d\mathbf{r}}{dt} = \sum_i \left(\frac{du_i}{dt}\right) h_i \mathbf{e}_i, \tag{1.130}$$

while $\mathbf{a} = d\mathbf{v}/dt$ can be obtained by direct differentiation of the several factors appearing in each term of Eq. (1.130).

In spherical coordinates, for example, Eq. (1.130) reads

$$\mathbf{v} = \dot{r}\mathbf{e}_r + \dot{\theta} r \mathbf{e}_\theta + \dot{\phi} r \sin\theta \, \mathbf{e}_\phi. \tag{1.131}$$

Since it is also true that

$$\mathbf{v} = \frac{d}{dt}(r\mathbf{e}_r) = \dot{r}\mathbf{e}_r + r\dot{\mathbf{e}}_r,$$

we see that

$$\dot{\mathbf{e}}_r = \dot{\theta}\mathbf{e}_\theta + \dot{\phi} \sin\theta \, \mathbf{e}_\phi, \tag{1.132a}$$

a result that can also be obtained by a direct differentiation of $\mathbf{e}_r$ from Eq. (1.118). Similarly, we find by a direct differentiation of $\mathbf{e}_\theta$

$$\begin{aligned} \dot{\mathbf{e}}_\theta &= \dot{\theta}(-\sin\theta \cos\phi \, \mathbf{e}_x - \sin\theta \sin\phi \, \mathbf{e}_y - \cos\theta \, \mathbf{e}_z) \\ &\quad + \dot{\phi}(-\cos\theta \sin\phi \, \mathbf{e}_x + \cos\theta \cos\phi \, \mathbf{e}_y) \\ &= -\dot{\theta} \, \mathbf{e}_r + \dot{\phi} \cos\theta \, \mathbf{e}_\phi, \end{aligned} \tag{1.132b}$$

$$\begin{aligned} \dot{\mathbf{e}}_\phi &= \dot{\phi}(-\cos\phi \, \mathbf{e}_x - \sin\phi \, \mathbf{e}_y) \\ &= -\dot{\phi}(\sin\theta \, \mathbf{e}_r + \cos\theta \, \mathbf{e}_\theta). \end{aligned} \tag{1.132c}$$

These are the basic relations that can be used to simplify the time rate of change of any vector field expressed in terms of spherical coordinates.

Problems

1.10.1. Obtain the unit tangent vectors $\mathbf{e}_i(\mathbf{r})$ for the cylindrical coordinate system and show that they are orthogonal to each other.

1.10.2. Obtain the scale factors and the unit tangent vectors, demonstrate the orthogonality, and describe the coordinate curves and surfaces of the generalized coordinate systems defined in Table 1.2 (p. 62). Plots of some of these curves and surfaces can be found in Morse and Feshbach, *Methods of Theoretical Physics,* pp. 657–66, and in Arfken, *Mathematical Methods for Physicists,* 2nd ed., pp. 95–120. The latter reference contains the answers to this question. The parameters a, b, and c in the table are constants having the dimension of length. They are not the generalized coordinates.

1.10.3. Calculate the differential elements ds_i, $(ds)^2$, $d\boldsymbol{\sigma}_{ij}$, and $d\tau$ for the curvilinear coordinate systems of Problem 1.10.2.

1.10.4. Express $\mathbf{v} = d\mathbf{r}/dt$ and $\mathbf{a} = d\mathbf{v}/dt$ in cylindrical coordinates.

1.10.5. Express $\mathbf{a} = d^2\mathbf{r}/dt^2$ in spherical coordinates.

1.10.6. Express $\mathbf{v} = d\mathbf{r}/dt$ and $\mathbf{a} = d\mathbf{v}/dt$ in the curvilinear coordinate systems of Problem 1.10.2.

1.11. Vector differential operators in orthogonal curvilinear coordinate systems

Since the generalized coordinates $\mathbf{r} = (u_1, u_2, u_3)$ define a point in space, scalar and vector fields in space can be written as $\phi(u_1, u_2, u_3)$ or $\mathbf{V}(u_1, u_2, u_3)$. In performing vector differential operations in such fields, it is of course possible to change variables back to the rectangular coordinates x, y, and z before applying the VDO in its familiar rectangular form. This procedure is often unwise, because the calculation may become rather complicated and the physical insights to the problem that might have motivated the use of a special system of generalized coordinates might be lost this way.

The purpose of this section is to show how such calculations can be made directly by using the generalized coordinates themselves. The procedures are not very complicated and are well worth mastering because they give insight to our understanding of generalized coordinates. We shall restrict ourselves to orthogonal curvilinear coordinates.

Let us start with $\nabla\phi(u_1, u_2, u_3)$. At every point $\mathbf{r}$, it describes a vector that can be decomposed into components along the local unit tangents $\mathbf{e}_i(\mathbf{r})$:

$$\nabla\phi(\mathbf{r}) = \sum_{i=1}^{3} \mathbf{e}_i(\mathbf{r})(\nabla\phi)_i,$$

Table 1.2. Orthogonal curvilinear coordinate systems

Coordinates	x	y	z
(a) Elliptic cylindrical	$a \cosh u \cos v$	$a \sinh u \sin v$	z
(b) Parabolic cylindrical	$\xi\eta$	$\frac{1}{2}(\eta^2 - \xi^2)$	z
(c) Bipolar (cylindrical)	$\dfrac{a \sinh \eta}{\cosh \eta - \cos \xi}$	$\dfrac{a \sin \xi}{\cosh \eta - \cos \xi}$	z
(d) Prolate spheroidal	$a \sinh u \sin v \cos \phi$	$a \sinh u \sin v \sin \phi$	$a \cosh u \cos v$
(e) Oblate spheroidal	$a \cosh u \cos v \cos \phi$	$a \cosh u \cos v \sin \phi$	$a \sinh u \cos v$
(f) Parabolic	$\xi\eta \cos \phi$	$\xi\eta \sin \phi$	$\frac{1}{2}(\eta^2 - \xi^2)$
(g) Toroidal	$\dfrac{a \sinh \eta \cos \phi}{\cosh \eta - \cos \xi}$	$\dfrac{a \sinh \eta \sin \phi}{\cosh \eta - \cos \xi}$	$\dfrac{a \sin \xi}{\cosh \eta - \cos \xi}$
(h) Bispherical	$\dfrac{a \sin \xi \cos \phi}{\cosh \eta - \cos \xi}$	$\dfrac{a \sin \xi \sin \phi}{\cosh \eta - \cos \xi}$	$\dfrac{a \sinh \eta}{\cosh \eta - \cos \xi}$
(i) Confocal ellipsoidal	$\left(\dfrac{(a^2 - \xi_1)(a^2 - \xi_2)(a^2 - \xi_3)}{(a^2 - b^2)(a^2 - c^2)}\right)^{1/2}$	$\left(\dfrac{(b^2 - \xi_1)(b^2 - \xi_2)(b^2 - \xi_3)}{(b^2 - a^2)(b^2 - c^2)}\right)^{1/2}$	$\left(\dfrac{(c^2 - \xi_1)(c^2 - \xi_2)(c^2 - \xi_3)}{(c^2 - a^2)(c^2 - b^2)}\right)^{1/2}$
(j) Conical	$\left(\dfrac{\xi_1 \xi_2 \xi_3}{bc}\right)^{1/2}$	$\left(\dfrac{\xi_1^2(b^2 - \xi_2^2)(b^2 - \xi_3^2)}{b^2(b^2 - c^2)}\right)^{1/2}$	$\left(\dfrac{\xi_1^2(c^2 - \xi_2^2)(c^2 - \xi_3^2)}{c^2(c^2 - b^2)}\right)^{1/2}$
(k) Confocal parabolic	$\left(\dfrac{(a^2 - \xi_1)(a^2 - \xi_2)(a^2 - \xi_3)}{(b^2 - a^2)}\right)^{1/2}$	$\left(\dfrac{(b^2 - \xi_1)(b^2 - \xi_2)(b^2 - \xi_3)}{(a^2 - b^2)}\right)^{1/2}$	$\frac{1}{2}(a^2 + b^2 - \xi_1 - \xi_2 - \xi_3)$

where

$$(\nabla\phi)_i = (\nabla\phi)\cdot \mathbf{e}_i(\mathbf{r})$$
$$= \left(\frac{\partial\phi}{\partial x}\mathbf{e}_x + \frac{\partial\phi}{\partial y}\mathbf{e}_y + \frac{\partial\phi}{\partial z}\mathbf{e}_z\right)\cdot \mathbf{e}_i(\mathbf{r}).$$

The direction cosines involved turn out to be just the partial derivatives

$$\mathbf{e}_x\cdot\mathbf{e}_i = \mathbf{e}_x\cdot\frac{\partial\mathbf{r}}{\partial s_i} = \mathbf{e}_x\cdot\frac{\partial}{\partial s_i}(x\mathbf{e}_x + y\mathbf{e}_y + z\mathbf{e}_z)$$
$$= \frac{\partial x}{\partial s_i}.$$

Hence

$$(\nabla\phi_i) = \frac{\partial\phi}{\partial x}\frac{\partial x}{\partial s_i} + \frac{\partial\phi}{\partial y}\frac{\partial y}{\partial s_i} + \frac{\partial\phi}{\partial z}\frac{\partial z}{\partial s_i}$$
$$= \frac{\partial\phi}{\partial s_i},$$

and

$$\nabla\phi(u_1,u_2,u_3) = \sum_{i=1}^{3}\mathbf{e}_i(\mathbf{r})\frac{\partial\phi}{\partial s_i}. \tag{1.133}$$

That is, u_i in $\nabla\phi(u_1,u_2,u_3)$ can be treated as if they were rectangular coordinates if the local tangents $\mathbf{e}_i(\mathbf{r})$ are used together with the dimensioned displacements $ds_i = h_i\, du_i$.

Example 1.11.1. Calculate $\nabla(r^2\sin\theta)$.
According to Eq. (1.119)

$$\nabla\phi(r,\phi,\phi) = \left(\mathbf{e}_r\frac{\partial}{\partial r} + \mathbf{e}_\theta\frac{1}{r}\frac{\partial}{\partial\theta} + \mathbf{e}_\phi\frac{1}{r\sin\theta}\frac{\partial}{\partial\phi}\right)\phi(r,\theta,\phi).$$

Hence

$$\nabla(r^2\sin\theta) = \mathbf{e}_r 2r\sin\theta + \mathbf{e}_\theta r\cos\theta.$$

Perhaps the most interesting gradients in a curvilinear coordinate system are those constructed from the curvilinear coordinates $u_j(\mathbf{r})$ themselves:

$$\nabla u_j(\mathbf{r}) = \sum_i \mathbf{e}_i(\mathbf{r})\frac{1}{h_i}\frac{\partial u_j}{\partial u_i} = \frac{\mathbf{e}_j(\mathbf{r})}{h_j(\mathbf{r})}. \tag{1.134}$$

This result expresses the fact that $\nabla \mathrm{u_j}(\mathbf{r})$ changes most rapidly along the u_j coordinate curve. It also shows that $\mathbf{e}_j$ is normal to the coordinate surface defined by the equation

$$u_j(\mathbf{r}) = \text{const}.$$

Now the gradient of a scalar field is always irrotational; hence

$$\nabla \times (\mathbf{e}_j/h_j) = 0. \tag{1.135}$$

Since

$$\nabla \times (\mathbf{e}_j/h_j) = (\nabla h_j^{-1}) \times \mathbf{e}_j + h_j^{-1}(\nabla \times \mathbf{e}_j),$$

this result states that a generalized coordinate curve has a natural curliness of

$$\nabla \times \mathbf{e}_j = -h_j(\nabla h_j^{-1}) \times \mathbf{e}_j. \tag{1.136}$$

One consequence of Eq. (1.136) is that the curl of a vector field

$$\mathbf{V}(\mathbf{r}) = \sum_j \mathbf{e}_j(\mathbf{r}) V_j(\mathbf{r}) \tag{1.137}$$

has contributions from both $\mathbf{e}_j$ and V_j. Both contributions can be combined into a simple expression with the help of Eqs. (1.55) and (1.135):

$$\begin{aligned}
\nabla \times \mathbf{V} &= \sum_j \nabla \times \left(\frac{\mathbf{e}_j}{h_j} h_j V_j\right) \\
&= \sum_j \left[\left(\nabla \times \frac{\mathbf{e}_j}{h_j}\right) h_j V_j + \nabla(h_j V_j) \times \frac{\mathbf{e}_j}{h_j}\right] \\
&= \sum_j \nabla(h_j V_j) \times \frac{\mathbf{e}_j}{h_j} \\
&= \sum_{i,j,k} \varepsilon_{ijk} \frac{1}{h_j} \frac{\partial}{\partial s_i}(h_j V_j)\, \mathbf{e}_k.
\end{aligned} \tag{1.138a}$$

The result shows that, besides the usual "Cartesian" contribution proportional to $\partial V_j/\partial s_i$, there is an additional effect proportional to $(V_j/h_j)(\partial h_j/\partial s_i)$ that arises from the curliness of a coordinate curve.

Another useful way of writing Eq. (1.138a) is

$$\begin{aligned}
\nabla \times \mathbf{V} &= \sum_{i,j,k} \varepsilon_{ijk} \frac{1}{h_i h_j} \frac{\partial}{\partial u_i}(h_j V_j)\, \mathbf{e}_k \\
&= \frac{1}{h_1 h_2 h_3} \begin{vmatrix} h_1 \mathbf{e}_1 & h_2 \mathbf{e}_2 & h_3 \mathbf{e}_3 \\ \dfrac{\partial}{\partial u_1} & \dfrac{\partial}{\partial u_2} & \dfrac{\partial}{\partial u_3} \\ h_1 V_1 & h_2 V_2 & h_3 V_3 \end{vmatrix}.
\end{aligned} \tag{1.138b}$$

To get a similarly compact and useful expression for $\nabla \cdot \mathbf{V}$, we start with the curl of $u_m \nabla u_n$:

$$\begin{aligned}
\nabla \times (u_m \nabla u_n) &= (\nabla u_m) \times (\nabla u_n) + u_m \nabla \times (\nabla u_n) \\
&= \left(\frac{\mathbf{e}_m}{h_m}\right) \times \left(\frac{\mathbf{e}_n}{h_n}\right) \\
&= \sum_k \varepsilon_{mnk} \frac{\mathbf{e}_k}{h_m h_n}.
\end{aligned} \tag{1.139}$$

Since this vanishes for $m = n$ for which the permutation symbol has repeated indices, the nonzero cases involve $m \neq n$. For example,

$$\nabla \times (u_1 \nabla u_2) = \frac{\mathbf{e}_3}{h_1 h_2} = -\nabla \times (u_2 \nabla u_1).$$

Now according to Eq. (1.51), the curl of any vector field is always solenoidal and free of divergence. This holds also for the curl of the last equation

$$\nabla \cdot (\mathbf{e}_i/p_i) = 0, \qquad \text{if} \qquad p_i = h_1 h_2 h_3/h_i. \tag{1.140}$$

Since

$$\nabla \cdot (\mathbf{e}_i/p_i) = (\nabla p_i^{-1}) \cdot \mathbf{e}_i + p_i^{-1} \nabla \cdot \mathbf{e}_i,$$

this means that

$$\nabla \cdot \mathbf{e}_i = -p_i(\nabla p_i^{-1}) \cdot \mathbf{e}_i = -p_i \frac{\partial}{\partial s_i}\left(\frac{1}{p_i}\right)$$

does not in general vanish. In other words, the curliness of the coordinate curve contributes to the divergence.

We are now in a position to put these results together to get the divergence of a vector field written in generalized coordinates

$$\begin{aligned}\nabla \cdot \mathbf{V} &= \sum_i \nabla \cdot (\mathbf{e}_i V_i) = \sum_i \nabla \cdot \left[\left(\frac{\mathbf{e}_i}{p_i}\right) p_i V_i\right] \\ &= \sum_i \left[\left(\nabla \cdot \frac{\mathbf{e}_i}{p_i}\right) p_i V_i + \nabla(p_i V_i) \cdot \frac{\mathbf{e}_i}{p_i}\right] \\ &= \sum_i \frac{1}{p_i} \frac{\partial}{\partial s_i} (p_i V_i).\end{aligned} \tag{1.141}$$

We have thus reduced $\nabla \times \mathbf{V}$ and $\nabla \cdot \mathbf{V}$ in the OCC system to something involving derivatives of scalar fields that are similar in structure to those in rectangular coordinates. All one has to do is to put in an appropriate scale function before a differentiation and to divide it out afterwards, as shown in both Eqs. (1.138) and (1.141).

It is also not accidental that the formula for $\nabla \times \mathbf{V}$ comes from ∇u_i, while that for $\nabla \cdot \mathbf{V}$ is related to $u_m \nabla u_n$. The reason is that the surface integral of $\nabla \times \mathbf{V}$ is related to a line integral (i.e., the circulation of $\mathbf{V}$) according to Stokes's theorem. Hence it depends on the one-dimensional changes ∇u_i. For $\nabla \cdot \mathbf{V}$, Gauss's theorem relates a volume integral to a surface integral in which the scale factor h_i appears quadratically. The derivation of Eqs. (1.138) and (1.141) directly from these integral theorems will be left as exercises (Problems 1.11.7 and 1.11.8).

Example 1.11.2. Calculate $\boldsymbol{\nabla} \cdot [(\mathbf{e}_r + \mathbf{e}_\theta + \mathbf{e}_\phi)/r^2]$.

The scale functions for $\boldsymbol{\nabla} \cdot \mathbf{V}$ in spherical coordinates are

$$p_r = h_r h_\theta h_\phi / h_r = r^2 \sin\theta$$
$$p_\theta = r \sin\theta$$
$$p_\phi = r.$$

Hence

$$\begin{aligned}
\boldsymbol{\nabla} \cdot (\mathbf{e}_r V_r + \mathbf{e}_\theta V_\theta + \mathbf{e}_\phi V_\phi) &= \frac{1}{p_r}\frac{\partial}{\partial s_r}(p_r V_r) + \frac{1}{p_\theta}\frac{\partial}{\partial s_\theta}(p_\theta V_\theta) + \frac{1}{p_\phi}\frac{\partial}{\partial s_\phi}(p_\phi V_\phi) \\
&= \frac{1}{r^2 \sin\theta}\frac{\partial}{\partial r}(r^2 \sin\theta\, V_r) \\
&\quad + \frac{1}{r\, \sin\theta}\frac{1}{r}\frac{\partial}{\partial\theta}(r \sin\theta\, V_\theta) \\
&\quad + \frac{1}{r}\frac{1}{r \sin\theta}\frac{\partial}{\partial\phi}(r V_\phi) \\
&= \frac{1}{r^2}\frac{\partial}{\partial r}(r^2 V_r) + \frac{1}{r \sin\theta}\frac{\partial}{\partial\theta}(\sin\theta\, V_\theta) \\
&\quad + \frac{1}{r \sin\theta}\frac{\partial}{\partial\phi} V_\phi.
\end{aligned}$$

Therefore

$$\boldsymbol{\nabla} \cdot \frac{\mathbf{e}_r + \mathbf{e}_\theta + \mathbf{e}_\phi}{r^2} = \frac{1}{r \sin\theta}\frac{\partial}{\partial\theta}\left(\frac{\sin\theta}{r^2}\right) = \frac{\cot\theta}{r^3}.$$

1.11.1. Repeated operations

To avoid possible confusion and error in calculating repeated VDOs such as $\nabla^2 = \boldsymbol{\nabla} \cdot \boldsymbol{\nabla}$, it is best to proceed one step at a time, because the $p_j(\mathbf{r})$ functions may differ in different operators. This stepwise procedure is illustrated by the following examples.

Example 1.11.3. Obtain $\nabla^2 = \boldsymbol{\nabla} \cdot \boldsymbol{\nabla}$ in an OCC system.

For definiteness we may consider $\nabla^2 \Phi(u_1, u_2, u_3) = \boldsymbol{\nabla} \cdot \mathbf{V}(u_1, u_2, u_3)$, where $\mathbf{V} = \boldsymbol{\nabla}\Phi$.

(1) $\mathbf{V} = \boldsymbol{\nabla}\Phi, \qquad v_i = \dfrac{\partial}{\partial s_i}\Phi = \dfrac{\partial}{h_i\, \partial u_i}\Phi, \qquad \mathbf{V} = \sum_i V_i \mathbf{e}_i.$

(2) $\boldsymbol{\nabla} \cdot \mathbf{V} \to \dfrac{\partial}{\partial s_1} V_1 + \dfrac{\partial}{\partial s_2} V_2 + \dfrac{\partial}{\partial s_3} V_3$ if Cartesian; hence

$$\boldsymbol{\nabla} \cdot \mathbf{V}(u_1, u_2, u_3) = \frac{1}{p_1}\frac{\partial}{\partial s_1}(p_1 V_1) + \frac{1}{p_2}\frac{\partial}{\partial s_2}(p_2 V_2) + \frac{1}{p_3}\frac{\partial}{\partial s_3}(p_3 V_3), \quad (1.142)$$

where $p_i = h_1 h_2 h_3 / h_i$, $ds_i = h_i\, du_i$.

Example 1.11.4. Calculate $\nabla^2\Phi$ in spherical coordinates.
We note that

	For gradient	For divergence
$h_r = 1$	$(\boldsymbol{\nabla}\Phi)_r = \dfrac{\partial\Phi}{\partial r}$	$p_r = h_\theta h_\phi = r^2 \sin\theta$
$h_\theta = r$	$(\boldsymbol{\nabla}\Phi)_\theta = \dfrac{1}{r}\dfrac{\partial\Phi}{\partial\theta}$	$p_\theta = h_\phi h_r = r\sin\theta$
$h_\phi = r\sin\theta$	$(\boldsymbol{\nabla}\Phi)_\phi = \dfrac{1}{r\sin\theta}\dfrac{\partial\Phi}{\partial\phi}$	$p_\phi = h_r h_\theta = r$

Hence

$$\begin{aligned}
\nabla^2\Phi &= \sum_{i=1}^{3}\frac{1}{p_i}\frac{\partial}{\partial s_i}(p_i V_i)\\
&= \left[\frac{1}{r^2\sin\theta}\frac{\partial}{\partial r}\left(r^2\sin\theta\frac{\partial}{\partial r}\right) + \frac{1}{r\sin\theta}\left(\frac{\partial}{r\,\partial\theta}\right)\left(r\sin\theta\frac{\partial}{r\,\partial\theta}\right)\right.\\
&\quad \left. + \frac{1}{r}\left(\frac{\partial}{r\sin\theta\,\partial\phi}\right)\left(r\frac{\partial}{r\sin\theta\,\partial\phi}\right)\right]\Phi(r,\theta,\phi)\\
&= \left(\frac{1}{r^2}\frac{\partial}{\partial r}r^2\frac{\partial}{\partial r} + \frac{1}{r^2\sin\theta}\frac{\partial}{\partial\theta}\sin\theta\frac{\partial}{\partial\theta} + \frac{1}{r^2\sin^2\theta}\frac{\partial^2}{\partial\phi^2}\right)\Phi(r,\theta,\phi).
\end{aligned}$$

In both examples, the additional factors between differential operators come from the additional scale functions appearing in OCC systems.

Problems

1.11.1. Show that $\partial/\partial x$, $\partial/\partial y$, $\partial/\partial z$ expressed in terms of spherical coordinates are:

$$\frac{\partial}{\partial x} = \mathbf{e}_x\cdot\boldsymbol{\nabla} = \sin\theta\cos\phi\frac{\partial}{\partial r} + \cos\theta\cos\phi\frac{1}{r}\frac{\partial}{\partial\theta} - \frac{\sin\phi}{r\sin\theta}\frac{\partial}{\partial\phi},$$

$$\frac{\partial}{\partial y} = \sin\theta\sin\phi\frac{\partial}{\partial r} + \cos\theta\sin\phi\frac{1}{r}\frac{\partial}{\partial\theta} + \frac{\cos\phi}{r\sin\theta}\frac{\partial}{\partial\phi},$$

$$\frac{\partial}{\partial z} = \cos\theta\frac{\partial}{\partial r} - \sin\theta\frac{1}{r}\frac{\partial}{\partial\theta}.$$

1.11.2. Calculate $\nabla^2\Phi(\rho,\phi,z)$ in cylindrical coordinates.

1.11.3. Calculate $\boldsymbol{\nabla}\Phi(\rho,\phi,z)$ and $\nabla^2\Phi$ in cylindrical coordinates for
(a) $\Phi = \rho$,
(b) $\Phi = \rho^2 + z^2$,
(c) $\Phi = \rho^2\tan\phi + \rho z\tan^2\phi$.

1.11.4. Calculate $\nabla\Phi(r,\theta,\phi)$ and $\nabla^2\Phi$ in spherical coordinates for
(a) $\Phi = \Phi(r)$,
(b) $\Phi = r^2(\sin\theta + \cos\phi)$,
(c) $\Phi = Ar\cos\theta\cos\phi + Br^3\cos^3\theta\sin\phi$, A and B being constants.

1.11.5. Calculate $\nabla\cdot\mathbf{V}(\rho,\theta,z)$ and $\nabla\times\mathbf{V}$ in cylindrical coordinates for
(a) $\mathbf{V} = \rho\mathbf{e}_\rho + z\mathbf{e}_z$,
(b) $\mathbf{V} = \mathbf{e}_\rho$,
(c) $\mathbf{V} = \mathbf{e}_\phi$,
(d) $\mathbf{V} = \ln\rho\ \mathbf{e}_z$,
(e) $\mathbf{V} = \ln\rho\ \mathbf{e}_\phi$.

1.11.6. Calculate $\nabla\cdot\mathbf{V}(r,\theta,\phi)$ and $\nabla\times\mathbf{V}$ in spherical coordinates for
(a) $\mathbf{V} = r\mathbf{e}_r$,
(b) $\mathbf{V} = f(r)\mathbf{e}_r$,
(c) $\mathbf{V} = \mathbf{e}_\theta$,
(d) $\mathbf{V} = -(\cot\theta/r)\mathbf{e}_\phi$.

1.11.7. Use Gauss's theorem in the form [of Eq. (1.84)]

$$\nabla\cdot\mathbf{V} = \lim_{\Delta\tau\to 0}\frac{1}{\Delta\tau}\oint_{\Delta S}\mathbf{V}\cdot d\boldsymbol{\sigma}$$

to derive the formula for $\nabla\cdot\mathbf{V}$ in an OCC system.

1.11.8. Use Stokes's theorem in the form of Eq. (1.100)

$$(\nabla\times\mathbf{V})_i = \lim_{\Delta\sigma_i\to 0}\frac{1}{\Delta\sigma_i}\oint_{c_i}\mathbf{V}\cdot d\mathbf{r}$$

to derive the formula for $\nabla\times\mathbf{V}$ in an OCC system.

1.11.9. Obtain a general expression for each of the following VDO in OCC systems:
(a) $\nabla(\nabla\cdot\mathbf{V})$,
(b) $\nabla\cdot(\nabla\times\mathbf{V})$,
(c) $\nabla\times(\nabla\times\mathbf{V})$,
(d) $\nabla\times(\nabla\Phi)$,

where $\mathbf{V} = \sum_i V_i(u_1,u_2,u_3)\mathbf{e}_i$, $\Phi = \Phi(u_1,u_2,u_3)$ are understood to be given functions of the generalized coordinates.

1.11.10. Obtain explicit expressions for $\nabla\cdot\mathbf{V}$, $\nabla\times\mathbf{V}$, and $\nabla^2\Phi$ for each of the orthogonal curvilinear coordinate systems of Problem 1.10.2.

Appendix 1. Tables of mathematical formulas

1. Vectors

$$\mathbf{A}\cdot\mathbf{B} = A_xB_x + A_yB_y + A_zB_z = \mathbf{B}\cdot\mathbf{A}$$

$$\mathbf{A}\times\mathbf{B} = \begin{vmatrix} \mathbf{i} & \mathbf{j} & \mathbf{k} \\ A_x & A_y & A_z \\ B_x & B_y & B_z \end{vmatrix} = \sum_{i,j,k}\varepsilon_{ijk}\mathbf{e}_iA_jB_k$$

$$\mathbf{A} \times (\mathbf{B} \times \mathbf{C}) = \mathbf{B}(\mathbf{A} \cdot \mathbf{C}) - \mathbf{C}(\mathbf{A} \cdot \mathbf{B})$$

$$\mathbf{A} \cdot (\mathbf{B} \times \mathbf{C}) = \mathbf{B} \cdot (\mathbf{C} \times \mathbf{A}) = \mathbf{C} \cdot (\mathbf{A} \times \mathbf{B}) = \begin{vmatrix} A_x & A_y & A_z \\ B_x & B_y & B_z \\ C_x & C_y & C_z \end{vmatrix}$$

$$(\mathbf{A} \times \mathbf{B}) \cdot (\mathbf{C} \times \mathbf{D}) = (\mathbf{A} \cdot \mathbf{C})(\mathbf{B} \cdot \mathbf{D}) - (\mathbf{A} \cdot \mathbf{D})(\mathbf{B} \cdot \mathbf{C})$$

$$\begin{aligned}(\mathbf{A} \times \mathbf{B}) \times (\mathbf{C} \times \mathbf{D}) &= -\mathbf{A}[\mathbf{B} \cdot (\mathbf{C} \times \mathbf{D})] + \mathbf{B}[\mathbf{A} \cdot (\mathbf{C} \times \mathbf{D})] \\ &= \mathbf{C}[(\mathbf{A} \times \mathbf{B}) \cdot \mathbf{D}] - \mathbf{D}[(\mathbf{A} \times \mathbf{B}) \cdot \mathbf{C}].\end{aligned}$$

2. Vector differential operators

$$\nabla(\phi\psi) = \phi\, \nabla\psi + \psi\, \nabla\phi$$

$$\nabla \cdot (\phi\mathbf{A}) = \phi\, \nabla \cdot \mathbf{A} + (\nabla\phi) \cdot \mathbf{A}$$

$$\nabla \times (\phi\mathbf{A}) = \phi\, \nabla \times \mathbf{A} + (\nabla\phi) \times \mathbf{A}$$

$$\nabla \cdot (\mathbf{A} \times \mathbf{B}) = \mathbf{B} \cdot (\nabla \times \mathbf{A}) - \mathbf{A} \cdot (\nabla \times \mathbf{B})$$

$$\nabla \times (\mathbf{A} \times \mathbf{B}) = (\mathbf{B} \cdot \nabla)\mathbf{A} - \mathbf{B}(\nabla \cdot \mathbf{A}) - (\mathbf{A} \cdot \nabla)\mathbf{B} + \mathbf{A}\nabla \cdot \mathbf{B}$$

$$\nabla(\mathbf{A} \cdot \mathbf{B}) = (\mathbf{B} \cdot \nabla)\mathbf{A} + (\mathbf{A} \cdot \nabla)\mathbf{B} + \mathbf{B} \times (\nabla \times \mathbf{A}) + \mathbf{A} \times (\nabla \times \mathbf{B})$$

$$\nabla \times \nabla\phi = 0$$

$$\nabla \cdot (\nabla \times \mathbf{A}) = 0$$

$$\nabla \times (\nabla \times \mathbf{A}) = \nabla(\nabla \cdot \mathbf{A}) - \nabla^2\mathbf{A}$$

$$\nabla \cdot \mathbf{r} = 3$$

$$\nabla \times \mathbf{r} = 0$$

$$\nabla r^2 = 2\mathbf{r}$$

$$\nabla^2 r^2 = 6.$$

3. Orthogonal curvilinear coordinates (u_1, u_2, u_3)

$$\frac{\partial \mathbf{r}}{\partial u_i} = h_i(\mathbf{r})\mathbf{e}_i(\mathbf{r})$$

$$h_i(\mathbf{r}) = \left[\left(\frac{\partial x}{\partial u_i}\right)^2 + \left(\frac{\partial y}{\partial u_i}\right)^2 + \left(\frac{\partial z}{\partial u_i}\right)^2\right]^{1/2}$$

$$\mathbf{e}_i(\mathbf{r}) \cdot \mathbf{e}_j(\mathbf{r}) = \delta_{ij}$$

$$d\mathbf{r} = \sum_i h_i(\mathbf{r})\mathbf{e}_i(\mathbf{r})\, du_i = \sum_i ds_i(\mathbf{r})\, \mathbf{e}_i(\mathbf{r})$$

$$ds_i = h_i\, du_i$$

$$d\boldsymbol{\sigma}_{ij} = ds_i\, ds_j(\mathbf{e}_i \times \mathbf{e}_j)$$

$$d\tau = ds_1\, ds_2\, ds_3 = h_1 h_2 h_3\, du_1\, du_2\, du_3$$

$$\nabla u_i(\mathbf{r}) = \mathbf{e}_i(\mathbf{r})/h_i(\mathbf{r}), \qquad \nabla \times (\mathbf{e}_i/h_i) = 0$$

$$\nabla \times (u_i\, \nabla u_j) = \sum_k \frac{\varepsilon_{ijk}\mathbf{e}_k}{h_i h_j}, \qquad \nabla \cdot (\nabla \times u_i\, \nabla u_j) = 0$$

$$\nabla\phi(u_1,u_2,u_3)=\sum_i \mathbf{e}_i\frac{\partial\phi}{\partial s_i}$$

$$\nabla\cdot(V_1\mathbf{e}_1+V_2\mathbf{e}_2+V_3\mathbf{e}_3)=\sum_i\frac{1}{p_i}\frac{\partial}{\partial s_i}(p_iV_i),\qquad p_i=\frac{h_1h_2h_3}{h_i}$$

$$\nabla\times(V_1\mathbf{e}_1+V_2\mathbf{e}_2+V_3\mathbf{e}_3)=\sum_{i,j,k}\varepsilon_{ijk}\mathbf{e}_i\frac{1}{h_k}\frac{\partial}{\partial s_j}(h_kV_k)$$

$$\nabla^2\phi(u_1,u_2,u_3)=\sum_i\frac{1}{p_i}\frac{\partial}{\partial s_i}\left(p_i\frac{\partial}{\partial s_i}\right)\phi.$$

4. Cylindrical coordinates (ρ,ϕ,z)

$$x=\rho\cos\phi,\qquad y=\rho\sin\phi,\qquad z=z$$

$$h_\rho=1,\qquad h_\phi=\rho,\qquad h_z=1,\qquad d\tau=\rho\,d\rho\,d\phi\,dz$$

$$\mathbf{e}_\rho=\cos\phi\,\mathbf{i}+\sin\phi\,\mathbf{j}$$

$$\mathbf{e}_\phi=-\sin\phi\,\mathbf{i}+\cos\phi\,\mathbf{j}$$

$$\mathbf{e}_z=\mathbf{k}$$

$$\mathbf{e}_\rho\times\mathbf{e}_\phi=\mathbf{e}_z,\qquad \mathbf{e}_\phi\times\mathbf{e}_z=\mathbf{e}_\rho,\qquad \mathbf{e}_z\times\mathbf{e}_\rho=\mathbf{e}_\phi$$

$$\mathbf{r}=\rho\mathbf{e}_\rho+z\mathbf{e}_z$$

$$\dot{\mathbf{e}}_\rho=\frac{d}{dt}\mathbf{e}_\rho=\dot\phi\mathbf{e}_\phi$$

$$\dot{\mathbf{e}}_\phi=-\dot\phi\mathbf{e}_\rho$$

$$\dot{\mathbf{e}}_z=0$$

$$\mathbf{v}=\dot{\mathbf{r}}=\dot\rho\mathbf{e}_\rho+\rho\dot\phi\mathbf{e}_\phi+\dot z\mathbf{e}_z$$

$$\nabla\cdot(A_\rho\mathbf{e}_\rho+A_\phi\mathbf{e}_\phi+A_z\mathbf{e}_z)=\frac{1}{\rho}\frac{\partial}{\partial\rho}(\rho A_\rho)+\frac{1}{\rho}\frac{\partial}{\partial\phi}A_\phi+\frac{\partial}{\partial z}A_z$$

$$\nabla\times(A_\rho\mathbf{e}_\rho+A_\phi\mathbf{e}_\phi+A_z\mathbf{e}_z)=\frac{1}{\rho}\begin{vmatrix}\mathbf{e}_\rho & \rho\mathbf{e}_\phi & \mathbf{e}_z\\ \dfrac{\partial}{\partial\rho} & \dfrac{\partial}{\partial\phi} & \dfrac{\partial}{\partial z}\\ A_\rho & \rho A_\phi & A_z\end{vmatrix}$$

$$\nabla^2u(\rho,\phi,z)=\frac{1}{\rho}\frac{\partial}{\partial\rho}\rho\frac{\partial u}{\partial\rho}+\frac{1}{\rho^2}\frac{\partial^2u}{\partial\phi^2}+\frac{\partial^2u}{\partial z^2}.$$

5. Spherical coordinates (r,θ,ϕ)

$$x=r\sin\theta\cos\phi,\qquad y=r\sin\theta\sin\phi,\qquad z=r\cos\theta$$

$$h_r=1,\qquad h_\theta=r,\qquad h_\phi=r\sin\theta,\qquad d\tau=r^2\sin\theta\,dr\,d\theta\,d\phi$$

$$\mathbf{e}_r=\sin\theta\cos\phi\mathbf{i}+\sin\theta\sin\phi\,\mathbf{j}+\cos\theta\,\mathbf{k}$$

$$\mathbf{e}_\theta=\cos\theta\cos\phi\,\mathbf{i}+\cos\theta\sin\phi\,\mathbf{j}-\sin\theta\,\mathbf{k}$$

$$\mathbf{e}_\phi = -\sin\phi\,\mathbf{i} + \cos\phi\,\mathbf{j}$$

$$\mathbf{e}_r \times \mathbf{e}_\theta = \mathbf{e}_\phi, \qquad \mathbf{e}_\theta \times \mathbf{e}_\phi = \mathbf{e}_r, \qquad \mathbf{e}_\phi \times \mathbf{e}_r = \mathbf{e}_\theta$$

$$\mathbf{r} = r\mathbf{e}_r$$

$$\dot{\mathbf{e}}_r = \dot{\theta}\mathbf{e}_\theta + \dot{\phi}\sin\theta\,\mathbf{e}_\phi$$

$$\dot{\mathbf{e}}_\theta = -\dot{\theta}\mathbf{e}_r + \dot{\phi}\cos\theta\,\mathbf{e}_\phi$$

$$\dot{\mathbf{e}}_\phi = -\dot{\phi}(\sin\theta\,\mathbf{e}_r + \cos\theta\,\mathbf{e}_\theta)$$

$$\mathbf{v} = \dot{\mathbf{r}} = \dot{r}\mathbf{e}_r + \dot{\theta}r\mathbf{e}_\theta + \dot{\phi}r\sin\theta\,\mathbf{e}_\phi$$

$$\nabla \cdot (A_r\mathbf{e}_r + A_\theta\mathbf{e}_\theta + A_\phi\mathbf{e}_\phi) = \frac{1}{r^2}\frac{\partial}{\partial r}(r^2A_r) + \frac{1}{r\sin\theta}\frac{\partial}{\partial\theta}(\sin\theta\,A_\theta) + \frac{1}{r\sin\theta}\frac{\partial}{\partial\phi}A_\phi$$

$$\nabla \times (A_r\mathbf{e}_r + A_\theta\mathbf{e}_\theta + A_\phi\mathbf{e}_\phi) = \frac{1}{r^2\sin\theta}\begin{vmatrix} \mathbf{e}_r & r\mathbf{e}_\theta & r\sin\theta\,\mathbf{e}_\phi \\ \dfrac{\partial}{\partial r} & \dfrac{\partial}{\partial\theta} & \dfrac{\partial}{\partial\phi} \\ A_r & rA_\theta & r\sin\theta\,A_\phi \end{vmatrix}$$

$$\nabla^2 u(r,\theta,\phi) = \frac{1}{r^2}\frac{\partial}{\partial r}\left(r^2\frac{\partial u}{\partial r}\right) + \frac{1}{r^2\sin\theta}\frac{\partial}{\partial\theta}\left(\sin\theta\frac{\partial u}{\partial\theta}\right) + \frac{1}{r^2\sin^2\theta}\frac{\partial^2 u}{\partial\phi^2}.$$

2

TRANSFORMATIONS, MATRICES, AND OPERATORS

2.1. Transformations and the laws of physics

Physics deals with physical events in space-time. Given a physical system at a given point in space-time, we would like to know how it looked in the past, and how it will evolve in the future. In addition, we wonder if it might be related to other relevant events in the past, in the future, and at other points and orientations in space. To answer these questions we need space-time machines to move us back and forth in space-time so that we can examine the situation ourselves. These space-time machines are called *space-time transformation operators.*

In this chapter, we show that the transformation operators in space-time can appear as square matrices and as differential operators. Square matrices can be multiplied into column vectors to give other column vectors; functions can be differentiated to yield other functions. If we represent the state of a physical system by a column vector or by a function (of position and/or time), its evolution in space-time can then be symbolized mathematically as the consequence of an operation by a square matrix or by a differential operator.

Transformation operators also appear in statements of physical laws. A law of physics codifies the behavior of a class of physical events. Since the evolution of physical events can be symbolized by transformation operators, a physical law is essentially a statement concerning one or more of these operators. For example, Newton's equation of motion in its simplest form

$$m\frac{d^2}{dt^2}x(t) = F(x,t)$$

describes how the position of a mass is changed, or transformed, in response to a driving force. An even simpler physical law is contained in the (one-dimensional) wave equation

$$\left(\frac{\partial^2}{\partial x^2} - \frac{1}{v^2}\frac{\partial^2}{\partial t^2}\right)u(x,t) = 0.$$

This states that wave motion will appear if the transformation operator $\partial^2/\partial x^2$ in space happens to be proportional to the transformation operator $\partial^2/\partial t^2$ in time. We thus see that a wave is a manifestation of a certain correlation between disturbances in space and those in time.

Besides simple space-time properties, a physical system may also have additional "internal" attributes that may or may not be describable directly in the space-time language. Their transformations are often represented in matrix form. Since successive transformations add up to a total transformation, matrices, like space-time transformations, form mathematical structures called *groups*. Modern physics is often a detective story in which an unknown group structure behind relevant experimental observations is to be identified. In this way, new physical laws can be established even before we can describe the internal attribute in space-time terms. To help the reader in his future adventures in group theory, this chapter closes with a short section describing briefly a number of common matrix groups.

2.2. Rotations in space: Matrices

Besides scalars and vectors, which are linear arrays of numbers, we also use rectangular arrays of numbers called *matrices*. Matrices are particularly useful in describing rotations in space.

By a *rotation in space* we mean *either* the rotation of a point in space about the origin of the coordinate system, the coordinate axes being fixed, *or* the rotation of the coordinate axes about the origin, the point in space being fixed. It is obvious that these two rotations are simply related to each other. In the present context, it is more convenient to consider the second type of rotation—the rotation of coordinate axes.

Let us therefore suppose that we are interested in the position $\mathbf{r}$ of a fixed point in space in two coordinate systems related by a rotation:

$$\begin{aligned}\mathbf{r} &= x_1\mathbf{e}_1 + x_2\mathbf{e}_2 + x_3\mathbf{e}_3 \\ &= x_1'\mathbf{e}_1' + x_2'\mathbf{e}_2' + x_3'\mathbf{e}_3'.\end{aligned} \tag{2.1}$$

We ask: What is the relation between the "new" coordinates (x_1',x_2',x_3') and the old coordinates (x_1,x_2,x_3)?

The answer can be obtained readily by writing the old unit vectors $\mathbf{e}_i$ in terms of the new vectors $\mathbf{e}_i'$ with the help of the completeness relation (Eq. 1.20):

$$\mathbf{e}_j = \mathbf{e}_1'(\mathbf{e}_1' \cdot \mathbf{e}_j) + \mathbf{e}_2'(\mathbf{e}_2' \cdot \mathbf{e}_j) + \mathbf{e}_3'(\mathbf{e}_3' \cdot \mathbf{e}_j) = \sum_{i=1}^{3} \mathbf{e}_i'\lambda_{ij}, \tag{2.2}$$

where

$$\lambda_{ij} = \mathbf{e}_i' \cdot \mathbf{e}_j = \mathbf{e}_i^{\text{new}} \cdot \mathbf{e}_j^{\text{old}} \tag{2.3}$$

is a direction cosine. Thus

$$\begin{aligned}\mathbf{r} = x_1&(\mathbf{e}_1'\lambda_{11} + \mathbf{e}_2'\lambda_{21} + \mathbf{e}_3'\lambda_{31}) \\ &+ x_2(\mathbf{e}_1'\lambda_{12} + \mathbf{e}_2'\lambda_{22} + \mathbf{e}_3'\lambda_{32}) \\ &+ x_3(\mathbf{e}_1'\lambda_{13} + \mathbf{e}_2'\lambda_{23} + \mathbf{e}_3'\lambda_{33}).\end{aligned}$$

That is,

$$x_i' = \lambda_{i1}x_1 + \lambda_{i2}x_2 + \lambda_{i3}x_3 = \sum_{j=1}^{3} \lambda_{ij}x_j, \tag{2.4}$$

with the indices j for the old coordinate system appearing together side by side in λ and x. The summations in Eqs. (2.2) and (2.4) are the same as those for matrix multiplications, as we shall see shortly. For this reason, Eq. (2.4) can be written conveniently in the following compact "matrix" form

$$\begin{pmatrix} x_1' \\ x_2' \\ x_3' \end{pmatrix} = \begin{pmatrix} \lambda_{11} & \lambda_{12} & \lambda_{13} \\ \lambda_{21} & \lambda_{22} & \lambda_{23} \\ \lambda_{31} & \lambda_{32} & \lambda_{33} \end{pmatrix} \begin{pmatrix} x_1 \\ x_2 \\ x_3 \end{pmatrix}, \tag{2.5}$$

or

$$\mathbf{x}' = \lambda \mathbf{x}, \tag{2.6}$$

where $\mathbf{x}$ and $\mathbf{x}'$ are the column matrices (also called *column vectors*). The 3×3 matrix λ is a tabular array contains the 9 direction cosines between the two sets of coordinate axes. It is called a *transformation matrix*, or more specifically a *rotation matrix*.

Example 2.2.1. The xy (or 12) axes are rotated by an angle θ in a positive (i.e., counterclockwise) direction, as shown in Fig. 2.1. What is the resulting rotation matrix for the transformation of the coordinates of a fixed point in space?

We note that

$$\lambda_{11} = \mathbf{e}_1' \cdot \mathbf{e}_1 = \cos\theta, \qquad \lambda_{12} = \mathbf{e}_1' \cdot \mathbf{e}_2 = \sin\theta,$$
$$\lambda_{13} = \mathbf{e}_1' \cdot \mathbf{e}_3 = 0, \qquad \text{etc.}$$

Therefore

$$\lambda(\theta) = \begin{pmatrix} \cos\theta & \sin\theta & 0 \\ -\sin\theta & \cos\theta & 0 \\ 0 & 0 & 1 \end{pmatrix} \tag{2.7}$$

is the *rotation matrix* describing the coordinate changes under a rotation.

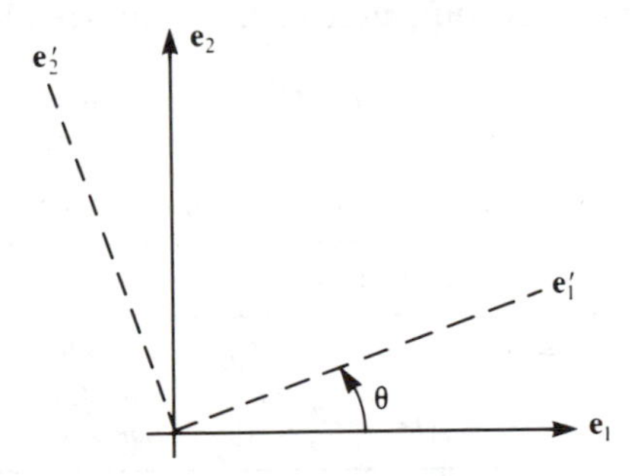

Fig. 2.1. A two-dimensional rotation.

Table 2.1. Matrix operations

Operation	Matrix element		A	B	C
Equality	$C = A$	$C_{ij} = A_{ij}$	$m \times n$	—	$m \times n$
Addition	$C = A + B$	$C_{ij} = A_{ij} + B_{ij}$	$m \times n$	$m \times n$	$m \times n$
Zero	$A = 0$	$A_{ij} = 0$	$m \times n$	—	—
Multiplication by a scalar	$C = \alpha A$	$C_{ij} = \alpha A_{ij}$	$m \times n$	—	$m \times n$
Matrix multiplication	$C = AB$				
	$\neq BA$	$C_{ij} = \sum_{k=1}^{n} A_{ik} B_{kj}$	$m \times n$	$n \times l$	$m \times l$

2.2.1. *Matrices*

It is worthwhile to review the algebra of matrices at this point. A rectangular array of numbers (called *elements* or *matrix elements*) A_{ij}

$$A = \begin{pmatrix} A_{11} & \cdots & A_{1n} \\ \vdots & & \\ A_{m1} & \cdots & A_{mn} \end{pmatrix} = (A_{ij})$$

containing m rows and n columns is called an $m \times n$ *matrix* if it satisfies the *matrix operations* of Table 2.1. The definition of a zero $m \times n$ matrix also implies the existence of a negative matrix

$$(-A)_{ij} = -A_{ij}.$$

Thus $m \times n$ matrices are "closed under addition," meaning that the result of an addition is also a matrix.

Table 2.2 shows the operations that can be performed on matrices. A matrix that remains unchanged under one of these operations is given a special name. For an operation involving a *transposition* (under which rows become columns), only square matrices (with $m = n$) can be invariant under the operation. A square matrix is *symmetric* if $A_{ji} = A_{ij}$; it is *antisymmetric* if $A_{ji} = -A_{ij}$.

The number of rows or columns of a square matrix is called its *order* or *degree*. A square matrix in which only diagonal matrix elements are nonzero is a *diagonal* matrix. A diagonal matrix with ones along the diagonal is the *identity*, or *unit*, matrix I. Its matrix elements are

$$I_{ij} = \delta_{ij}. \tag{2.8}$$

Table 2.2. Operations on matrices

Operation	Matrix element		A	C	if $C = A$
Transposition	$C = A^{\mathrm{T}}$	$C_{ij} = A_{ji}$	$m \times n$	$n \times m$	Symmetric[a]
Complex conjugation	$C = A^*$	$C_{ij} = A_{ij}^*$	$m \times n$	$m \times n$	Real
Hermitian conjugation	$C = A^\dagger = A^{\mathrm{T}*}$	$C_{ij} = A_{ji}^*$	$m \times n$	$n \times m$	Hermitian[a,b]

[a] For square matrices only.

[b] Named after Hermite.

The square matrix A^{-1} that satisfies the properties

$$A^{-1}A = AA^{-1} = I$$

is called the *inverse* of the square matrix A. A square matrix A that does not have an inverse is a *singular matrix*. The sum of the diagonal matrix elements of a square matrix is called its *trace*:

$$\operatorname{Tr} A = \sum_{i=1}^{n} A_{ii}. \tag{2.9}$$

An *orthogonal* matrix is one whose transpose is its inverse. Thus

$$(O^{\mathrm{T}}O)_{ij} = \delta_{ij} = (OO^{\mathrm{T}})_{ij},$$

that is,

$$\sum_k O_{ki}O_{kj} = \delta_{ij} = \sum_k O_{ik}O_{jk}. \tag{2.10}$$

These are called *orthogonality relations.* A *unitary* matrix is one whose *Hermitian conjugate* is its inverse. Thus

$$(U^{\dagger}U)_{ij} = \delta_{ij} = (UU^{\dagger})_{ij},$$

that is,

$$\sum_k U^*_{ki}U_{kj} = \delta_{ij} = \sum_k U_{ik}U^*_{jk}. \tag{2.11}$$

These are called *unitarity relations.*

Finally, we note that the order of the matrix multiplication is important. For example, if AB is defined, as shown in Table 2.1, the product BA does not even exist, unless $m = l$. For square matrices $(m = n = l)$, BA is always defined, but may not be equal to AB.

Example 2.2.2.

$$A = \begin{pmatrix} 1 & 2 \\ 2 & 4 \end{pmatrix} \quad \text{is symmetric, } \operatorname{Tr} A = 5.$$

$$B = \begin{pmatrix} 0 & 2 \\ -2 & 0 \end{pmatrix} \quad \text{is antisymmetric.}$$

$$C = \begin{pmatrix} 0 & i \\ -i & 0 \end{pmatrix} \quad \text{is Hermitian because}$$

$$C^{\mathrm{T}} = \begin{pmatrix} 0 & -i \\ i & 0 \end{pmatrix}, \quad \text{and} \quad C^* = \begin{pmatrix} 0 & -i \\ i & 0 \end{pmatrix} = C^{\mathrm{T}}.$$

$$AC = i\begin{pmatrix} -2 & 1 \\ -4 & 2 \end{pmatrix}, \quad \text{while}$$

$$CA = i\begin{pmatrix} 2 & 4 \\ -1 & -2 \end{pmatrix} \neq AC.$$

The simplest matrices are 2×2 matrices, each of which has four matrix elements. It is convenient to express them in terms of the four linearly independent 2×2 matrices

$$I = \begin{pmatrix} 1 & 0 \\ 0 & 1 \end{pmatrix}, \qquad \sigma_1 = \begin{pmatrix} 0 & 1 \\ 1 & 0 \end{pmatrix},$$
$$\sigma_2 = \begin{pmatrix} 0 & -i \\ i & 0 \end{pmatrix}, \qquad \sigma_3 = \begin{pmatrix} 1 & 0 \\ 0 & -1 \end{pmatrix} \tag{2.12}$$

so that

$$a_1\sigma_1 + a_2\sigma_2 + a_3\sigma_3 + bI = \mathbf{a} \cdot \boldsymbol{\sigma} + bI$$
$$= \begin{pmatrix} b + a_3 & a_1 - ia_2 \\ a_1 + ia_2 & b - a_3 \end{pmatrix} = \begin{pmatrix} A_{11} & A_{12} \\ A_{21} & A_{22} \end{pmatrix} = A, \tag{2.13}$$

where

$$\mathbf{a} = a_1\mathbf{e}_1 + a_2\mathbf{e}_2 + a_3\mathbf{e}_3,$$
$$\boldsymbol{\sigma} = \sigma_1\mathbf{e}_1 + \sigma_2\mathbf{e}_2 + \sigma_3\mathbf{e}_3.$$

The matrices σ_i were used by Pauli to describe the three components of the spin (a vector) of the electron. They are called *Pauli spin matrices*. One can verify by direct calculation that they satisfy the relation

$$\sigma_i\sigma_j = \delta_{ij} + i \sum_k \varepsilon_{ijk}\sigma_k, \tag{2.14a}$$

where ε_{ijk} is a permutation symbol. Equation (2.14a) is equivalent to the relations

$$\sigma_i^2 = I, \qquad \sigma_i\sigma_j = i\sigma_k, \tag{2.14b}$$

where (i,j,k) is a cyclic permutation of (1,2,3). It is also useful to think of $\pm I$, $\pm\sigma_1$, $\pm\sigma_2$, $\pm\sigma_3$ as eight of the square roots of the 2×2 matrix I. Unlike simple numbers, these matrices do not commute, and are said to make up a *noncommuting algebra*.

The expression (2.13) of 2×2 matrices in terms of Pauli matrices may appear very complicated, but it is actually no worse than the representation of vectors by their components. Vector components are extracted from a given vector by taking scalar products. The components a_i and b in Eq. (2.13) can be calculated from a given matrix A by taking traces. This is done by first noting that

$$\operatorname{Tr} I = 2, \qquad \operatorname{Tr} \sigma_i = 0.$$

As a consequence

$$\operatorname{Tr} A = a_1 \operatorname{Tr} \sigma_1 + a_2 \operatorname{Tr} \sigma_2 + a_3 \operatorname{Tr} \sigma_3 + b \operatorname{Tr} I = 2b$$
$$\operatorname{Tr}(A\sigma_1) = \operatorname{Tr}(a_1 + a_2\sigma_2\sigma_1 + a_3\sigma_3\sigma_1 + b\sigma_1)$$
$$= \operatorname{Tr}[a_1 + a_2(-i\sigma_3) + a_3(i\sigma_2) + b\sigma_1]$$
$$= \operatorname{Tr} a_1 = 2a_1, \qquad \text{etc.}$$

Hence

$$A = \sum_i \sigma_i \tfrac{1}{2} \operatorname{Tr}(A\sigma_i) + (\tfrac{1}{2} \operatorname{Tr} A)I.$$

2.2.2. *Transformation matrices*

Let us return to matrices for coordinate transformations. Since coordinate transformations are reversible by interchanging old and new indices, we must have

$$(\lambda^{-1})_{ij} = \hat{e}_i^{\text{old}} \cdot \hat{e}_j^{\text{new}} = \hat{e}_j^{\text{new}} \cdot \hat{e}_i^{\text{old}} = \lambda_{ji} = (\lambda^{\text{T}})_{ij}. \tag{2.15}$$

Hence transformation matrices are orthogonal matrices.

Example 2.2.3. The rotation matrix of Eq. (2.7) is orthogonal because

$$\lambda^{\text{T}}(\theta)\lambda(\theta) = \begin{pmatrix} \cos\theta & -\sin\theta & 0 \\ \sin\theta & \cos\theta & 0 \\ 0 & 0 & 1 \end{pmatrix}\begin{pmatrix} \cos\theta & \sin\theta & 0 \\ -\sin\theta & \cos\theta & 0 \\ 1 & 0 & 1 \end{pmatrix} = I.$$

A rotation matrix such as that given in Eq. (2.7) is a continuous function of its argument θ. As a result, its determinant is also a continuous function. In fact, it is equal to 1 for any θ. In addition, there are matrices of coordinate transformations with a determinant of -1. These transformations change the handedness of the coordinate system. Examples of such *parity transformations* are

$$P_1 = \begin{pmatrix} -1 & 0 & 0 \\ 0 & 1 & 0 \\ 0 & 0 & 1 \end{pmatrix}, \qquad P_3 = \begin{pmatrix} -1 & 0 & 0 \\ 0 & -1 & 0 \\ 0 & 0 & -1 \end{pmatrix}, \qquad P_i^2 = I.$$

They change the signs of an odd number of coordinates of a fixed point $\mathbf{r}$ in space, and therefore must involve a change in handedness of the coordinate system, as shown in Fig. 2.2. A change of handedness, unlike a simple rotation, cannot be made in a continuous manner.

We note finally that a rotation matrix such as that given in Eq. (2.7) can often be obtained by a geometrical method, but it is easier to use the simple algebraic (i.e., abstract) method described here. All we need here is the ability to calculate the 9 direction cosines of Eq. (2.3).

2.2.3. *Successive transformations*

One advantage of the matrix method for coordinate transformations is that successive transformations $1, 2, \ldots, m$ of the coordinate axes about the origin are described by successive matrix multiplications as far as their effects on the coordinates of a fixed point are concerned. This is because, if $\mathbf{x}^{(1)} = \lambda_1\mathbf{x}$, $\mathbf{x}^{(2)} = \lambda_2\mathbf{x}^{(1)}, \ldots,$ then

$$\mathbf{x}^{(m)} = \lambda_m\mathbf{x}^{(m-1)} = (\lambda_m\lambda_{m-1}\cdots\lambda_1)\mathbf{x} = L\mathbf{x},$$

where

$$L = \lambda_m\lambda_{m-1}\cdots\lambda_1 \tag{2.16}$$

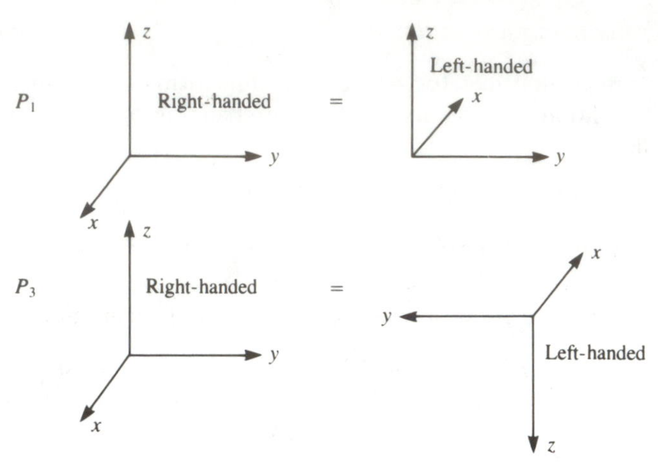

Fig. 2.2. Parity transformations of the coordinate system.

is the net transformation matrix for the m successive transformations taken in the specified manner. Note that λ_k acts on the kth coordinate system obtained after $k-1$ rotations.

Example 2.2.4. Consider a rotation of the xy axes about the z axis by an angle θ. If this rotation is followed by a back-rotation of the same angle θ in the opposite direction, that is, by $-\theta$, we recover the original coordinate system. Thus

$$R(-\theta)R(\theta) = I = \begin{pmatrix} 1 & 0 & 0 \\ 0 & 1 & 0 \\ 0 & 0 & 1 \end{pmatrix} = R^{-1}(\theta)R(\theta).$$

Hence

$$R^{-1}(\theta) = R(-\theta) = \begin{pmatrix} \cos\theta & -\sin\theta & 0 \\ \sin\theta & \cos\theta & 0 \\ 0 & 0 & 1 \end{pmatrix} = R^{\mathrm{T}}(\theta). \tag{2.17}$$

Example 2.2.5. Use rotation matrices to show that

$$\sin(\theta_1 + \theta_2) = \sin\theta_1 \cos\theta_2 + \cos\theta_1 \sin\theta_2.$$

We make two successive two-dimensional rotations of angles θ_1 and θ_2

$$R(\theta_2)R(\theta_1) = R(\theta_1 + \theta_2),$$

that is

$$\begin{pmatrix} \cos\theta_2 & \sin\theta_2 \\ -\sin\theta_2 & \cos\theta_2 \end{pmatrix}\begin{pmatrix} \cos\theta_1 & \sin\theta_1 \\ -\sin\theta_1 & \cos\theta_1 \end{pmatrix} = \begin{pmatrix} \cos(\theta_1+\theta_2) & \sin(\theta_1+\theta_2) \\ -\sin(\theta_1+\theta_2) & \cos(\theta_1+\theta_2) \end{pmatrix}.$$

A direct matrix multiplication of the LHS gives

$$\begin{pmatrix} \cos\theta_1\cos\theta_2-\sin\theta_1\sin\theta_2 & \sin\theta_1\cos\theta_2+\cos\theta_1\sin\theta_2 \\ -\sin\theta_1\cos\theta_2-\cos\theta_1\sin\theta_2 & -\sin\theta_1\sin\theta_2+\cos\theta_1\cos\theta_2 \end{pmatrix},$$

in which the 12 matrix element represents the desired result.

Two-dimensional rotations (i.e., successive rotations about the same axis $\mathbf{e}_i$) are *Abelian,* that is,

$$R_i(\theta_2)R_i(\theta_1)=R_i(\theta_1)R_i(\theta_2).$$

However, rotations about different axes do not generally commute. Thus

$$\begin{aligned} R_z(\alpha)R_y(\beta) &= \begin{pmatrix} \cos\alpha & \sin\alpha & 0 \\ -\sin\alpha & \cos\alpha & 0 \\ 0 & 0 & 1 \end{pmatrix}\begin{pmatrix} \cos\beta & 0 & -\sin\beta \\ 0 & 1 & 0 \\ \sin\beta & 0 & \cos\beta \end{pmatrix} \\ &= \begin{pmatrix} \cos\alpha\cos\beta & \sin\alpha & -\cos\alpha\sin\beta \\ -\sin\alpha\cos\beta & \cos\alpha & \sin\alpha\sin\beta \\ \sin\beta & 0 & \cos\beta \end{pmatrix}, \end{aligned}$$

whereas

$$\begin{aligned} R_y(\beta)R_z(\alpha) &= \begin{pmatrix} \cos\alpha\cos\beta & \sin\alpha\cos\beta & -\sin\beta \\ -\sin\alpha & \cos\alpha & 0 \\ \cos\alpha\sin\beta & \sin\alpha\sin\beta & \cos\beta \end{pmatrix} \\ &\neq R_z(\alpha)R_y(\beta). \end{aligned}$$

This distinction is illustrated in Fig. 2.3.

Example 2.2.6. The orientation of a rigid body about its center of mass is specified by three angles, not two. The reason is that, while an axis rigidly attached to the body can be specified by only two angles (e.g., the longitude and latitude on a sphere), a third angle is needed to specify the orientation of the rigid body about this axis. A convenient choice of the three orientational angles is given by the *Euler angles* (α, β, γ) defined by first rotating the body about a z axis by an angle α, then about the new x

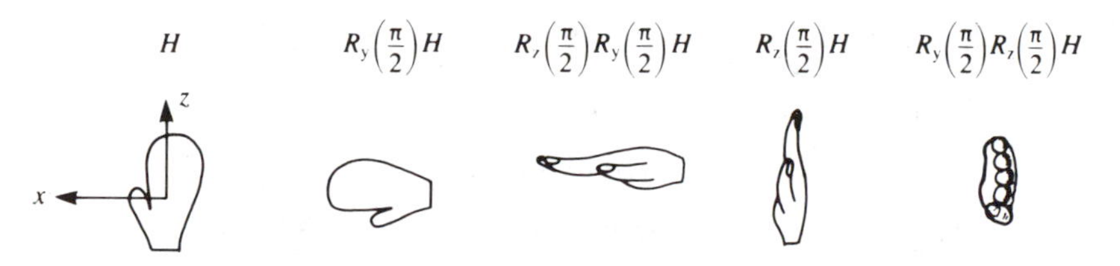

Fig. 2.3. Rotations of the right hand about its center of mass.

axis by an angle β, and finally about the latest z axis by an angle γ:

$$R(\alpha,\beta,\gamma)=R_z(\gamma)R_x(\beta)R_z(\alpha). \tag{2.18}$$

We should note the following: (1) An earlier rotation appears further to the right in the matrix multiplication. (2) The rotational axes of two successive rotations cannot be chosen the same, otherwise the rotation angles are not independent.

Problems

2.2.1. (a) Show that any square matrix B can be written in the form $B=S+A$, where S is symmetric and A is antisymmetric.
(b) Obtain the symmetric and antisymmetric parts of

$$B=\begin{pmatrix} 1 & 2 & 3i \\ 4 & 5i & 6 \\ 7i & 8 & 9 \end{pmatrix}.$$

2.2.2. Use rotation matrices to obtain explicit expressions for $\sin(\theta_1-\theta_2)$ and $\cos(\theta_1-\theta_2)$.

2.2.3. Show explicitly that $R^n(\theta)=R(n\theta)$.

2.2.4. There is a constant vector $\mathbf{V}=(V_x,V_y)=(2,1)$ at the point $(x,y)=(5,10)$ in a two-dimensional space. If the coordinate axes are rotated by 30° in the positive sense in the xy plane (i.e., about the z axis), calculate the components of $\mathbf{V}$ in the new (rotated) coordinate system.

2.2.5. Obtain explicitly the rotation matrices $R_x(\beta)$ and $R(\alpha,\beta,\gamma)$ of Eq. (2.18).

2.2.6. For any matrix A, show that $A+A^{\mathrm{T}}$ is symmetric, while $A-A^{\mathrm{T}}$ is *skew-symmetric* (or antisymmetric).

2.2.7. Show that $\mathrm{Tr}\,AB=\mathrm{Tr}\,BA$.

2.2.8. Show that two $n\times n$ square matrices do not necessarily commute even when one is diagonal.

2.2.9. (a) Show that $(AB)^\dagger=B^\dagger A^\dagger$.
(b) Show that the commutator

$$[H_1, H_2]=H_1H_2-H_2H_1$$

of two Hermitian matrices is *skew-Hermitian* or *anti-Hermitian*. (A square matrix A is skew-Hermitian if $A^\dagger=-A$).

2.2.10. Verify that the 2×2 Pauli matrices defined in Eq. (2.12) satisfy the multiplicative relations shown in Eqs. (2.14a) and (2.14b).

2.2.11. With the help of Eq. (2.13), show that

$$\mathrm{Tr}[(\boldsymbol{\sigma}\cdot\mathbf{a})(\boldsymbol{\sigma}\cdot\mathbf{b})]=2\mathbf{a}\cdot\mathbf{b},$$
$$\mathrm{Tr}[(\boldsymbol{\sigma}\cdot\mathbf{a})(\boldsymbol{\sigma}\cdot\mathbf{b})(\boldsymbol{\sigma}\cdot\mathbf{c})]=2i\mathbf{a}\cdot(\mathbf{b}\times\mathbf{c}).$$

2.2.12. With the help of Eq. (2.13), show that

$$(\boldsymbol{\sigma}\cdot\mathbf{a})(\boldsymbol{\sigma}\cdot\mathbf{b})=\mathbf{a}\cdot\mathbf{b}+i\boldsymbol{\sigma}\cdot(\mathbf{a}\times\mathbf{b}).$$

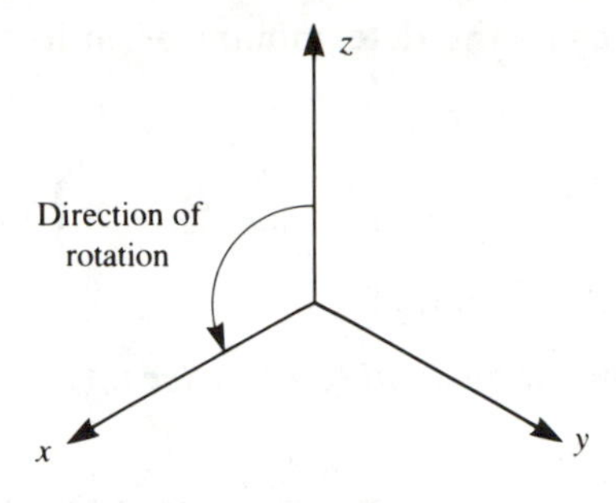

Fig. 2.4.

2.2.13. A rectangular coordinate system for the earth has its origin at the center of the earth, its z axis passing through the North Pole, and its x axis passing through the point 0°N 0°E. Its z axis is then rotated from the North Pole to Los Angeles (≈30°N 120°W) along a longitude, so that the final z axis goes from the center of the earth to Los Angeles. Show that the resulting transformation matrix can be expressed in terms of three successive rotations $R_z(-\alpha)R_x(\beta)R_z(\alpha)$, where $R_z(\alpha)$ rotates the x axis along the equator to the point 0°N 30°W. [Hint: The first rotation $R_z(\alpha)$ transforms to a certain coordinate system, while the last rotation $R_z(-\alpha)$ transforms back from that coordinate system.)

2.2.14. Find the transformation matrix that rotates a rectangular coordinate system through an angle of 120° about an axis through the origin, making equal angles with the original three coordinate axes. When viewed along the axis of rotation, the coordinate axes look like Fig. 2.4.

2.3. Determinant and matrix inversion

To facilitate the use of matrices, we need two additional mathematical objects that can be calculated from a square matrix: its determinant and its inverse matrix.

2.3.1. Determinant

The determinant of small square matrices are well known:

$$\det\begin{pmatrix} A_{11} & A_{12} \\ A_{21} & A_{22} \end{pmatrix} = A_{11}A_{22} - A_{21}A_{12} = \begin{vmatrix} A_{11} & A_{12} \\ A_{21} & A_{22} \end{vmatrix}, \tag{2.19}$$

$$\det\begin{pmatrix} A_{11} & A_{12} & A_{13} \\ A_{21} & A_{22} & A_{23} \\ A_{31} & A_{32} & A_{33} \end{pmatrix} = A_{11}\begin{vmatrix} A_{22} & A_{23} \\ A_{32} & A_{33} \end{vmatrix} - A_{21}\begin{vmatrix} A_{12} & A_{13} \\ A_{32} & A_{33} \end{vmatrix} + A_{31}\begin{vmatrix} A_{12} & A_{13} \\ A_{22} & A_{23} \end{vmatrix}. \tag{2.20}$$

More generally, we *define* the determinant of an $n \times n$ matrix to be

$$\det A = \begin{vmatrix} A_{11} \cdots A_{1n} \\ A_{21} \\ \vdots \\ A_{n1} \cdots A_{nn} \end{vmatrix} = \sum_{ij\cdots l} \varepsilon_{ij\ldots l} A_{i1} A_{j2} \cdots A_{ln}, \tag{2.21}$$

where $\varepsilon_{ij\ldots l}$ is a *permutation symbol of order n* (i.e., with n indices). It is defined as follows:

$$\varepsilon_{ij\ldots l} = \begin{cases} +1, & \text{if } ij\cdots l \text{ is an even permutation of } 12\cdots n, \\ -1, & \text{if } ij\cdots l \text{ is an odd permutation of } 12\cdots n, \\ 0, & \text{if any index is repeated.} \end{cases} \tag{2.22}$$

This defines n^n permutation symbols. Of these only $n!$ symbols are nonzero, since there are exactly $n!$ permutations of the n distinct characters $1, 2, \ldots, n$. Thus Eq. (2.21) contains $n!$ terms. The order n of the matrix is called the *order* of the determinant.

Using the definition given by Eq. (2.21), we write directly

$$\det\begin{pmatrix} A_{11} & A_{12} \\ A_{21} & A_{22} \end{pmatrix} = \not{\varepsilon}_{11} A_{11} A_{12} + \varepsilon_{12} A_{11} A_{22} + \varepsilon_{21} A_{21} A_{12} + \not{\varepsilon}_{22} A_{21} A_{22}$$
$$= A_{11} A_{22} - A_{12} A_{21}.$$

Similarly,

$$\det\begin{pmatrix} A_{11} & A_{12} & A_{13} \\ A_{21} & A_{22} & A_{23} \\ A_{31} & A_{32} & A_{33} \end{pmatrix} = A_{11}A_{22}A_{33} - A_{11}A_{32}A_{23} + A_{21}A_{32}A_{13}$$
$$- A_{21}A_{12}A_{33} + A_{31}A_{12}A_{23} - A_{31}A_{22}A_{13}, \tag{2.23}$$

where the six terms correspond to the six permutations (three even, three odd) of the indices 123. These expressions are equal to those shown in Eqs. (2.19) and (2.20), respectively.

Properties of determinants that can be proved directly from the definition in Eq. (2.21) are

1. $\det A^{\mathrm{T}} = \det A$ (2.24)

2. $$\det\begin{pmatrix} kA_{11} & A_{12} & \cdots & A_{1n} \\ kA_{21} & A_{22} & & \vdots \\ \vdots & \vdots & & \\ kA_{n1} & A_{n2} & & A_{nn} \end{pmatrix} = k \det A, \qquad \det(kA) = k^n \det A. \tag{2.25}$$

3. If A and B differ only in that two rows or two columns are interchanged, then

$$\det B = -\det A. \tag{2.26}$$

In particular, if A contains two identical rows or columns, then

$$\det A = 0. \tag{2.27}$$

4. $$\det(AB) = (\det A)(\det B) = \det(BA). \tag{2.28}$$

The proofs of these results will be left as exercises.

The expression of a determinant as a sum of determinants of a lower order, as shown for example in Eq. (2.20), is called a *Laplace development*. To obtain a Laplace development from Eq. (2.21), we factor out any one matrix element, say A_{kq}, so that

$$\det A = \sum_k A_{kq} C_{kq}, \tag{2.29}$$

where the *cofactor*

$$C_{kq} = \sum_{\substack{ij\cdots l\\(\neq k)}}{}' \varepsilon_{ij\ldots l}(A_{i1}\cdots A_{ln})' \tag{2.30}$$

of the element A_{kq} does not contain A_{kq}, while the sum does not involve k. These restrictions are denoted by the primes in Eq. (2.30).

Consider next the "reduced" determinant of order $n-1$

$$M_{kq} = \sum_{\substack{ij\cdots l\\(\neq k)}}{}' \varepsilon'_{ij\ldots l}(A_{i1}\cdots A_{ln})' \tag{2.31}$$

obtained from the matrix in which the kth row and the qth column are simply removed. It is identical to C_{kq} except that the permutation symbol $\varepsilon_{ij\ldots l}$ of n indices is replaced by the reduced permutation symbol $\varepsilon_{ij}'{}_{\ldots l(\neq k)}$ of $n-1$ indices obtained from it by dropping q from the standard arrangement $12\cdots n$, and dropping k from the permutation $ij\cdots l$. If $q=1$ and $k=1$ are both in the first position, they can both be dropped from C_{kq} to give M_{kq}. If q is in the qth position, it takes $q-1$ transpositions to move it to the first position. Similarly it takes $k-1$ transpositions to move k from the kth position to the first. Each transposition changes the "parity" of the permutation (i.e., even$\rightleftarrows$odd permutation). Hence

$$C_{kq} = (-1)^{k+q} M_{kq}, \tag{2.32}$$

and

$$\det A = \sum_{k=1}^{n} A_{kq}(-1)^{k+q} M_{kq}. \tag{2.33}$$

Going back to the expression in Eq. (2.31), it is easy to see that M_{kq} is the determinant of the reduced matrix constructed from A by removing the kth row and qth column. We call M_{kq} the *minor* of the element A_{kq}, and the expansion in Eq. (2.33) the Laplace development of the qth column. Equation (2.20) is thus a Laplace development of the first column.

Since rows and columns of a matrix can be interchanged without changing the value of the determinant, as stated in Eq. (2.24), we can also obtain the Laplace development of the kth row:

$$\det A = \sum_{q=1}^{n} A_{kq}(-1)^{k+q}M_{kq}. \tag{2.34}$$

The Laplace development of the second row, for a 3×3 matrix, is

$$\det A = -A_{21}\begin{vmatrix} A_{12} & A_{13} \\ A_{32} & A_{33} \end{vmatrix} + A_{22}\begin{vmatrix} A_{11} & A_{13} \\ A_{31} & A_{33} \end{vmatrix} - A_{23}\begin{vmatrix} A_{11} & A_{12} \\ A_{31} & A_{32} \end{vmatrix}.$$

All six Laplace developments of a third-order determinant (three by column and three by row) are contained in Eq. (2.23). They are just different regroupings of the same six terms.

2.3.2. *Matrix inversion*

An important application of the theory of determinants is concerned with the calculation of the inverse of a square matrix.

Let us first recall that a 2×2 matrix

$$A = \begin{pmatrix} A_{11} & A_{12} \\ A_{21} & A_{22} \end{pmatrix}$$

has the inverse

$$A^{-1} = \begin{pmatrix} A_{22} & -A_{12} \\ -A_{21} & A_{11} \end{pmatrix} \Big/ (A_{11}A_{22} - A_{12}A_{21}) = \begin{pmatrix} A_{22} & -A_{12} \\ -A_{21} & A_{11} \end{pmatrix} \Big/ \det a. \tag{2.35}$$

This result can be verified directly by showing that the inversion relations $A^{-1}A = I = AA^{-1}$ are indeed satisfied. Equation (2.35) shows clearly that the inverse exists only if $\det A \neq 0$.

The inverse matrix of a general $n \times n$ matrix A can be constructed readily by using a slightly more general form of Eq. (2.34) for the Laplace development of the ith row

$$\delta_{ik} \det A = \sum_{q=1}^{n} A_{iq}C_{kq} = (AC^{\mathrm{T}})_{ik} = (CA^{\mathrm{T}})_{ki}, \tag{2.36a}$$

while a Laplace development of a column gives

$$\delta_{ik} \det A = (C^{\mathrm{T}}A)_{ki}. \tag{2.36b}$$

(The proof of these results is left as an exercise.) Since $\delta_{ik} = I_{ik} = I_{ki}$, we find that

$$I = \frac{AC^{\mathrm{T}}}{\det A} = \frac{C^{\mathrm{T}}A}{\det A}.$$

Thus

$$A^{-1} = C^{\mathrm{T}}/\det A, \tag{2.37}$$

where C is the $n \times n$ *cofactor matrix* whose matrix elements are the cofactors C_{kj}. Since the cofactors [see Eq. (2.30)] always exist, it follows that A^{-1} exists if $\det A \neq 0$. If $\det A \neq 0$, or if A^{-1} exists, the matrix A is said to be *nonsingular* or *invertible.*

Example 2.3.1. Invert the matrix

$$A = \begin{pmatrix} 1 & 1 & 1 \\ 1 & 2 & 1 \\ 1 & 1 & 3 \end{pmatrix}.$$

First calculate $\det A$ to *determine* if A is invertible. Suppose we do a Laplace development of the first row. The required minors are

$$M_{11} = \begin{vmatrix} 2 & 1 \\ 1 & 3 \end{vmatrix} = 5, \qquad M_{12} = \begin{vmatrix} 1 & 1 \\ 1 & 3 \end{vmatrix} = 2,$$

$$M_{13} = \begin{vmatrix} 1 & 2 \\ 1 & 1 \end{vmatrix} = -1.$$

Therefore

$$\det A = A_{11}M_{11} - A_{12}M_{12} + A_{13}M_{13} = 2,$$

and A is invertible. The cofactor matrix is constructed from all the signed minors, that is,

$$C = \begin{pmatrix} M_{11} & -M_{12} & M_{13} \\ -M_{21} & M_{22} & -M_{23} \\ M_{31} & -M_{32} & M_{33} \end{pmatrix} = \begin{pmatrix} 5 & -2 & -1 \\ -2 & 2 & 0 \\ -1 & 0 & 1 \end{pmatrix}$$

after calculating the remaining minors. Hence

$$A^{-1} = \frac{1}{2}\begin{pmatrix} 5 & -2 & -1 \\ -2 & 2 & 0 \\ -1 & 0 & 1 \end{pmatrix}.$$

Finally, it is necessary to check that $A^{-1}A$ (or AA^{-1}) $= I$ to ensure against numerical mistakes.

2.3.3. *Simultaneous equations*

Determinants are also involved in the solution of simultaneous algebraic equations. Suppose we have n algebraic equations in n unknowns $x_1, \ldots, x_n$:

$$\begin{aligned} &A_{11}x_1 + A_{12}x_2 + \cdots + A_{1n}x_n = c_1 \\ &\vdots \\ &A_{n1}x_1 + A_{n2}x_2 + \cdots + A_{nn}x_n = c_n. \end{aligned}$$

These equations can be written very compactly in matrix form as

$$Ax = c, \tag{2.38}$$

where A is an $n \times n$ matrix and x and c are n-dimensional column vectors. Left multiplication with A^{-1} then yields the unique solution

$$x = A^{-1}c \tag{2.39}$$

if A^{-1} exists. This requires that $\det A \neq 0$, that is, that A be nonsingular. Equation (2.39) gives for the unknown

$$x_i = (A^{-1}c)_i = \sum_j \frac{(C^{\mathrm{T}})_{ij}c_j}{\det A} = \sum_j \frac{c_j C_{ji}}{\det A}. \tag{2.40}$$

The numerator of this expression looks like a Laplace development of the ith column of a determinant. This is indeed the case. The determinant is that of A modified by replacing its ith column by c. Equation (2.40) is called *Cramer's rule.*

Example 2.3.2. Solve the matrix equation

$$\begin{pmatrix} 1 & 1 & 1 \\ 1 & 2 & 1 \\ 1 & 1 & 3 \end{pmatrix} \begin{pmatrix} x_1 \\ x_2 \\ x_3 \end{pmatrix} = \begin{pmatrix} 1 \\ 2 \\ 3 \end{pmatrix}.$$

Since $\det A = 2$, we have by Cramer's rule

$$x_1 = \frac{1}{2} \begin{vmatrix} 1 & 1 & 1 \\ 2 & 2 & 1 \\ 3 & 1 & 3 \end{vmatrix} = -1$$

$$x_2 = \frac{1}{2} \begin{vmatrix} 1 & 1 & 1 \\ 1 & 2 & 1 \\ 1 & 3 & 3 \end{vmatrix} = 1$$

$$x_3 = \frac{1}{2} \begin{vmatrix} 1 & 1 & 1 \\ 1 & 2 & 2 \\ 1 & 1 & 3 \end{vmatrix} = 1.$$

That these are the solutions can easily be verified by doing the matrix multiplication Ax.

Problems

2.3.1. Verify Eqs. (2.24)–(2.27).

2.3.2. Show that $\det(A^{-1}) = (\det A)^{-1}$ if A is nonsingular.

2.3.3. Calculate $\begin{vmatrix} 1 & 2 & 3 \\ 4 & 5 & 6 \\ 7 & 8 & 9 \end{vmatrix}$ by Laplace developments of

(a) The second row, and

(b) The third column.

2.3.4. Verify Eqs. (2.36a) and (2.36b).

2.3.5. Show that the inverse of the inverse of a matrix is the original.

2.3.6. Invert the matrix $\begin{pmatrix} 1 & 2 & 3 \\ 4 & 5 & 6 \\ 7 & 8 & 1 \end{pmatrix}$.

2.3.7. Prove the relation $\det(AB) = (\det A)(\det B)$ for
(a) 2×2 matrices, and
(b) 3×3 matrices.

2.3.8. Use Cramer's rule to solve the matrix equation

$$\begin{pmatrix} 1 & 2 & 3 \\ 4 & 5 & 6 \\ 7 & 8 & 1 \end{pmatrix} \begin{pmatrix} x_1 \\ x_2 \\ x_3 \end{pmatrix} = \begin{pmatrix} 1 \\ 2 \\ 3 \end{pmatrix}.$$

2.4. Homogeneous equations

It is also possible to solve a set of n *homogeneous* algebraic equations written in matrix form as

$$Ax = 0, \tag{2.41}$$

that is, Eq. (2.38) with $c = 0$. It is obvious that there is a unique solution $x = A^{-1}c = 0$ if A is nonsingular. We call this a trivial case. A nontrivial solution is one with $x \neq 0$. This can only occur if $\det A = 0$, for then we have the quotient of two zeros, which can be finite.

To see what we are getting into, let us first consider the simple case of 2×2 matrices. Then $Ax = 0$ means the two simultaneous algebraic equations

$$A_{11}x_1 + A_{12}x_2 = 0$$
$$A_{21}x_1 + A_{22}x_2 = 0.$$

These equations state that $\mathbf{x} = (x_1, x_2)$ is perpendicular to the row vectors

$$\mathbf{a}_1 = (A_{11}, A_{12}), \qquad \mathbf{a}_2 = (A_{21}, A_{22})$$

contained in A. That is, $\mathbf{x}$ is a vector, any vector, perpendicular simultaneously to both $\mathbf{a}_1$ and $\mathbf{a}_2$. Its length is of no consequence in the homogeneous equation; only its direction $\mathbf{e}(\mathbf{x})$ or $\mathbf{e}(-\mathbf{x})$ can be determined.

If the two row vectors are nonparallel, $\det A$ does not vanish. There is then no nontrivial solution. The reason is that $\mathbf{a}_1$ and $\mathbf{a}_2$ already span the entire two-dimensional plane of the problem. An $\mathbf{x}$ perpendicular to this plane cannot also lie in it. There is thus no solution x in the plane. This is the reason why $\det A = 0$ is required.

Det $A = 0$ can be obtained by making $\mathbf{a}_1$ and $\mathbf{a}_2$ parallel. They then occupy only one of the two dimensions of the problem, leaving the other dimension for $\mathbf{x}$. The direction of $\mathbf{x}$ is then uniquely determined up to a sign.

For 3×3 matrices, $Ax = 0$ means that $\mathbf{x}$ must be perpendicular to all three row vectors of A:

$$\mathbf{a}_i \cdot \mathbf{x} = 0. \tag{2.42a}$$

Suppose two of these row vectors are not parallel to each other. To get $\det A = 0$, the third row vector, say $\mathbf{a}_3$, must be made up of a linear combination of these nonparallel vectors

$$\mathbf{a}_3 = c_1\mathbf{a}_1 + c_2\mathbf{a}_2. \tag{2.43a}$$

Equation (2.43a) states that $\mathbf{a}_3$ lies in the plane containing $\mathbf{a}_1$ and $\mathbf{a}_2$. Thus the three row vectors span only a two-dimensional plane. Equation (2.42a) now shows that $\mathbf{x}$ is the normal to this plane through the origin. (All vectors are measured from the origin.) Again, only its direction is determined, not its sense or length.

Det $A = 0$ is also realized when all three row vectors are parallel to one another. These vectors then span only the one-dimensional subspace specified by the common direction $\mathbf{e}(\mathbf{a}_1)$. Any vector in that plane normal to $\mathbf{e}(\mathbf{a}_1)$ that contains the origin is an acceptable solution. That is, the solutions $\mathbf{x}$ span the two-dimensional plane perpendicular to $\mathbf{e}(\mathbf{a}_1)$.

In both cases, the solutions span the subspace not spanned by the row vectors of A.

These considerations can immediately be generalized to $n \times n$ matrices. The row vectors $\mathbf{a}_i$ of A span an r-dimensional subspace S_a, with $r < n$ if $\det A = 0$ is to be realized. The solutions x of the homogeneous matrix equation $Ax = 0$ span the remaining $(n - r)$-dimensional subspace S_x. The subspace S_x is called the *orthogonal complement* of the subspace S_a. These subspaces are orthogonal to each other. Together they make up the entire n-dimensional space.

If $r = n - 1$, $n - r = 1$. Then $Ax = 0$ determines one direction only. More generally $Ax = 0$ determines the $n - r$ linearly independent directions that specify the subspace S_x.

Example 2.4.1. The homogeneous equation

$$\begin{pmatrix} 1 & 1 & 1 \\ 1 & 1 & 1 \\ 1 & 1 & 1 \end{pmatrix} \begin{pmatrix} x_1 \\ x_2 \\ x_3 \end{pmatrix} = 0$$

has solutions which satisfy the equation

$$x_1 + x_2 + x_3 = 0.$$

They are vectors lying in the plane perpendicular to the row vectors

$$\mathbf{a}_1 = \mathbf{a}_2 = \mathbf{a}_3 = (1,1,1).$$

Example 2.4.2. The homogeneous equation

$$\begin{pmatrix} -1 & 1 & 1 \\ 1 & 1 & 1 \\ 1 & 1 & 1 \end{pmatrix} \begin{pmatrix} x_1 \\ x_2 \\ x_3 \end{pmatrix} = 0$$

can be solved by examining the two distinct algebraic equations

$$-x_1 + x_2 + x_3 = 0$$
$$x_1 + x_2 + x_3 = 0.$$

The solutions can be extracted by taking their sum and difference

$$x_1 = 0, \qquad x_2 + x_3 = 0.$$

Thus the solutions are vectors along the line $\mathbf{x} = x_2(0,1,-1)$ in the yz plane.

Example 2.4.3. Solve the homogeneous equation

$$\begin{pmatrix} -1 & 1 & 1 \\ 1 & 1 & 1 \\ 0 & 2 & 2 \end{pmatrix} \begin{pmatrix} x_1 \\ x_2 \\ x_3 \end{pmatrix} = 0.$$

First verify that

$$\det A = -0 - 0 + 0 = 0$$

by a Laplace development of the first column. Then note that

$$\mathbf{a}_3 = \mathbf{a}_1 + \mathbf{a}_2$$

so that the three vectors are linearly dependent. Finally we throw out $\mathbf{a}_1$ to find the two independent equations

$$x_1 + x_2 + x_3 = 0$$
$$2(x_2 + x_3) = 0.$$

Again the solutions are

$$x_1 = 0, \qquad x_2 + x_3 = 0,$$

which are identical to the solutions of the last problem. This is not unexpected, since both problems have the same linearly independent row vectors $\mathbf{a}_1$ and $\mathbf{a}_2$.

The solution can be obtained even more readily by using the result that $\mathbf{x} = (x_1, x_2, x_3)$ is perpendicular to both $\mathbf{a}_1$ and $\mathbf{a}_2$. Hence

$$\mathbf{x} = \text{const} \times (\mathbf{a}_1 \times \mathbf{a}_2) = \text{const} \times (0,2,-2).$$

2.4.1. Linear independence of vectors and the rank of a matrix

The solution of the problem of homogeneous matrix equations given so far is adequate but quite sketchy for the general case. The inquisitive reader who must know how things work out in detail is urged to read on.

Suppose only the first r row vectors $\mathbf{a}_i$ of A are linearly independent. The remaining $n-r$ vectors $\mathbf{a}_k$ are linearly dependent on the former in the sense that they can be expressed as the linear combinations

$$\mathbf{a}_k = \sum_{i=1}^{r} c_i \mathbf{a}_i. \tag{2.43b}$$

This means that if a vector $\mathbf{x}$ is orthogonal to the $\mathbf{a}_i$s,

$$\mathbf{a}_i \cdot \mathbf{x} = 0, \qquad i = 1, \ldots, r, \tag{2.42b}$$

it is automatically orthogonal to the $\mathbf{a}_k$s as well. Equation (2.42b) can be written more compactly in the matrix form

$$Bx = 0, \tag{2.44a}$$

where B is an $r \times n$ nonsquare submatrix of A containing the r linearly independent row vectors $\mathbf{a}_i$, and x is an $n \times 1$ column matrix.

The $r \times n$ nonsquare matrix B contains n column vectors $\mathbf{b}_j$. Not all of these are linearly independent of one another, because an arbitrary r-dimensional vector has at most r independent components. Consequently, no more than r of these column vectors are linearly independent.

A column vector, say $\mathbf{b}_n$, linearly dependent on the other column vectors

$$\mathbf{b}_n = \sum_{j=1}^{n-1} d_j \mathbf{b}_j,$$

can be dropped from the matrix B without affecting the linear independence of the remaining $(n-1)$-dimensional *row* vectors. This is because $\mathbf{b}_n$ does not contain any information not already contained in the remaining column vectors $\mathbf{b}_j$.

There are exactly $n-r$ of these linearly dependent column vectors that can be discarded for the time being. We now have an $r \times r$ square submatrix R of B made up of the remaining columns of B. If we try to discard any more columns from R, we will end up with a nonsquare matrix in which the row vectors have fewer than r components each. Should this happen, not all the r row vectors still in the matrix can be linearly independent.

Thus by construction, we have obtained an $r \times r$ square matrix R containing r linearly independent column vectors. Its determinant does not vanish. Hence R is a nonsingular invertible matrix.

For example, the 2×3 nonsquare matrix

$$B = \begin{pmatrix} 1 & 2 & 3 \\ 4 & 5 & 6 \end{pmatrix}$$

is made up of two linearly independent row vectors. As a result, only two of the three column vectors are linearly independent. The third column vector

$$\mathbf{b}_3 = \begin{pmatrix} 3 \\ 6 \end{pmatrix} = -\begin{pmatrix} 1 \\ 4 \end{pmatrix} + 2\begin{pmatrix} 2 \\ 5 \end{pmatrix} = -\mathbf{b}_1 + 2\mathbf{b}_2$$

can be expressed in terms of the remaining two. It may be discarded for certain purposes, leaving a square invertible matrix

$$R = \begin{pmatrix} 1 & 2 \\ 4 & 5 \end{pmatrix}.$$

In this way we have isolated the largest submatrix R of A (or B) that has a nonzero determinant. The order r of this submatrix R is called the *rank* of A or B. For example, the rank of the matrix B in our example is 2, while that of $\begin{pmatrix} 0 & 0 & 1 \\ 0 & 0 & 2 \end{pmatrix}$ is 1.

Now the matrix equation (2.44a).

$$(Bx)_i = \sum_{j=1}^{n} B_{ij} x_j = 0,$$

can be written in the "column" form

$$\sum_{j=1}^{n} \mathbf{b}_j x_j = 0,$$

if $\mathbf{b}_j$ is the jth column of B. Keeping the r linearly independent vectors on the left, we have

$$\sum_{j=1}^{r} \mathbf{b}_j x_j = -\sum_{k=r+1}^{n} \mathbf{b}_k x_k. \tag{2.44b}$$

Any solution of this equation is guaranteed to be a solution of the original matrix equation $Ax = 0$, because B contains all the linearly independent row vectors of A.

Equation (2.44b) can be solved by exhausting all possibilities. Suppose all the components x_k appearing on the right-hand side are zero. Then the equation has the matrix representation $Rx' = 0$, where x' is the column vector containing the first r components of x. Since R is invertible, x' has the unique solution $x' = 0$. With the remaining components x_k of x already taken to be zero, we have constructed the trivial solution $x = 0$.

We next try a solution with only one nonzero component, x_k, $r + 1 \leqslant k \leqslant n$, on the right-hand side of Eq. (2.44b). This equation can now be written in the matrix form

$$R\mathbf{y}_k = -\mathbf{b}_k, \tag{2.45}$$

where $\mathbf{y}_k$ is an r-dimensional column vector whose jth component is x_j/x_k. Equation (2.45) has the unique solution

$$\mathbf{y}_k = -R^{-1}\mathbf{b}_k,$$

defining a unique direction

$$\mathbf{e}_k = \mathbf{e}(\mathbf{y}_k). \tag{2.46}$$

By taking successively $k = r+1, \ldots, k = n$, we can construct in this way $n-r$ linearly independent directions $\mathbf{e}_k$, $k = r+1, \ldots, n$. (See Problem 2.4.3.) These vectors span the entire $(n-r)$-dimensional space S_x that is the orthogonal complement of the space S_a containing the r linearly independent row vectors of B. Any other solution x can be expressed as a linear combination of the $\mathbf{e}_k$s. This completes the solution of the problem of homogeneous matrix equations.

Problems

2.4.1. Solve the homogeneous equation

$$\begin{pmatrix} 1 & 2 & 3 \\ 4 & 5 & 6 \\ 5 & 7 & 9 \end{pmatrix} \begin{pmatrix} x \\ y \\ z \end{pmatrix} = 0.$$

2.4.2. By expressing the last row vector in terms of the others, determine the rank of the following matrices

(a) $\begin{pmatrix} 1 & 2 & 3 & 4 \\ 5 & 6 & 7 & 8 \end{pmatrix}$;

(b) $\begin{pmatrix} 1 & 2 & 3 \\ 4 & 5 & 6 \\ 7 & 8 & 9 \end{pmatrix}$;

(c) $\begin{pmatrix} 1 & 2 & 3 & 4 \\ 5 & 6 & 7 & 8 \\ 9 & 10 & 11 & 12 \end{pmatrix}$.

2.4.3. Show that the $n-r$ directions $\mathbf{e}_k$ of Eq. (2.46) are linearly independent of one another.

2.5. The matrix eigenvalue problem

It is an experimental fact that the linear momentum of a massive object $\mathbf{p} = m\mathbf{v}$ is always parallel to its linear velocity $\mathbf{v}$. In contrast, the angular momentum $\mathbf{L}$ of a rigid body is not always parallel to its angular velocity

$\boldsymbol{\omega}$. A classic illustration of this observation is the rotating dumbbell shown in Fig. 2.5. It is made up of two point masses m separated by a massless rigid rod of length $2r$. It is rotated about its midpoint 0 about the z axis with angular speed ω. The angular momentum $\mathbf{L} = \mathbf{r} \times \mathbf{p}$ is always perpendicular to the dumbbell axis (if $\theta \neq 0$). Consequently, it is not always parallel to $\boldsymbol{\omega}$.

However, both $\mathbf{L}$ and $\boldsymbol{\omega}$ are three-dimensional column vectors when expressed in matrix notation. Hence there exists a 3×3 square matrix $\mathscr{I}$ that can connect them:

$$\mathbf{L} = \mathscr{I}\boldsymbol{\omega}, \qquad \text{or} \qquad L_i = \sum_j \mathscr{I}_{ij}\omega_j. \tag{2.47}$$

$\mathscr{I}$ is called the *inertial matrix*. For the special case of a point mass, we have with the help of the *BAC* rule,

$$\mathbf{L} = m\mathbf{r} \times \mathbf{v} = m\mathbf{r} \times (\boldsymbol{\omega} \times \mathbf{r}) = m[r^2\boldsymbol{\omega} - \mathbf{r}(\mathbf{r} \cdot \boldsymbol{\omega})].$$

Thus

$$\mathscr{I} = m(r^2 - \mathbf{rr}\cdot)$$

in vector notation. To write the relation in matrix notation, we note that

$$L_i = m\left(r^2\omega_i - x_i \sum_j x_j\omega_j\right) = \sum_j [m(r^2\delta_{ij} - x_ix_j)]\omega_j.$$

Comparison with Eq. (2.47) gives

$$\mathscr{I}_{ij} = m(r^2\delta_{ij} - x_ix_j)$$

in matrix notation.

Having established that $\mathbf{L}$ is in general not parallel to $\boldsymbol{\omega}$, we may ask next if occasionally they might be. It is easy to see that for the rotating dumbbell there are indeed three distinct directions along which this occurs. One of these $\boldsymbol{\omega}$s is parallel to the dumbbell axis (i.e., $\theta = 0$), while the remaining two are perpendicular to the dumbbell axis (i.e., $\theta = \pi/2$). These directions are called the *principal axes of inertia*. The corresponding inertial parameters (the proportionality constants $\lambda =$

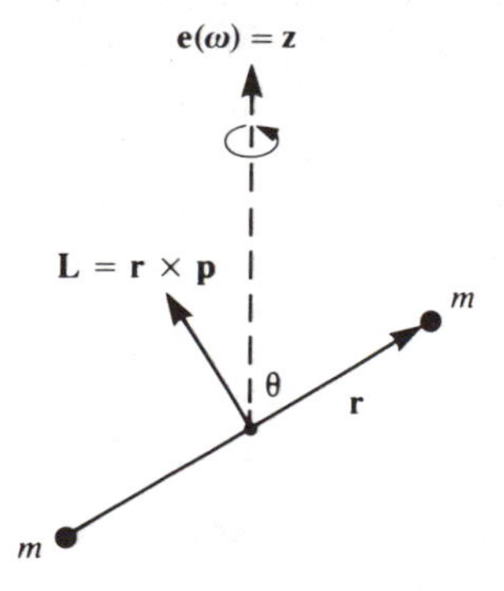

Fig. 2.5. A rotating dumbbell.

L/ω) are called the *principal moments of inertia.* The existence of three principal axes of inertia turns out to be a universal property of rigid bodies, as we shall now show.

The parallelism between $\mathbf{L}$ and $\boldsymbol{\omega}$ along these special directions is expressed by the condition

$$\mathscr{I}\boldsymbol{\omega} = \lambda\boldsymbol{\omega}, \tag{2.48}$$

where the scalar proportionality constant λ is called an *eigenvalue* (or *characteristic value*) of $\mathscr{I}$. (*Eigen* is the German word for "own, proper, characteristic, special.") Equation (2.48) itself is called an *eigenvalue equation.* It is an interesting mathematical observation that this equation can be solved for these special directions $\boldsymbol{\omega}$ and for the eigenvalue λ associated with each of these directions.

2.5.1. Solution of the eigenvalue problem

We shall write the general matrix eigenvalue equation as the homogeneous equation

$$(B - \lambda I)\mathbf{u} = 0. \tag{2.49}$$

It is clear that the matrix B should be square, otherwise $B\mathbf{u}$ will not have the same dimension as $\mathbf{u}$ and cannot possibly satisfy the equation. It also follows from the results of the last section that nontrivial solutions with $\mathbf{u} \neq 0$ can appear only if the determinant

$$\phi(\lambda) = \det(B - \lambda I) = \begin{vmatrix} B_{11} - \lambda & B_{12} & \cdots \\ B_{21} & B_{22} - \lambda & \\ \vdots & \vdots & \end{vmatrix} \tag{2.50}$$

vanishes. $\phi(\lambda)$ is called a *characteristic* (or *secular*) *determinant.* If matrix B is $n \times n$, $\phi(\lambda)$ is a polynomial of degree n in λ. The determinantal condition

$$\phi(\lambda) = 0, \tag{2.51}$$

called the *characteristic* (or *secular*) *equation,* therefore has n roots $\lambda_1, \lambda_2, \ldots, \lambda_n$, which are not necessarily different. When $\lambda = \lambda_i$, $\det(B - \lambda I)$ vanishes; Eq. (2.49) can then have a nontrivial solution $\mathbf{u}_i$. Thus the n roots of Eq. (2.51) are the only possible eigenvalues of Eq. (2.49) when the matrix B is $n \times n$.

To illustrate the calculation of eigenvalues, let us consider a general 2×2 matrix

$$B = \begin{pmatrix} a & c \\ d & b \end{pmatrix}.$$

The associated secular equation is quadratic in λ

$$\phi(\lambda) = \det(B - \lambda I) = \lambda^2 - (a + b)\lambda + ab - cd = 0.$$

It has two roots

$$\lambda_{1,2} = \tfrac{1}{2}[(a+b) \pm \sqrt{(a-b)^2 + 4cd}], \tag{2.52a}$$

which satisfy the useful identity

$$\lambda_1 + \lambda_2 = a + b = \mathrm{Tr}\, B.$$

This result turns out to be a special case of the general theorem that states that the sum of the eigenvalues of a matrix of any order is equal to its trace

$$\sum_i \lambda_i = \mathrm{Tr}\, b. \tag{2.53}$$

The proof of this general theorem is left as an exercise (Problem 2.5.6).

As we have discussed in the last section, only the directions $\mathbf{e}_i = \mathbf{e}(\mathbf{u}_i)$ of the nontrivial solutions $\mathbf{u}_i$ of the eigenvalue equation (2.49) are determined, not their lengths. Thus, the matrix eigenvalue equation may be written more precisely as

$$(B - \lambda_i I)\mathbf{e}_i = 0, \qquad \text{or} \qquad B\mathbf{e}_i = \lambda_i \mathbf{e}_i. \tag{2.54}$$

We call $\mathbf{e}_i$ the *eigenvector* "belonging" to the eigenvalue λ_i.

Since we already know how to solve homogeneous matrix equations, the eigenvalue equation (2.54) can be solved to obtain $\mathbf{e}_i$. The procedure is illustrated in the following examples.

Example 2.5.1. A uniform solid sphere has an inertial matrix (in suitable units) of

$$\begin{pmatrix} 1 & 0 & 0 \\ 0 & 1 & 0 \\ 0 & 0 & 1 \end{pmatrix}.$$

The characteristic equation is $\phi(\lambda) = (1-\lambda)^3$ so that the three roots are $\lambda_1 = \lambda_2 = \lambda_3 = 1$. Identical eigenvalues like these are said to be *degenerate*. The eigenvectors are to be calculated from the homogeneous equation $A\omega = 0$, where

$$A = \mathcal{I} - \lambda I = \begin{pmatrix} 0 & 0 & 0 \\ 0 & 0 & 0 \\ 0 & 0 & 0 \end{pmatrix}.$$

This shows that $\boldsymbol{\omega}$ can be any vector in space. Since there are three linearly independent vectors in space, we may choose $\boldsymbol{\omega}_1 = \mathbf{e}_x$, $\boldsymbol{\omega}_2 = \mathbf{e}_y$, $\boldsymbol{\omega}_3 = \mathbf{e}_z$.

Example 2.5.2. The matrix

$$S = \begin{pmatrix} 1 & 1 & 1 \\ 1 & 1 & 1 \\ 1 & 1 & 1 \end{pmatrix}$$

already has zero determinant. Hence it must have at least one zero eigenvalue, $\lambda_1 = 0$. The associated eigenvector $\boldsymbol{\omega}$ satisfies the homogeneous equation

$$0 = A\boldsymbol{\omega}_1 = (S - \lambda_1 I)\boldsymbol{\omega}_1 = \begin{pmatrix} 1 & 1 & 1 \\ 1 & 1 & 1 \\ 1 & 1 & 1 \end{pmatrix} \boldsymbol{\omega}_1.$$

Therefore $\boldsymbol{\omega}_1$ is *any* vector on the plane perpendicular to the vector $\mathbf{a} = (1,1,1)$. Once a vector $\boldsymbol{\omega}_1$ on this plane is chosen, we can find another vector $\boldsymbol{\omega}_2$ perpendicular to $\boldsymbol{\omega}_1$ on the plane. Since $S\boldsymbol{\omega}_2 = 0$, too, it is clear that there is a second eigenvalue $\lambda_2 = 0$ degenerate with λ_1.

The third and last eigenvalue can be found from Eq. (2.53)

$$\lambda_3 = \operatorname{Tr} S = 3.$$

For this

$$A = S - \lambda_3 I = \begin{pmatrix} -2 & 1 & 1 \\ 1 & -2 & 1 \\ 1 & 1 & -2 \end{pmatrix}.$$

Hence the eigenvector $\boldsymbol{\omega}_3 = (x, y, z)$ satisfies the homogeneous equations

$$-2x + y + z = 0$$
$$x - 2y + z = 0$$
$$x + y - 2z = 0.$$

Using any two of these equations, we obtain the solutions

$$\frac{y}{x} = 1, \qquad \frac{z}{x} = 1.$$

The normalized eigenvector is therefore $\mathbf{e}_3 = (1/\sqrt{3}) \begin{pmatrix} 1 \\ 1 \\ 1 \end{pmatrix}$. This turns out to be the direction of $\mathbf{a}$ itself.

Example 2.5.3. Obtain the eigenvalues and eigenvectors of the real, symmetric 2×2 matrix

$$S = \begin{pmatrix} a & c \\ c & b \end{pmatrix}, \qquad a \leq b.$$

According to Eq. (2.52a)

$$\lambda_{1,2} = \tfrac{1}{2}(a + b) \mp \tfrac{1}{2}\sqrt{(b - a)^2 + 4c^2}. \qquad (2.52b)$$

That is, λ_1 is below the mean $(a + b)/2$ of the diagonal matrix elements by an amount

$$\Delta = \tfrac{1}{2}\sqrt{(b - a)^2 + 4c^2},$$

and below a itself by $\delta = \Delta + \frac{1}{2}(a-b)$, while λ_2 is above the mean by an amount Δ and above b by an amount δ.

Any normalized two-dimensional vector can be written in the form

$$\mathbf{e} = \begin{pmatrix} \cos\theta \\ \sin\theta \end{pmatrix},$$

since $\cos^2\theta + \sin^2\theta = 1$ gives its normalization. The eigenvector $\mathbf{e}_i$ satisfies the homogeneous equation

$$(S - \lambda_i I)\mathbf{e}_i = 0.$$

That is,

$$\begin{pmatrix} a-\lambda_i & c \\ c & b-\lambda_i \end{pmatrix}\begin{pmatrix} \cos\theta_i \\ \sin\theta_i \end{pmatrix} = 0,$$

or

$$\tan\theta_i = \frac{\lambda_i - a}{c} = \frac{c}{\lambda_i - b}.$$

More specifically,

$$\tan\theta_1 = \frac{\lambda_1 - a}{c} = -\frac{\delta}{c},$$

$$\tan\theta_2 = \frac{c}{\lambda_2 - b} = \frac{c}{\delta} = -\cot\theta_1.$$

Hence

$$\cos\theta_2 = -\sin\theta, \qquad \sin\theta_2 = \cos\theta, \qquad \text{if} \qquad \theta = \theta_1.$$

This shows that

$$\mathbf{e}_2 = \begin{pmatrix} -\sin\theta \\ \cos\theta \end{pmatrix} \tag{2.55}$$

is perpendicular to $\mathbf{e}_1$.

Example 2.5.4. Obtain the eigenvalues and eigenvectors of the Hermitian matrix

$$B = \begin{pmatrix} \gamma & i\beta\gamma \\ -i\beta\gamma & \gamma \end{pmatrix}$$

with real constants β and γ. (This is a transformation matrix for the Lorentz transformation.)

From Eq. (2.52a), the eigenvalues are found to be

$$\lambda_1 = \gamma(1-\beta), \qquad \lambda_2 = \gamma(1+\beta).$$

The first eigenvector $\mathbf{e}_1$ is obtained from the equation

$$\begin{pmatrix} \gamma-\lambda_1 & i\beta\gamma \\ -i\beta\gamma & \gamma-\lambda_1 \end{pmatrix}\begin{pmatrix} q_1 \\ q_2 \end{pmatrix} = 0. \tag{2.56}$$

That is,

$$(\gamma - \lambda_1)q_1 + i\beta\gamma q_2 = 0,$$

or

$$\frac{q_2}{q_1} = \frac{\gamma - \lambda_1}{-i\beta\gamma} = i, \qquad \text{and} \qquad \mathbf{e}_1 = c_1\begin{pmatrix} 1 \\ i \end{pmatrix}.$$

The question now arises as to how the normalization constant c_1 is to be chosen. The usual scalar product

$$\mathbf{e}_1^{\mathrm{T}} \cdot \mathbf{e}_1 = c_1^2(1 \;\; i)\begin{pmatrix} 1 \\ i \end{pmatrix} = 0$$

seems to suggest that $\mathbf{e}_1$ might have no length. This cannot be correct, since it has nonzero components. We recall that the length (or absolute value) of a complex number $z = x + iy$ is not the square root of $z^2 = x^2 - y^2 + 2ixy$, but that of the real non-negative number $z^*z = x^2 + y^2$. For the same reason, the length of $\mathbf{e}_1$ should be calculated from the *Hermitian scalar product* (usually referred to as just the scalar product)

$$\mathbf{e}_1^{\dagger} \cdot \mathbf{e}_1 = \mathbf{e}_1^{\mathrm{T}*} \cdot \mathbf{e}_1 = |c_1|^2 (1 \;\; -i)\begin{pmatrix} 1 \\ i \end{pmatrix} = 2\,|c_1|^2. \tag{2.57a}$$

This gives

$$|c_1|^2 = \tfrac{1}{2}, \qquad \text{or} \qquad |c_1| = \frac{1}{\sqrt{2}}.$$

It is clear that c_1 itself can be a complex number of phase ϕ_1. Hence

$$\mathbf{e}_1 = \frac{1}{\sqrt{2}} e^{i\phi_1}\begin{pmatrix} 1 \\ i \end{pmatrix}. \tag{2.58}$$

Any choice of ϕ_1 is acceptable, since Eq. (2.57) is satisfied for any choice of ϕ_1. We refer to this undetermined degree of freedom as a *phase*, or *gauge, transformation* of the eigenvector. (In advanced physics, gauge degrees of freedom can be used to describe certain internal properties of the system. For our purposes here, however, this guage transformation is as yet undetermined, and therefore unimportant.)

In a similar way, we can determine the eigenvector belonging to λ_2:

$$\mathbf{e}_2 = \frac{1}{\sqrt{2}} e^{i\phi_2}\begin{pmatrix} 1 \\ -i \end{pmatrix}. \tag{2.59}$$

The orthogonality between $\mathbf{e}_1$ and $\mathbf{e}_2$ can be checked by examining the scalar product

$$\begin{aligned} \mathbf{e}_2^{\dagger} \cdot \mathbf{e}_1 &= \tfrac{1}{2} e^{i(\phi_1 - \phi_2)}(1 \;\; i)\begin{pmatrix} 1 \\ i \end{pmatrix} = 0 \\ &= \mathbf{e}_1^{\dagger} \cdot \mathbf{e}_2. \end{aligned} \tag{2.57b}$$

Because this scalar product vanishes, these complex vectors are said to be orthogonal to each other.

The eigenvalues of a 3×3 matrix are the roots of a cubic equation in λ. They can also be expressed in closed form with the help of Cardan's formula for these roots. Unfortunately, the formula is rather long, and it involves cube roots as well as square roots. It is more profitable to restrict ourselves to simple special cases. All the eigenvalues in the following problems can be obtained by simple considerations without using general formulas. The eigenvalues and eigenvectors of more complicated matrices are usually calculated by using a computer.

Problems

2.5.1. Calculate the eigenvalues and eigenvectors of the following matrices:

(a) $\sigma_1 = \begin{pmatrix} 0 & 1 \\ 1 & 0 \end{pmatrix}$;

(b) $\sigma_2 = \begin{pmatrix} 0 & -i \\ i & 0 \end{pmatrix}$;

(c) $\begin{pmatrix} 0 & 0 & 1 \\ 0 & 0 & 1 \\ 1 & 1 & 0 \end{pmatrix}$;

(d) $\begin{pmatrix} 0 & 0 & 1 & 1 \\ 0 & 0 & 1 & 1 \\ 1 & 1 & 0 & 0 \\ 1 & 1 & 0 & 0 \end{pmatrix}$;

(e) $\begin{pmatrix} 1 & 1 & 1 & 1 \\ 1 & 1 & 1 & 1 \\ 1 & 1 & 1 & 1 \\ 1 & 1 & 1 & 1 \end{pmatrix}$.

2.5.2. Obtain the eigenvalues and eigenvectors of an $n \times n$ diagonal matrix.

2.5.3. Three unequal masses are connected by massless rigid rods of unequal lengths into a rigid right triangle. Obtain the principal axes of inertia for rotation about the vertex at the right angle.

2.5.4. Show that if B is singular it has one or more zero eigenvalues.

2.5.5. By equating the coefficients of the λ^{n-1} terms in the equation

$$|B - \lambda I| = (\lambda_1 - \lambda)(\lambda_2 - \lambda) \cdots (\lambda_n - \lambda)$$

for an arbitrary $n \times n$ matrix B, show that

$$\sum_{i=1}^{n} \lambda_i = \operatorname{Tr} B$$

holds for the sum of the eigenvalues λ_i. (Hint: We may want to do this first for $n = 2$ and 3.)

2.6. Generalized matrix eigenvalue problems

Figure 2.6 shows a system of two masses m_1 and m_2 connected by three springs with spring constants k_1, k_{12}, and k_2. Their motion is described by Newton's equations of motion:

$$\begin{aligned} m_1\ddot{q}_1 &= -k_1 q_1 - k_{12}(q_1 - q_2), \\ m_2\ddot{q}_2 &= -k_2 q_2 - k_{12}(q_2 - q_1), \end{aligned} \tag{2.60}$$

where q_i is the displacement of mass m_i from its position of equilibrium.

This system of two coupled differential equations can be written in matrix form

$$M\ddot{\mathbf{q}}(t) = -K\mathbf{q}(t), \tag{2.61a}$$

where

$$\mathbf{q}(t) = \begin{pmatrix} q_1(t) \\ q_2(t) \end{pmatrix}, \qquad M = \begin{pmatrix} m_1 & 0 \\ 0 & m_2 \end{pmatrix}, \qquad K = \begin{pmatrix} k_1 + k_{12} & -k_{12} \\ -k_{12} & k_2 + k_{12} \end{pmatrix}.$$

The equation can be solved by first asking the following question: Is it at all possible that $K\mathbf{q}$ is parallel not to $\mathbf{q}$ itself, but to $M\mathbf{q}$

$$K\mathbf{q} = \lambda M\mathbf{q}, \tag{2.62a}$$

where λ is a scalar proportionality constant? It turns out not to be important (in principle) whether M is diagonal. It can be a general square matrix with nonzero off-diagonal matrix elements. Equation (2.62a) will be referred to as the generalized matrix eigenvalue problem. The special values of λ for which Eq. (2.62a) is satisfied are called its eigenvalues. The corresponding vector $\mathbf{q}$ is its eigenvector.

Writing Eq. (2.62a) as a homogeneous matrix equation

$$(K - \lambda M)\mathbf{q} = 0, \tag{2.62b}$$

we see that nontrivial solutions $\mathbf{q}$ exist only if

$$\det(K - \lambda M) = \phi(\lambda) = 0. \tag{2.63}$$

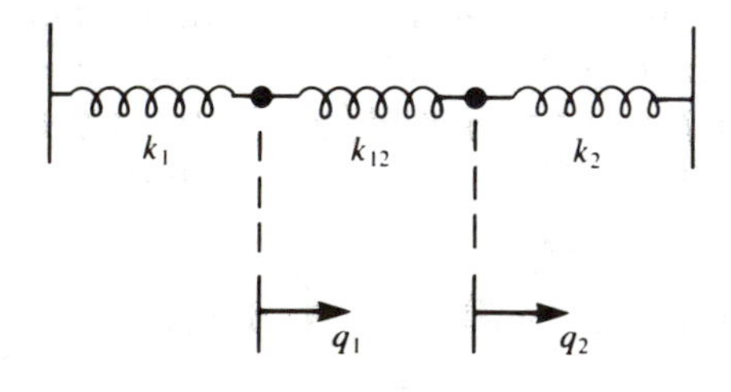

Fig. 2.6. Two masses connected by three springs.

With 2×2 square matrices K and M, the determinant is a quadratic function of λ. Hence the secular equation (2.63) has two roots, the eigenvalues λ_1 and λ_2, for which Eq. (2.62a) can be satisfied. The actual vector $\mathbf{q}^{(r)}$ along which Eq. (2.62a) is realized

$$(K-\lambda_r M)\mathbf{q}^{(r)}=0 \tag{2.62c}$$

is just the eigenvector "belonging" to λ_r. We thus see that the mathematics of this generalized matrix eigenvalue problem is essentially the same as that of the simple eigenvalue problem of the last section.

We should add that, in the problem of coupled oscillators, the mass matrix M is diagonal if $\mathbf{q}$ contains actual coordinates. M can be off diagonal if $\mathbf{q}$ contains generalized coordinates such as angles.

Example 2.6.1. Solve Eq. (2.62a) when

$$M=m\begin{pmatrix}1 & 1\\ 1 & 6\end{pmatrix},\qquad K=k\begin{pmatrix}2 & -1\\ -1 & 2\end{pmatrix}.$$

$$\det(K-\lambda M)=m\begin{vmatrix}2\omega_0^2-\lambda & -\omega_0^2-\lambda\\ -\omega_0^2-\lambda & 2\omega_0^2-6\lambda\end{vmatrix}$$

$$=m(5\lambda^2-16\omega_0^2\lambda+3\omega_0^4)=0,$$

where $\omega_0^2=k/m$. The roots are

$$\lambda_1=\omega_0^2/5,\qquad \lambda_2=3\omega_0^2.$$

The corresponding eigenvectors satisfy the homogeneous equation

$$(K-\lambda_r M)\mathbf{q}^{(r)}=0.$$

They can be shown to be

$$\mathbf{q}^{(1)}=\text{const}\times\begin{pmatrix}2\\ 3\end{pmatrix}=\eta_1\mathbf{e}_1,\qquad \mathbf{q}^{(2)}=\text{const}\times\begin{pmatrix}4\\ -1\end{pmatrix}=\eta_2\mathbf{e}_2, \tag{2.64a}$$

where $\mathbf{e}_r$ is the unit vector along $\mathbf{q}^{(r)}$. These vectors are not orthogonal to each other, but they are not parallel to each other either; they are linearly independent of each other.

The reason why we look for these special directions is that along each of these directions Eq. (2.61a) has the simple form

$$M\ddot{\mathbf{q}}^{(r)}(t)=-\lambda_r M\mathbf{q}^{(r)}(t). \tag{2.61b}$$

Each $\mathbf{q}^{(r)}$ has a time-independent direction represented by $\mathbf{e}_r$. Therefore its time dependence resides solely in its length $\eta_r(t)$:

$$\mathbf{q}^{(r)}(t)=\eta_r(t)\mathbf{e}_r. \tag{2.64b}$$

As a result, we may write Eq. (2.61b) as

$$[\ddot{\eta}_r(t)+\lambda_r\eta_r(t)]M\mathbf{e}_r=0. \tag{2.61c}$$

Thus the length $\eta_r(t)$, called a *normal* or *natural coordinate,* satisfies the differential equation for simple harmonic oscillations:

$$\ddot{\eta}_r(t) + \lambda_r \eta_r(t) = 0. \tag{2.61d}$$

The solutions are known to be

$$\eta_r(t) = C_r e^{i\omega_r t}, \qquad \text{or} \qquad D_r e^{-i\omega_r t}, \tag{2.65}$$

a result that can be verified by substitution into Eq. (2.61d). They describe a normal vibration of normal frequency

$$\omega_r = \lambda_r^{1/2}. \tag{2.66}$$

If $\lambda_r > 0$, ω_r is real. The amplitude of the vibration, $|\eta_r(t)|$ as defined in Eq. (2.65), remains constant in time, and the vibration is said to be stable. If $\lambda_r < 0$, ω_r is purely imaginary. $|\eta_r(t)|$ then changes with time, and the vibration is unstable.

If $\mathbf{q}^{(1)}(t)$ is a solution of Eq. (2.61a) and $\mathbf{q}^{(2)}(t)$ is another solution, we can see by direct substitution that their sum

$$\mathbf{q}(t) = \sum_{r=1}^{2} \mathbf{q}^{(r)}(t) \tag{2.67}$$

is also a solution. This is referred to as a *superposition principle.*

The generalization of these results to $n \times n$ matrices is straightforward, and will be left to the reader.

It turns out that the generalized matrix eigenvalue equation (2.62) is mathematically equivalent to the simple matrix eigenvalue equation (2.49) only when the eigenvalues of both K and M are positive definite. If M itself has m zero eigenvalues, the number of solutions λ will be only $n - m$, where n is the dimension of the matrices (Problem 2.6.2). The interested reader will find a number of other interesting possibilities in Problem 2.6.3.

Problems

2.6.1. Calculate the eigenvalues and eigenvectors of the generalized eigenvalue problem Eq. (2.62a) for the following matrices:

(a) $M = \begin{pmatrix} 1 & 0 \\ 0 & 3 \end{pmatrix}, \qquad K = \begin{pmatrix} 3 & 1 \\ 1 & 1 \end{pmatrix}$

(*b*) $M = \begin{pmatrix} 1 & 1 \\ 1 & 3 \end{pmatrix}, \qquad K = \begin{pmatrix} 3 & 1 \\ 1 & 1 \end{pmatrix}.$

2.6.2. Show that if the matrix M itself has m zero eigenvalues, then Eq. (2.62a) for $n \times n$ matrices has at most $n - m$ eigenvalues.

2.6.3. Calculate the eigenvalues and eigenvectors of Eq. (2.62a) for the following matrices:

$$\text{(a)}\ M=\begin{pmatrix}1 & 2\\ 2 & 3\end{pmatrix},\qquad K=\begin{pmatrix}3 & 2\\ 2 & 1\end{pmatrix}$$

$$\text{(b)}\ M=\begin{pmatrix}1 & 0\\ 0 & 0\end{pmatrix},\qquad K=\begin{pmatrix}1 & 1\\ 1 & 6\end{pmatrix}$$

$$\text{(c)}\ M=\begin{pmatrix}1 & 0\\ 0 & -1\end{pmatrix},\qquad K=\begin{pmatrix}3 & 3\\ 3 & -13\end{pmatrix}.$$

2.7. Eigenvalues and eigenvectors of Hermitian matrices

Many matrices appearing in eigenvalue problems in physics are Hermitian matrices for which

$$H^{\dagger} = H^{*\mathrm{T}} = H. \tag{2.68}$$

Hermitian matrices can be real or complex. If real, they are symmetric ($S^{\mathrm{T}} = S$).

Hermitian matrices appear frequently in physics because of the following properties of their eigenvalues and eigenvectors:

1. Their eigenvalues are real.
2. Eigenvectors belonging to distinct eigenvalues are orthogonal.
3. Eigenvectors belonging to degenerate eigenvalues (e.g., $\lambda_1 = \lambda_2$) can be orthogonalized.

Property (1) is appropriate to physics because all measurable quantities are real. Properties (2) and (3) show that the eigenvectors of a Hermitian matrix define a Cartesian coordinate system in the space of n-dimensional vectors. This is a feature of considerable convenience in describing the physics of a system in which the original matrix H plays a significant role.

To derive these properties, we first note that a Hermitian matrix is in general a complex matrix. Its eigenvectors $\mathbf{e}_i$ in the simple eigenvalue equation

$$H\mathbf{e}_i = \lambda_i \mathbf{e}_i \tag{2.69}$$

can be complex; that is, they can have complex components. The scalar product $\mathbf{e}_i^{\dagger}\mathbf{e}_i = \mathbf{e}_i^{\mathrm{T}*}\mathbf{e}_i$ is always non-negative. Two complex vectors $\mathbf{e}_i, \mathbf{e}_j$ are said to be *orthogonal* if

$$\mathbf{e}_i^{\dagger}\mathbf{e}_j = \mathbf{e}_i^{\mathrm{T}*}\mathbf{e}_j = \delta_{ij} = \mathbf{e}_j^{\dagger}\mathbf{e}_i. \tag{2.70}$$

Coming back to our eigenvalue problem in Eq. (2.69), we see that

$$\mathbf{e}_j^{\dagger}H\mathbf{e}_i = \lambda_i \mathbf{e}_j^{\dagger}\mathbf{e}_i. \tag{2.71}$$

Since H is Hermitian, the left-hand side can also be written as

$$\begin{aligned}(\mathbf{e}_j^\dagger H\mathbf{e}_i)^{\dagger\dagger} &= \mathbf{e}_i^\dagger H^\dagger \mathbf{e}_j)^\dagger = (\mathbf{e}_i^\dagger H\mathbf{e}_j)^\dagger \\ &= (\lambda_j \mathbf{e}_i^\dagger \mathbf{e}_j)^\dagger = \lambda_j^* \mathbf{e}_j^\dagger \mathbf{e}_i.\end{aligned} \tag{2.72}$$

The difference between the last two equations is

$$O = (\lambda_i - \lambda_j^*)\mathbf{e}_j^\dagger \mathbf{e}_i. \tag{2.73}$$

Thus, if $j = i$, $\mathbf{e}_i^\dagger \mathbf{e}_j > 0$; hence $\lambda_i^* = \lambda_i = \text{real}$. If $\lambda_j \neq \lambda_i$, then $\mathbf{e}_j^\dagger \mathbf{e}_i = 0$ is required to satisfy Eq. (2.73). Hence the corresponding eigenvectors are orthogonal.

Equation (2.72) does not exclude the possibility that two or more linearly independent eigenvectors may belong to the same eigenvalue. Since any linear combination of these eigenvectors also belongs to the same eigenvalue, we are free to choose linear combinations that lead to orthogonal eigenvectors.

That there are indeed distinct orthogonal eigenvectors belonging to degenerate eigenvalues can be seen as follows: Suppose the Hermitian matrix H_0 has two degenerate eigenvalues $\lambda_1 = \lambda_2$, while the eigenvalues of the Hermitian matrix $H_0 + H_1$ are all distinct. We now consider the Hermitian matrix

$$H(\varepsilon) = H_0 + \varepsilon H_1,$$

and decrease ε from 1 to 0. The eigenvalues $\lambda_1(\varepsilon)$ and $\lambda_2(\varepsilon)$ are distinct for finite ε, but coincide in the limit $\varepsilon \to 0$. When they are distinct, the corresponding eigenvectors are orthogonal. Since there is no abrupt change in the mathematical structure of the problem as $\varepsilon \to 0$, the orthogonal eigenvectors remain orthogonal as $\varepsilon \to 0$.

2.7.1. Matrix diagonalization

Consider the square matrix

$$U = \begin{pmatrix} e_{11} & e_{21} & \cdots & e_{n1} \\ e_{12} & & & \\ \vdots & & & \\ e_{1n} & & \cdots & e_{nn} \end{pmatrix} \tag{2.74}$$

made up of the orthonormalized eigenvectors $\mathbf{e}_j = (e_{j1}, e_{j2}, \ldots, e_{jn})$ of a Hermitian matrix H stored columnwise. The matrix element $(U^\dagger U)_{ij}$ involves the row vector $\mathbf{e}_i$ making up the ith row of $U^\dagger$ and the column vector $\mathbf{e}_j$ making up the jth column of U. It is just the scalar product

$$(U^\dagger U)_{ij} = \mathbf{e}_i^\dagger \mathbf{e}_j = \delta_{ij}. \tag{2.75}$$

Hence $U^\dagger U = I$, and U must be a unitary matrix.

Consider next the matrix element

$$D_{ij} = (U^\dagger HU)_{ij} = \sum_{k,l} (U^\dagger)_{ik} H_{kl} U_{lj}.$$

This also involves the ith row of $U^\dagger$ and the jth column of U. Hence

$$D_{ij} = \mathbf{e}_i^\dagger H \mathbf{e}_j = \lambda_j \mathbf{e}_i^\dagger \mathbf{e}_j = \lambda_j \delta_{ij}, \qquad (2.76)$$

and D is a diagonal matrix containing the eigenvalues on its diagonal.

The matrix $U^\dagger HU$ is called a *unitary transformation* of H. We see that H is diagonalized if the unitary matrix U of the transformation contains the orthonormal eigenvectors of H stored columnwise.

The eigenvalue equation itself also changes under U:

$$\begin{aligned}(H - \lambda_\alpha)\mathbf{e}_\alpha = 0 \\ &= U^\dagger (H - \lambda_\alpha)\mathbf{e}_\alpha \\ &= U^\dagger (H - \lambda_\alpha) UU^\dagger \mathbf{e}_\alpha \\ &= (D - \lambda_\alpha)\mathbf{e}'_\alpha.\end{aligned}$$

Since D in the transformed eigenvalue equation is diagonal, the new eigenvectors

$$\mathbf{e}'_\alpha = U^\dagger \mathbf{e}_\alpha$$

must be unit vectors along the new coordinate axes. This result can be deduced directly from the last equation itself, for the components of the new eigenvector are

$$(\mathbf{e}'_\alpha)_\beta = (U^\dagger \mathbf{e}_\alpha)_\beta = \mathbf{e}_\beta^\dagger \mathbf{e}_\alpha = \delta_{\alpha\beta}. \qquad (2.77)$$

From the perspective of the coordinate transformation of Section 2.2, the eigenvalue system has not changed. Rather it is the coordinate system that has rotated. The transformation $U^\dagger$ has rotated the coordinate system so that the eigenvectors have become the new coordinate axes. In physics, the eigenvectors are often called the *principal axes* of H. The unitary transformation U is then referred to as a *principal-axes transformation.*

Example 2.7.1. Diagonalize the matrices $H = \begin{pmatrix} 1 & 1 \\ 1 & 1 \end{pmatrix}$, H^2, H^n, and e^H.

By using the method of Section 2.5, we can show that the eigenvalues of H are $\lambda_1 = 0$ and $\lambda_2 = 2$. The corresponding eigenvectors are

$$\mathbf{e}_1 = \frac{1}{\sqrt{2}}\begin{pmatrix} 1 \\ -1 \end{pmatrix}, \qquad \mathbf{e}_2 = \frac{1}{\sqrt{2}}\begin{pmatrix} 1 \\ 1 \end{pmatrix}.$$

The unitary matrix $U = \frac{1}{\sqrt{2}}\begin{pmatrix} 1 & 1 \\ -1 & 1 \end{pmatrix}$ containing these *normalized* eigenvectors columnwise will therefore diagonalize H through the unitary transformation

$$U^\dagger HU = \frac{1}{\sqrt{2}}\begin{pmatrix} 1 & -1 \\ 1 & 1 \end{pmatrix} \frac{1}{\sqrt{2}}\begin{pmatrix} 0 & 2 \\ 0 & 2 \end{pmatrix} = \begin{pmatrix} 0 & 0 \\ 0 & 2 \end{pmatrix} = D.$$

The diagonalized matrix D contains the eigenvalues 0 and 2 along its diagonal.

The same unitary transformation applied to H^2 yields

$$U^\dagger H^2 U = U^\dagger HU^\dagger UHU = D^2.$$

Since the product of two diagonal matrices AB is a diagonal matrix whose diagonal matrix elements are the products of corresponding matrix elements, that is, $(AB)_{ij} = A_{ii}B_{ii}\delta_{ij}$, we have

$$D^2 = \begin{pmatrix} 0 & 0 \\ 0 & 4 \end{pmatrix}.$$

Similarly,

$$U^\dagger H^n U = D^n = \begin{pmatrix} 0 & 0 \\ 0 & 2^n \end{pmatrix}.$$

We come finally to the *matrix function* e^H of a matrix H. It may be defined by the usual power series for the exponential function

$$e^H = \sum_{n=0}^{\infty} H^n/n!,$$

now applied to the matrix H. Since each term on the right-hand side is a 2×2 matrix, e^H itself must also be a 2×2 matrix. Its diagonalization can now be achieved as follows:

$$\begin{aligned} U^\dagger e^H U &= U^\dagger\left(\sum_{n=0}^{\infty} \frac{H^n}{n!}\right)U \\ &= \sum_{n=0}^{\infty} \frac{D^n}{n!} = e^D. \end{aligned}$$

Since the product of diagonal matrices is diagonal, e^D is diagonal. Its matrix elements are

$$(e^D)_{ii} = \sum_{n=0}^{\infty} \frac{D_{ii}^n}{n!} = e^{D_{ii}}.$$

Hence

$$e^D = \begin{pmatrix} 1 & 0 \\ 0 & e^2 \end{pmatrix}.$$

Exercise 2.7.1. Show explicitly that the matrix

$$U = \frac{1}{\sqrt{2}}\begin{pmatrix} 1 & 1 \\ i & -i \end{pmatrix}$$

containing the eigenvectors of Eq. (2.56) columnwise will also diagonalize the matrix $\begin{pmatrix} 0 & i \\ -i & 0 \end{pmatrix}$. Is U unitary?

2.7.2. The generalized problem

Let us next turn our attention to the generalized matrix eigenvalue problem of Eq. (2.62a)

$$K\mathbf{e}_i = \lambda_i M\mathbf{e}_i.$$

This turns out to be equivalent to Eq. (2.69) if M is nonsingular. To see this, we first note that the matrix inverse square root $M^{-1/2}$, which is itself a square matrix, exists if M is nonsingular. If it is left multiplied into Eq. (2.62a), we obtain

$$M^{-1/2}K\mathbf{e}_i = M^{-1/2}KM^{-1/2}(M^{1/2}\mathbf{e}_i) = \lambda_i(M^{1/2}\mathbf{e}_i), \tag{2.78a}$$

or

$$K'\mathbf{b}_i = \lambda_i \mathbf{b}_i,$$

where

$$\begin{aligned} K' &= M^{-1/2}KM^{-1/2}, \\ \mathbf{b}_i &= M^{1/2}\mathbf{e}_i. \end{aligned} \tag{2.79}$$

As usual, the eigenvalue Eq. (2.78a) defines only the direction $\mathbf{e}'_i$ of $\mathbf{b}_i$

$$K'\mathbf{e}'_i = \lambda_i \mathbf{e}'_i. \tag{2.78b}$$

It is clear that $\mathbf{e}'_i$ differs in general from $\mathbf{e}_i$ (the direction of $\mathbf{q}^{(i)}$) unless $M^{1/2}$ is a scalar. In that case, Eq. (2.62a) is identical to the simpler matrix eigenvalue problem of Eq. (2.49).

Let us examine the matrix K' when both M and K are general Hermitian matrices. Since the inverse and the square root of Hermitian matrices are also Hermitian, K' is Hermitian. The situation is therefore the same as in Eq. (2.69): Eigenvalues are real, and eigenvectors are, or can be made, orthogonal. The only difference from Eq. (2.69) is that, while the $\mathbf{e}'_i$ are orthogonal, the $\mathbf{e}_i$ are not, although they are necessarily linearly independent of one another. (A proof of this statement is requested in Problem 2.7.6.) An example of such nonorthogonal eigenvectors has been given in Eq. (2.64a) of the preceding section.

We note finally that if M is Hermitian, it can be diagonalized. Hence the *weight matrix* M may be taken to be diagonal without any loss of generality. The orthogonality relation then has the explicit form

$$\mathbf{b}_i^* \cdot \mathbf{b}_j = \mathbf{e}_i^\dagger M \mathbf{e}_j = \delta_{ij} \sum_k |e_{ik}|^2 M_{kk}, \tag{2.80}$$

where e_{ik} is the kth component of $\mathbf{e}_i$.

2.7.3. Matrix transformations

We shall conclude this section by introducing some mathematical terms describing matrix transformations. Two square matrices A and B are said to be *equivalent* to each other if there exist *non*singular matrices P and Q such that

$$B = PAQ. \tag{2.81}$$

(A useful result: Every matrix is equivalent to some diagonal matrix.) The tranformation PAQ of A is called an *equivalent transformation*. The important special cases are:

1. If $P = Q^{-1}$, then $Q^{-1}AQ$ is a *similarity transformation* of A;
2. If $P = Q^{\mathrm{T}}$, then $Q^{\mathrm{T}}AQ$ is a *congruent transformation* of A;
3. If $P = Q^{\dagger}$, then $Q^{\dagger}AQ$ is a *conjunctive transformation* of A;
4. If $P = Q^{-1} = Q^{\mathrm{T}}$, it is an *orthogonal transformation*;
5. If $P = Q^{-1} = Q^{\dagger}$, it is a *unitary transformation*.

Finally, we mention three useful mathematical results:

1. The trace of a matrix is unchanged under a similarity transformation:
$$\mathrm{Tr}\, Q^{-1}AQ = \mathrm{Tr}\, QQ^{-1}A = \mathrm{Tr}\, A.$$
2. The trace of a Hermitian matrix is real and is equal to the sum of its eigenvalues:
$$\mathrm{Tr}\, H = \mathrm{Tr}\, U^{\dagger}HU = \mathrm{Tr}\, D = \sum_i \lambda_i.$$
3. A matrix A is *normal* if $[A, A^{\dagger}] = 0$. A normal matrix can be diagonalized by a unitary transformation. Hermitian, anti-Hermitian, unitary, and antiunitary matrices are examples of normal matrices.

Problems

2.7.1. Show that two commuting Hermitian matrices can be diagonalized by the same unitary transformation.

2.7.2. Show that the eigenvalues of an anti-Hermitian matrix are purely imaginary.

2.7.3. Show that the eigenvalues of a unitary matrix are complex numbers of unit amplitude.

2.7.4. If H is a Hermitian matrix, show that
   (a) $\mathrm{Tr}\, H = \mathrm{Tr}\, D$, where $D = U^{\dagger}HU$ is its diagonalized form;
   (b) $\det H = \det D$.

2.7.5. Show that the determinant of a Hermitian matrix H is the product of its eigenvalues.

2.7.6. Given n orthonormal unit vectors $\mathbf{e}'_i$, show that the n vectors $Q\mathbf{e}'_i$ are linearly independent if Q is a nonsingular Hermitian matrix.

2.7.7. Discuss the solution of the generalized matrix eigenvalue problem described by Eq. (2.62a) if one of the eigenvalues of M is zero. (You may assume that both M and K are Hermitian.)

2.7.8. (a) A function $F(x) = f_0 + f_1 x + f_2 x^2 + \cdots$ is defined by its Taylor series. Show that this also defines a matrix function $f(B)$ of an $n \times n$ matrix B and that $f(B)$ is also an $n \times n$ matrix.

(b) If matrix B has the eigenvalues $b_1, b_2, \ldots, b_n$, show that $F(B)$ has the eigenvalues $f(b_1), f(b_2), \ldots, f(b_n)$.

(c) If H is a Hermitian matrix, show that $\det(e^H) = e^{\mathrm{Tr}\, H}$.

2.8. The wave equation

One of the most important equations of motion in classical physics is the wave equation. To facilitate subsequent discussions, it is profitable to describe briefly a simple physical system whose motion satisfies this equation. It is hoped that this will provide a meaningful physical context in which the often abstract considerations of several subsequent sections can be embedded. The reader who already understands the wave equation can move immediately to the next section.

The simple system in question is that of n identical masses m connected by identical massless springs each of length d and spring constant k. The displacement $q_i(t)$ of the ith mass from its position of equilibrium satisfies the Newtonian equation of the form of Eq. (2.60)

$$\frac{m}{k}\ddot{q}_i = -(q_i - q_{i-1}) - (q_i - q_{i+1}) = q_{i+1} - 2q_i + q_{i-1},$$

$$i = 1, \ldots, n. \tag{2.82}$$

Here we have chosen $q_0 = q_{n+1} = 0$ as the (zero) displacements of the fixed ends. We would like to go to the limit $n, k \to \infty$, $m, d \to 0$ in such a way that

$$M = nm, \qquad L = (n+1)d, \qquad \tau = kd$$

are all finite. In this "continuum" limit, it is more convenient to replace the position index i of the ith mass along the length of the spring by the finite variable $x_i = id$:

$$q_i(t) = q(x_i, t).$$

Dividing Eq. (2.82) by d^2 and taking this limit of $n \to \infty$, we find

$$\lim \frac{m}{kd^2} = \lim \frac{mn}{knd^2} = \frac{M}{\tau L} \equiv \frac{1}{v^2} = \text{finite},$$

and

$$\lim \frac{1}{d^2}[q(x_{i+1}, t) - 2q(x, t) + q(x_{i-1}, t)] = \frac{\partial^2}{\partial x^2} q(x,t).$$

Hence Eq. (2.82) now reads

$$\left(\frac{1}{v^2}\frac{\partial^2}{\partial t^2} - \frac{\partial^2}{\partial x^2}\right) q(x,t) = 0, \tag{2.83}$$

where the parameter v has the dimension of a speed (i.e., distance divided by time). This is called a one-dimensional *wave equation*. It describes the longitudinal vibrations of a continuous, uniform, *massive* spring. (The mass comes from the point masses, not from the originally massless springs.)

The solution of Eq. (2.83) can be written in the form

$$q(x,t) = f(x - vt) + g(x + vt), \tag{2.84}$$

where f and g are completely arbitrary functions of their arguments. To see this, first set $g = 0$, and denote $x - vt$ by s. Then

$$\frac{\partial}{\partial x} f(s) = \frac{d}{ds} f(s), \qquad \frac{1}{v}\frac{\partial}{\partial t} f(s) = -\frac{d}{ds} f(s).$$

The left-hand side of Eq. (2.83) is now

$$\frac{d^2}{ds^2} f(s) - \frac{d^2}{ds^2} f(s),$$

which must necessarily be zero. Hence any function $f(x - vt)$ is a solution. In a similar way one can show that any function $g(x + vt)$ is also a solution. Finally their sum, Eq. (2.84), is also a solution, since each term contributes nothing on the left-hand side of Eq. (2.83). The fact that a sum of solutions is also a solution is referred to as a *superposition principle.*

Although $f(s)$ does not change when s is constant, the wave does move in general in space and in time because s itself is a function of x and t. In particular, $ds = dx - v\,dt$. Hence when $ds = 0$, we must have

$$\frac{dx}{dt} = v.$$

Thus the point of constant s, and constant $f(s)$, must move along the $+x$ axis with speed v. We call this point of constant s a *wave front* of constant phase (which refers to s itself). The speed v is called the *phase velocity* of the wave.

In a similar way, one can show that in $g(x + vt)$ the wave front of constant phase $x + vt$ moves with a velocity

$$\frac{dx}{dt} = -v$$

along the $-x$ axis.

The relative negative sign between the two partial differential operators in Eq. (2.83) is of importance in wave propagation. It is quite conceivable that there exist physical phenomena that can be described by the differential equation

$$\left(\frac{1}{v^2}\frac{\partial^2}{\partial t^2} + \frac{\partial^2}{\partial x^2}\right)\Pi(s,t) = 0, \tag{2.85}$$

with the opposite (i.e., positive) relative sign. If so, its solution must be expressible in the form

$$\Pi(x,t) = f(x - ivt) + g(x + ivt)$$

with a purely imaginary speed of propagation iv. All measurable physical quantities are real. An imaginary speed is not a speed at all. It simply gives a position divided by time scale over which the physical phenomenon in question is appearing and disappearing. In fact, we do not know of any important physical phenomenon that is described by Eq. (2.85).

Let us now return to the wave equation, Eq. (2.83). It could also describe a wave propagating along the $\pm x$ axis, but in a three-dimensional space. (Obviously, we are no longer considering the problem of the massive linear spring.) Such a wave is called a *plane wave,* since the wave front is the entire yz plane at x. If we had chosen the coordinate axes along different directions, the same plane wave propagating along the same direction would have involved disturbances along the new y and z coordinates as well. Thus the equation of motion should contain $\partial^2/\partial y^2$ and $\partial^2/\partial z^2$ in a way symmetrical with $\partial^2/\partial x^2$. This consideration leads to the three-dimensional wave equation

$$\left(\frac{1}{v^2}\frac{\partial^2}{\partial t^2} - \nabla^2\right)u(\mathbf{r},t) = 0. \tag{2.86}$$

For this equation, the general solution is of the form

$$u(\mathbf{r},t) = f(\mathbf{r} - \mathbf{v}t) + g(\mathbf{r} + \mathbf{v}t).$$

2.9. Displacement in time and translation in space: Infinitesimal generators

The wave equation [Eq. (2.83)] describes a certain relation between the acceleration of masses and the restoring force responsible for it. More specifically, it states that wave propagation is the consequence of a simple correlation between the curvature (i.e., the second derivative) of the wave function $q(x,t)$ in time to its curvature in space. Like other laws of nature, it makes a nontrivial statement relating these apparently distinct physical properties. In particular, the relationship is that between a differential operation in time and those in space, again like many other laws of nature.

One might well wonder why so many laws of nature can be expressed as differential equations in space and in time. The answer, which we would like to discuss in more detail in this section, is that the differential operators in space and in time are related to time displacements and space translations. Physics is concerned not with a dead and static universe, but with one that is dynamic and changing both in space and in time. Physical laws relate properties at different points of space and at different times.

Let us consider time translation first, because it is simpler. We are familiar with the idea that, under an infinitesimal time displacement dt, a differentiable function $f(t)$ of time t changes to

$$f(t+dt)=f(t)+dt\frac{df}{dt}=\left(1+dt\frac{d}{dt}\right)f(t). \tag{2.87}$$

After another time displacement dt, we have

$$f(t+2\,dt)=\left(1+dt\frac{d}{dt}\right)f(t+dt)=\left(1+dt\frac{d}{dt}\right)^2 f(t).$$

Since a finite time displacement τ can be constructed from an infinite number of consecutive infinitesimal time displacements, we have

$$\begin{aligned} f(t+\tau) &= \lim_{n\to\infty}\left(1+\frac{\tau}{n}\frac{d}{dt}\right)^n f(t)=\exp\left(\tau\frac{d}{dt}\right)f(t) \\ &= T_t(\tau)f(t), \end{aligned} \tag{2.88}$$

where

$$T_t(\tau)=\exp(-i\tau W), \qquad W=i\frac{d}{dt}. \tag{2.89}$$

Thus the increase by τ of the time argument t in $f(t)$ can be expressed formally as the result of a left multiplication of $f(t)$ by a *time displacement operator* $T_t(\tau)$. The operator W itself is called the *infinitesimal generator* of time displacement.

To displace the time of an arbitrary function $f(t)$ by the amount τ, it might be necessary to compute the operator algebraic series

$$\begin{aligned} f(t+\tau) &= \exp\left(\tau\frac{d}{dt}\right)f(t) \\ &= \left[1+\tau\frac{d}{dt}+\frac{1}{2!}\left(\tau\frac{d}{dt}\right)^2+\cdots\right]f(t) \end{aligned} \tag{2.90}$$

term by term. This is of course just the Taylor expansion of $f(t+\tau)$ about $f(t)$, as one might suspect.

Example 2.9.1. Calculate $\sin \omega(t+\tau)$

$$\begin{aligned}\sin \omega(t+\tau) &= \left[1+\tau\frac{d}{dt}+\frac{1}{2!}\left(\tau\frac{d}{dt}\right)^2+\frac{1}{3!}\left(\tau\frac{d}{dt}\right)^3+\cdots\right]\sin \omega t \\ &= \sin \omega t+\omega\tau\cos \omega t+\frac{1}{2!}(\omega\tau)^2(-\sin \omega t) \\ &\quad +\frac{1}{3!}(\omega\tau)^3(-\cos \omega t)+\cdots \\ &= \sin \omega t\left(1-\frac{1}{2!}(\omega\tau)^2+\cdots\right) \\ &\quad +\cos \omega t\left(\omega\tau-\frac{1}{3!}(\omega\tau)^3+\cdots\right) \\ &= \sin \omega t\cos \omega\tau+\cos \omega t\sin \omega\tau.\end{aligned}$$

There are certain advantages to using the exponential time displacement operator of Eq. (2.89) to represent a Taylor expansion of a function of time. Two successive time displacements

$$T_t(\tau_1)T_t(\tau_2)=\exp[-i(\tau_1+\tau_2)W]=T_t(\tau_1+\tau_2)$$

is equivalent to a single composite displacement in the same way that two rotation matrices can be multiplied into a single rotation matrix. Perhaps more important, the time displacement operator takes on a very simple form for those functions that satisfy the equation

$$Wf(t)=\omega f(t), \tag{2.91}$$

with a constant ω. This is called a *differential eigenvalue equation,* in which $f(t)$ is the eigenfunction belonging to the eigenvalue ω. For such a function, Eq. (2.90) simplifies to

$$\begin{aligned}f(t+\tau) &= \left(1+\tau(-i\omega)+\frac{1}{2!}(-i\omega\tau)^2+\cdots\right)f(t) \\ &= \exp(-i\omega\tau)f(t).\end{aligned} \tag{2.92}$$

In particular, if at $t=0$, $f(t=0)=1$, we find

$$f(\tau)=\exp(-i\omega\tau), \tag{2.93}$$

that is, $f(t)=\exp(-i\omega t)$, which is obviously a solution of Eq. (2.91).

This eigenfunction

$$\exp(-i\omega t)=\cos \omega t-i\sin \omega t \tag{2.94}$$

should be familiar to us as the function describing the time dependence of a system oscillating with frequency ω. This suggests that time displacement is related to the frequencies ω with which the system can oscillate. We may therefore call W a *frequency operator.*

Another interesting feature of W arises from the fact that it is a differential operator:

$$\frac{d}{dt}[tf(t)] = t\frac{d}{dt}f(t) + f(t).$$

That is, it satisfies the *commutation relation*

$$[W,t] \equiv Wt - tW = i. \tag{2.95}$$

This is a direct consequence of the fact that time is changed by the time displacement operation.

In a similar way, an infinitesimal translation $d\mathbf{r}$ in space results in a change of a scalar field (or a component of a vector field) $\phi(\mathbf{r})$ to

$$\phi(\mathbf{r} + d\mathbf{r}) = \phi(\mathbf{r}) + d\mathbf{r} \cdot \nabla\phi(\mathbf{r}) = (1 + d\mathbf{r} \cdot \nabla)\phi(\mathbf{r}).$$

A finite translation $\boldsymbol{\rho}$ *of the position operator* $\mathbf{r}$ can be constructed from an infinite number of successive infinitesimal space translations. The resulting scalar field is

$$\begin{aligned}\phi(\mathbf{r} + \boldsymbol{\rho}) &= \lim_{n\to\infty}\left(1 + \frac{\boldsymbol{\rho}}{n} \cdot \nabla\right)^n \phi(\mathbf{r}) = \exp(\boldsymbol{\rho} \cdot \nabla)\phi(\mathbf{r}) \\ &= T_{\mathbf{r}}(\boldsymbol{\rho})\phi(\mathbf{r}),\end{aligned} \tag{2.96}$$

where

$$T_{\mathbf{r}}(\boldsymbol{\rho}) = \exp(i\boldsymbol{\rho} \cdot \mathbf{K}), \qquad \mathbf{K} = \frac{1}{i}\nabla. \tag{2.97}$$

Thus the space coordinate in $\phi(\mathbf{r})$ is increased by $\boldsymbol{\rho}$ as the result of a left multiplication of $\phi(\mathbf{r})$ by the *space translation operator* $T_{\mathbf{r}}(\boldsymbol{\rho})$, which is an exponential function of the generator $\mathbf{K}$ of space translation. Its eigenfunction belonging to the eigenvalue $\mathbf{k}$ is proportional to $\exp(i\mathbf{k} \cdot \mathbf{r})$. The generator $\mathbf{K}$ satisfies the commutation relations

$$[K_i, x_j] = -i\delta_{ij}, \tag{2.98}$$

where x_j is the jth component of $\mathbf{r}$.

Example 2.9.2. Calculate the electrostatic potential of a point charge at $\mathbf{r}_s = (0,0,d)$

$$\begin{aligned}\Phi(\mathbf{r} - \mathbf{r}_s) &= \frac{\text{const}}{|\mathbf{r} - \mathbf{r}_s|} = \exp(-i\mathbf{r}_s \cdot \mathbf{K})\Phi(\mathbf{r}) = \exp(-idK_z)\Phi(\mathbf{r}) \\ &= \left[1 + \left(-d\frac{\partial}{\partial z}\right) + \frac{1}{2!}\left(-d\frac{\partial}{\partial z}\right)^2 + \cdots\right]\Phi(\mathbf{r}).\end{aligned}$$

These derivatives can be calculated with the help of the identity

$$\begin{aligned}\frac{\partial}{\partial z}\frac{1}{r^m} &= 2z\frac{\partial}{\partial r^2}\frac{1}{(r^2)^{m/2}} = -\frac{mz}{r^{m+2}} \\ &= -\frac{m\cos\theta}{r^{m+1}}.\end{aligned}$$

Thus

$$\frac{\partial}{\partial z}\frac{1}{r} = -\frac{z}{r^3} = -\frac{\cos\theta}{r^2},$$

$$\frac{\partial^2}{\partial z^2}\frac{1}{r} = \frac{3\cos^2\theta - 1}{r^3}.$$

Hence

$$\Phi(\mathbf{r} - \mathbf{r}_s) = \text{const} \times \left(\frac{1}{r} + \frac{d\cos\theta}{r^2} + \frac{d^2}{2}\frac{3\cos^2\theta - 1}{r^3} + \cdots\right).$$

These terms are referred to as the *monopole, dipole,* and *quadrupole terms,* respectively. The expansion is called a *multipole expansion.*

2.9.1. Transformation groups

The exponentiated form of these space-time transformation operators show that they satisfy the following multiplicative properties:

1. Closure under multiplication:

$$T_t(\tau_2)T_t(\tau_1) = T_t(\tau_2 + \tau_1),$$
$$T_{\mathbf{r}}(\boldsymbol{\rho}_2)T_{\mathbf{r}}(\boldsymbol{\rho}_1) = T_{\mathbf{r}}(\boldsymbol{\rho}_2 + \boldsymbol{\rho}_1). \tag{2.99}$$

   That is, the product of operators is itself an operator of the same type.

2. Associative property:

$$T_t(\tau_3)[T_t(\tau_2)T_t(\tau_1)] = [T_t(\tau_3)T_t(\tau_2)]T_t(\tau_1). \tag{2.100}$$

3. Identity operator:

$$T_t(0) = 1, \qquad T_{\mathbf{r}}(\mathbf{0}) = 1. \tag{2.101}$$

4. Inverse operators:

$$T_t^{-1}(\tau)T_t(\tau) = 1 = T_t(\tau)T_t^{-1}(\tau),$$
$$\text{i.e.,} \qquad T_t^{-1}(\tau) = T_t(-\tau). \tag{2.102}$$

A collection or set of objects (called *elements*) satisfying the algebraic properties (1)–(4) is said to form a *group* with respect to the specified multiplication rule between group elements. For this reason, we talk about the group of time displacements and the group of space translations. In addition, our group multiplications are commutative:

$$T_t(\tau_1)T_t(\tau_2) = T_t(\tau_2)T_t(\tau_1),$$
$$T_{\mathbf{r}}(\boldsymbol{\rho}_1)T_{\mathbf{r}}(\boldsymbol{\rho}_2) = T_{\mathbf{r}}(\boldsymbol{\rho}_2)T_{\mathbf{r}}(\boldsymbol{\rho}_1). \tag{2.103}$$

We call groups with commutative products *Abelian groups,* after Abel, a mathematician of the nineteenth century.

Rotation matrices also form a group, but they do not in general commute. Hence they form a non-Abelian group.

2.9.2. *Infinitesimal generators, invariance principles, and physical laws*

A study of the powerful mathematics of group theory is outside the scope of this book; so let us return to the space-time transformation operators themselves. Their importance arises from that basic postulate of physics that states that physical laws should remain essentially unchanged in all space and for all time. If this were not true, physics would be just a part of history, which involves a study of the past, and never an *objective prediction* of the future. Unlike historical events, there is every reason to believe that the fall of an apple here tomorrow can be predicted with the same precision using Newton's laws as that of an apple Newton saw in England, or of the balls Galileo dropped from the Leaning Tower in Pisa at a time when Newton's laws of motion and gravitation had not yet been discovered.

These nice invariant properties of physical laws can readily be ascertained with the help of the space and time displacement operators. Let us suppose that the state of motion of a system here and now is described by a function $f(\mathbf{r},t)$ that satisfies the equation of state

$$\mathscr{D}f(\mathbf{r},t)=0. \tag{2.104}$$

where $\mathscr{D}$ is an appropriate operator. Under a time displacement it is changed into

$$\begin{aligned}0 &= T_t(\tau)[\mathscr{D}f(\mathbf{r},t)] = T_t(\tau)\mathscr{D}T_t^{-1}[T_t(\tau)f(\mathbf{r},t)]\\ &= \mathscr{D}_t(\tau)f(\mathbf{r},t+\tau).\end{aligned}$$

Since the choice of the starting time $t=0$ is arbitrary, we see that the invariance of the equation of state (2.104) under time displacement requires that

$$\mathscr{D}_t(\tau) = T_t(\tau)\mathscr{D}T_t^{-1}(\tau) = \mathscr{D}. \tag{2.105}$$

That is, the operator must commute with $T_t(\tau)$, or equivalently with the frequency operator W. In a similar way, invariance under space translation requires that $\mathscr{D}$ commutes with the space translation operator $T_{\mathbf{r}}(\boldsymbol{\rho})$, or equivalently with its generator $\mathbf{K}$.

One operator that will commute with $\mathbf{K}$ is $\mathbf{K}$ itself. It changes sign when the coordinate $\mathbf{r}$ changes sign. Its vector components also change when the coordinate system is rotated. If the physical description is to be independent of these changes, only scalar products such as $K^2=\mathbf{K}\cdot\mathbf{K}$ should appear. The simplest choice satisfying these *invariance principles* is

$$\mathscr{D} = K^2 - k^2. \tag{2.106}$$

Its use in Eq. (2.104) gives the three-dimensional *Helmholtz equation*

$$(K^2-k^2)u(\mathbf{r})=0, \tag{2.107}$$

which describes the spatial part of wave motion.

The Helmholtz equation can be solved as follows. The infinitesimal generator $\mathbf{K}$ of space translation has eigenvalues $\pm\mathbf{k}$ and eigenfunctions $\exp(\pm i\mathbf{k}\cdot\mathbf{r})$, the latter being the spatial version of Eq. (2.93):

$$\mathbf{K}\exp(\pm i\mathbf{k}\cdot\mathbf{r}) = \pm\mathbf{k}\exp(\pm i\mathbf{k}\cdot\mathbf{r}). \tag{2.108}$$

Applying $\mathbf{K}$ twice in the form

$$\begin{aligned}\mathbf{K}\cdot\mathbf{K}\exp(\pm i\mathbf{k}\cdot\mathbf{r}) &= \pm\mathbf{k}\cdot\mathbf{K}\exp(\pm i\mathbf{k}\cdot\mathbf{r})\\ &= k^2\exp(\pm i\mathbf{k}\cdot\mathbf{r}),\end{aligned}$$

we get just the Helmholtz equation. Thus the two eigenfunctions of $\mathbf{K}$ belonging to the eigenvalues $\pm\mathbf{k}$ are also the linearly independent eigenfunctions of K^2 belonging to the "doubly degenerate" eigenvalue k^2. We thus see that the eigenfunctions themselves appear in the description of physical disturbances for the simple reason that they describe states of the system in which these physical disturbances appear in their simplest form.

In a similar way, the time part of $u(\mathbf{r},t)$ at fixed $\mathbf{r}$ will satisfy a differential operator even in the frequency operator W:

$$(W^2 - \omega^2)u(\mathbf{r} = \text{fixed},t) = 0.$$

This shows that the complete time-dependent wave function $u(\mathbf{r},t)$ satisfies the partial differential equation (i.e., a differential equation in both $\mathbf{r}$ and t)

$$\left(K^2 - \frac{W^2}{v^2}\right)u(\mathbf{r},t) = \left(k^2 - \frac{\omega^2}{v^2}\right)u(\mathbf{r},t) = 0, \tag{2.109}$$

if $k = \pm\omega/v$. This is just the wave equation, Eq. (2.86),

$$\left(-\nabla^2 + \frac{1}{v^2}\frac{\partial^2}{\partial t^2}\right)u(\mathbf{r},t) = 0,$$

where v is the wave speed.

If space (or $\mathbf{r}$) and time are originally independent of each other, the eigenfunctions $u(\mathbf{r},t)$ must be products (actually linear combinations of products) of the eigenfunctions for $\mathbf{K}$ (for the spatial part) and W (for the time part). There are four independent products with the same eigenvalues k^2 for K^2 and ω^2 for W^2:

$$\begin{aligned}u_1(\mathbf{r},t) &= \exp i(\mathbf{k}\cdot\mathbf{r} - \omega t), & u_2(\mathbf{r},t) &= \exp i(-\mathbf{k}\cdot\mathbf{r} - \omega t),\\ u_3(\mathbf{r},t) &= \exp i(\mathbf{k}\cdot\mathbf{r} + \omega t), & u_4(\mathbf{r},t) &= \exp i(-\mathbf{k}\cdot\mathbf{r} + \omega t).\end{aligned} \tag{2.110}$$

Two of these solutions, u_1 and u_4, describe waves moving with speed v along the direction $\mathbf{k}$, while the remaining two solutions, u_2 and u_3, describe waves propagating along $-\mathbf{k}$, as the reader is asked to confirm in Problem 2.9.2. It is the wave equation itself that sets up the specific correlation between changes in position and changes in time, which leads to the propagation of a wave along $\mathbf{k}$ or $-\mathbf{k}$. For this reason, $\mathbf{K}$ may be called a *propagation operator*.

2.9.3. Wave equation in quantum mechanics

One of the greatest discoveries of physics in this century is that matter in motion shows wave properties, an observation that is of utmost importance in describing the physics of atomic phenomena. The mathematical description of matter waves, like that of any other wave motion, requires a wave equation. This is achieved in quantum mechanics with the help of the following "quantization rules" deduced from experimental facts: The concepts of momentum and energy for a material particle must be generalized to the *momentum* and the *Hamiltonian operator,* respectively, involving Planck's constant $\hbar$:

$$\mathbf{p}_{\text{op}} = \hbar \mathbf{K}, \qquad \text{and} \qquad H = \hbar W. \tag{2.111}$$

The classical energy–momentum relation, such as the

$$E^2 = p^2c^2 + m_0^2c^4 \tag{2.112}$$

relation of Einstein, should next be generalized to its operator equivalent, $H^2 = p_{\text{op}}^2c^2 + m_0^2c^4$ in our example. But since H and $\mathbf{p}_{\text{op}}$ are differential operators of independent variables $\mathbf{r}$ and t, this operator equivalent cannot be taken literally as an operator equation. It is conceivable, however, that H^2 operating on a special class of functions $u(\mathbf{r},t)$ gives a result exactly equal to $p_{\text{op}}^2c^2 + m_0^2c^4$ operating on the same function. That is,

$$H^2u(\mathbf{r},t) = (p_{\text{op}}^2c^2 + m_0^2c^4)u(\mathbf{r},t). \tag{2.113}$$

When the differential forms of H and $\mathbf{p}_{\text{op}}$ are used, this equation can be simplified to the form of a Schrödinger wave equation

$$\left(-\nabla^2 + \frac{m^2c^2}{\hbar^2} + \frac{1}{c^2}\frac{\partial^2}{\partial t^2}\right)u(\mathbf{r},t) = 0, \tag{2.114}$$

known as the *Klein–Gordon equation.*

Problems

2.9.1. Calculate the electrostatic potential at large distances of
(a) An electric dipole with charges Q at $\mathbf{s}/2$ and $-Q$ at $-\mathbf{s}/2$;
(b) A linear electric quadrupole with charges Q at $\pm\mathbf{s}/2$ and $-2Q$ at the origin.

2.9.2. Determine the direction of propagation of the wave described by each of the four wave functions shown in Eq. (2.110).

2.9.3. Show that all $m \times n$ matrices form a group under matrix *addition* as the group multiplication. What is the identity element of the group? What is the inverse operation?

2.9.4. Show that all $n \times n$ square nonsingular matrices form a group under matrix multiplication as the group multiplication.

2.9.5. Show that all 3×3 rotation matrices form a group under matrix multiplication.

2.9.6. Show that the four 2×2 matrices

$$I = \begin{pmatrix} 1 & 0 \\ 0 & 1 \end{pmatrix}, \qquad A = \begin{pmatrix} 0 & i \\ i & 0 \end{pmatrix},$$

$$B = \begin{pmatrix} -1 & 0 \\ 0 & -1 \end{pmatrix}, \qquad C = \begin{pmatrix} 0 & -i \\ -i & 0 \end{pmatrix}$$

are closed under matrix multiplication. (That is, their products with each other can be expressed in terms of the matrices themselves.) Show that these matrices form a group. Show that this group has the same mathematical structure (i.e., group multiplicative properties) as any one of the following groups of four elements:

(a) $I = 1, \quad A = i, \quad B = -1, \quad C = -i;$

(b) $I = R(0), \quad A = R(\pi/2), \quad B = R(\pi), \quad C = R(3\pi/2).$

2.10. Rotation operators

Let us now return to rotation matrices, because our earlier discussion is somewhat incomplete. We obtained in Section 2.2 the matrices for finite rotations. According to the ideas discussed in Section 2.9, it should be possible to generate finite rotations from infinitesimal rotations. The associated infinitesimal generators are in a sense more interesting than the finite rotation matrices themselves, since the former may appear as operators in terms of which the states of rotating systems can be described. In this section, we concentrate on these generators of infinitesimal rotations.

2.10.1. Infinitesimal rotations

The rotation matrix for an infinitesimally small rotation of amount $d\theta_3$ about the 3 axis is from Eq. (2.7)

$$R(d\theta) = \begin{pmatrix} 1 & d\theta & 0 \\ -d\theta & 1 & 0 \\ 0 & 0 & 1 \end{pmatrix} = 1 + i\, d\theta_3\, J_3, \tag{2.115}$$

where

$$J_3 = i \begin{pmatrix} 0 & -1 & 0 \\ 1 & 0 & 0 \\ 0 & 0 & 0 \end{pmatrix}. \tag{2.116a}$$

Similarly, for rotations about the 1 and 2 axes separately, we have

$1+i\,d\theta_1\,J_1$ and $1+i\,d\theta_2\,J_2$, with

$$J_1 = i\begin{pmatrix} 0 & 0 & 0 \\ 0 & 0 & -1 \\ 0 & 1 & 0 \end{pmatrix}, \qquad J_2 = i\begin{pmatrix} 0 & 0 & 1 \\ 0 & 0 & 0 \\ -1 & 0 & 0 \end{pmatrix}. \tag{2.116b}$$

The matrices J_i, $i=1,2,3$, are the *generators* of infinitesimal rotations.

By direct manipulation, one can show that these matrices satisfy the following properties:

1. They are Hermitian matrices. (2.117)
2. $$J_1^2 = \begin{pmatrix} 0 & 0 & 0 \\ 0 & 1 & 0 \\ 0 & 0 & 1 \end{pmatrix}, \quad J_2^2 = \begin{pmatrix} 1 & 0 & 0 \\ 0 & 0 & 0 \\ 0 & 0 & 1 \end{pmatrix}, \quad J_3^2 = \begin{pmatrix} 1 & 0 & 0 \\ 0 & 1 & 0 \\ 0 & 0 & 0 \end{pmatrix}; \tag{2.118}$$

   hence
   $$J^2 = \mathbf{J}\cdot\mathbf{J} = J_1^2 + j_2^2 + J_3^2 = 2I. \tag{2.119}$$
3. They do not commute with one another. More specifically, their *commutators* are
   $$[J_i, J_j] \equiv J_iJ_j - J_jJ_i = i\sum_k \varepsilon_{ijk}J_k. \tag{2.120}$$

2.10.2. Finite rotations

Let us next rotate our coordinate system about an arbitrary direction $\mathbf{e}_\theta$ by an infinitesimal amount $d\theta$. The appropriate rotation matrix is

$$R(d\theta) = 1 + i\,d\boldsymbol{\theta}\cdot\boldsymbol{J}, \qquad d\boldsymbol{\theta} = d\theta\,\mathbf{e}_\theta. \tag{2.121}$$

This can also be written as

$$R(\boldsymbol{\theta}+d\boldsymbol{\theta}) = R(d\boldsymbol{\theta})R(\boldsymbol{\theta}) = (1+i\,d\boldsymbol{\theta}\cdot\mathbf{J})R(\boldsymbol{\theta}).$$

That is,

$$dR(\boldsymbol{\theta}) = R(\boldsymbol{\theta}+d\boldsymbol{\theta}) - R(\boldsymbol{\theta}) = i\,d\boldsymbol{\theta}\cdot\mathbf{J}R(\boldsymbol{\theta}). \tag{2.122}$$

For repeated rotations about the same direction $\mathbf{e}_\theta$, Eq. (2.122) can be integrated to give

$$R(\boldsymbol{\theta}) = \exp(i\boldsymbol{\theta}\cdot\mathbf{J}), \qquad \boldsymbol{\theta} = \theta\mathbf{e}_\theta. \tag{2.123}$$

This is called a *rotation operator*.

The rotation operator, being a function of the 3×3 matrix operator $\mathbf{J}$, is also a 3×3 matrix operator. This can be seen readily in the power-series expansion

$$R(\boldsymbol{\theta}) = \exp(i\boldsymbol{\theta}\cdot\mathbf{J}) = \sum_{n=0}^{\infty}\frac{(i\boldsymbol{\theta}\cdot\mathbf{J})^n}{n!}. \tag{2.124}$$

The series expansion is also useful in understanding the properties of $R(\boldsymbol{\theta})$. For example, one can show in this way that $R_3(\theta) = \exp(i\theta J_3)$ is identical to the rotation matrix of Eq. (2.7). (See Problem 2.10.2.) The form of Eq. (2.124) is the more flexible because the direction $\mathbf{e}(\boldsymbol{\theta})$ can be oriented along any direction in space. It is also more compact.

Various matrix operations can be performed on functions of matrices. For example,

$$R^{\dagger}(\boldsymbol{\theta}) = \exp(-i\boldsymbol{\theta} \cdot \mathbf{J}^{\dagger}) = \exp(-i\boldsymbol{\theta} \cdot \mathbf{J}) = R^{-1}(\boldsymbol{\theta}). \tag{2.125}$$

This shows that the orthogonal matrix $R(\boldsymbol{\theta})$ is also unitary. [$R(\theta)$ is orthogonal by virtue of Eq. (2.15).]

2.10.3. *Vector algebra in matrix form*

There is a useful connection between the generators J_i of infinitesimal rotations and the cross product of vector algebra that can be brought out by writing vector algebra in a matrix form.

We first point out that the scalar product may be written as the matrix product of a row vector and a column vector

$$\mathbf{A} \cdot \mathbf{B} = (A_1 A_2 A_3) \begin{pmatrix} B_1 \\ B_2 \\ B_3 \end{pmatrix} = A^{\mathrm{T}} B. \tag{2.126}$$

It is unchanged by a coordinate transformation, because in the new coordinate system we have

$$\mathbf{A}' \cdot \mathbf{B}' = (\lambda A)^{\mathrm{T}} \lambda B = A^{\mathrm{T}} (\lambda^{\mathrm{T}} \lambda) B = \mathbf{A} \cdot \mathbf{B}. \tag{2.127}$$

This invariant property can also be seen in the familiar cosine form $\mathbf{A} \cdot \mathbf{B} = AB \cos \theta_{AB}$ of the scalar product, since neither A, B, nor θ_{AB} depends on the choice of the coordinate axes. We refer to this invariance property by calling scalar products *rotational scalars*.

The matrix form of the vector product is

$$\mathbf{A} \times \mathbf{B} = \begin{pmatrix} 0 & -A_3 & A_2 \\ A_3 & 0 & -A_1 \\ -A_2 & A_1 & 0 \end{pmatrix} \begin{pmatrix} B_1 \\ B_2 \\ B_3 \end{pmatrix} = -i(\mathbf{A} \cdot \mathbf{J})\mathbf{B}, \tag{2.128}$$

where

$$\mathbf{A} \cdot \mathbf{J} = A_1 J_1 + A_2 J_2 + A_3 J_3 = \mathbf{J} \cdot \mathbf{A} \tag{2.129}$$

is an antisymmetric matrix. In this way, we can see explicitly that the cross product is related to rotation.

The result shown in Eq. (2.128) has some useful applications. For example, the change in the coordinates of a constant vector $\mathbf{r}$ under an infinitesimal rotation of the coordinate axes is from Eq. (2.122).

$$d\mathbf{r} = \mathbf{r}' - \mathbf{r} = i\, d\boldsymbol{\theta} \cdot \mathbf{J}\mathbf{r} = -d\boldsymbol{\theta} \times \mathbf{r}. \tag{2.130}$$

As a result

$$\mathbf{v} = \frac{d\mathbf{r}}{dt} = -\boldsymbol{\omega} \times \mathbf{r}, \qquad \boldsymbol{\omega} = \frac{d\boldsymbol{\theta}}{dt}. \tag{2.131}$$

If the coordinate axes are fixed, but the point is rotated, its linear velocity $\mathbf{v}_{\text{fixed}}$ in the fixed coordinate system is the reverse of this

$$\mathbf{v}_{\text{fixed}} = -\mathbf{v} = \boldsymbol{\omega} \times \mathbf{r}. \tag{2.132}$$

This result is useful in describing the kinematics of rotating objects.

The purpose of the above discussion is not the derivation of simple results by a complicated procedure. It is rather to show how pervasive the generators J_i are in the description of rotations, although we may not be aware of them.

Problems

2.10.1. Obtain the matrix that generates infinitesimal rotations about an axis making equal angles with the three coordinate axes.

2.10.2. Prove directly the identity

$$R_3(\theta) = e^{i\theta J_3} = \begin{pmatrix} \cos\theta & \sin\theta & 0 \\ -\sin\theta & \cos\theta & 0 \\ 0 & 0 & 1 \end{pmatrix}.$$

2.10.3. Verify Eq. (2.120): $[J_i, J_j] = i \sum_k \varepsilon_{ijk} J_k$.

2.10.4. Show that

(a) $[\exp(A)]^\dagger = \exp(A^\dagger)$;

(b) If H is Hermitian, $\exp(iaH)$ is unitary.

2.10.5. If the commutator $[B,A] = \lambda I$ of two $n \times n$ square matrices A and B is proportional to the identity matrix I, show that

(a) $[B, A^n] = n\lambda A^{n-1}$;

(b) $[B, f(A)] = \lambda I \dfrac{\partial}{\partial A} f(A)$;

(c) $e^{x(A+B)} = e^{xA} e^{xB} e^{(1/2)x^2\lambda I}$ (by showing that both sides satisfy a first-order differential equation in x), and hence that $e^{A+B} = e^A e^B e^{(1/2)[B,A]}$;

(d) $e^A e^B = e^B e^A e^{-[B,A]}$.

2.10.6. In deriving Eq. (2.123), why is it all right not to worry about the fact that the three components of $\mathbf{J}$ do not commute?

2.10.7. A point mass m is moving with velocity $\mathbf{v}_{\text{rot}}$ relative to a rotating coordinate system that is rotating with constant angular velocity $\boldsymbol{\omega}$ relative to a fixed Newtonian inertial frame. Show that

(a) Its velocity in the fixed frame is

$$\mathbf{v}_{\text{fixed}} = \mathbf{v}_{\text{rot}} + \boldsymbol{\omega} \times \boldsymbol{r};$$

(b) Its acceleration in the fixed frame is

$$\begin{aligned} \mathbf{a}_{\text{fixed}} &= \left(\frac{d}{dt}\mathbf{v}_{\text{rot}}\right)_{\text{fixed}} + \boldsymbol{\omega} \times \mathbf{v}_{\text{fixed}} \\ &= \mathbf{a}_{\text{rot}} + 2\boldsymbol{\omega} \times \boldsymbol{v}_{\text{rot}} + \boldsymbol{\omega} \times (\boldsymbol{\omega} \times \mathbf{r}). \end{aligned}$$

Thus the effective force on the point mass differs from the Newtonian force $m\mathbf{a}_{\text{fixed}}$ by

$$\Delta\mathbf{F} = m(\mathbf{a}_{\text{rot}} - \mathbf{a}_{\text{fixed}}) = -m[\boldsymbol{\omega} \times (\boldsymbol{\omega} \times \mathbf{r}) + 2\boldsymbol{\omega} \times \mathbf{v}_{\text{rot}}].$$

The first term is called the *centrifugal force,* while the second term is called the *Coriolis force*.

2.10.8. Prove the Jacobi identity for the commutators of matrices

$$[A,[B,C]] + [B,[C,A]] + [C,[A,B]] = 0.$$

2.11. Matrix groups

Matrices are also important in physics because they are the simplest mathematical objects that form groups under noncommutative multiplications. The definition of a group has been given in Section 2.9 in connection with the study of transformations. The use of group properties of matrices is indeed related to the transformation and classification of physical properties. Although it is not our intention to study group theory in this book, it is nevertheless useful to recognize the more common matrix groups.

All invertible (i.e., nonsingular) complex $n \times n$ matrices form a group under matrix multiplication. This group is called the *complex linear group* of order n and is denoted by the symbol GL(n,c). It is a very large group, and it contains many *subgroups,* each of which satisfies all group properties.

Interesting subgroups of GL(n,c) with complex matrices are (1) the *special linear group* SL(n,c) of unimodular matrices (i.e., matrices with determinants of value 1 only), (2) the *unitary group* U(n) of unitary matrices, and (3) the *special unitary group* SU(n) of unitary unimodular matrices.

Interesting subgroups of GL(n,c) with real matrices are (1) the *linear group* GL(n) of all real invertible $n \times n$ matrices, (2) the *special linear group* SL(n) of unimodular matrices, and (3) the *orthogonal group* O(n) of orthogonal matrices.

Some of these groups are of particular interest in physics. We shall discuss a few of these.

2.11.1. *The group* O(n)

The group O(n) is the group of coordinate transformation matrices λ in n-dimensional space. Under the O(n) transformation, the scalar product

$$\mathbf{A}' \cdot \mathbf{B}' = (\lambda A)^{\mathrm{T}}\lambda B = A^{\mathrm{T}}(\lambda^{\mathrm{T}}\lambda)B = \mathbf{A} \cdot \mathbf{B}$$

is invariant. Matrices of O(n) are characterized by $\frac{1}{2}n(n-1)$ real parameters, because there are n^2 real matrix elements that satisfy

$\frac{1}{2}n(n+1)$ orthogonality relations $\lambda^T\lambda = I$. (See Problem 2.11.1.) Thus the group O(3) of rotation matrices in space is characterized by three parameters. These may be taken to be the Euler angles α, β, γ defined in Eq. (2.18) in Section 2.2,

$$R(\alpha, \beta, \gamma) = R_z(\gamma)R_x(\beta)R_z(\alpha),$$

where $R_i(\theta)$ is the matrix for rotation about the axis $\mathbf{e}_i$ by an angle θ. Alternatively, we may use Eq. (2.123)

$$R(\boldsymbol{\theta}) = \exp(i\boldsymbol{\theta} \cdot \mathbf{J}),$$

where $\boldsymbol{\theta} = \theta_1\mathbf{e}_1 + \theta_2\mathbf{e}_2 + \theta_3\mathbf{e}_3$ contains three real parameters.

2.11.2. *The groups* U(*n*) *and* SU(*n*)

The group U(n) contains all complex unitary matrices of order n. The n^2 complex matrix elements are equivalent to $2n^2$ real parameters, but the unitarity relation $U^\dagger U = I$ imposes n^2 constraints. (Prove this statement.) Hence the matrices of U(n) are characterized by n^2 real parameters.

A unitary matrix of order n can always be written in the form

$$U = \exp(iH), \tag{2.133}$$

where H is an $n \times n$ Hermitian matrix. One can show directly that $n \times n$ Hermitian matrices are characterized by n^2 real parameters.

The determinant of a unitary matrix can be calculated readily with the help of the result of Problem 2.7.8:

$$\det(\exp A) = \exp(\operatorname{Tr} A).$$

Thus

$$\det U = \det[\exp(iH)] = \exp(i \operatorname{Tr} H) = \exp(i\alpha), \tag{2.134}$$

where

$$\alpha = \sum_i \lambda_i$$

is the sum of the eigenvalues λ_i of H. Since λ_i and therefore α are real, we see that $\det U$ is a complex number of unit magnitude. It is also clear that the group $U(1)$ is made up of the complex numbers $\exp(i\beta)$, where β is a real number. That is, $U(1)$ is the group of complex phase factors.

Unitary matrices with $\det U = 1$ (i.e., $\alpha = 0$) form the group SU(n). If these matrices are denoted U_0, then

$$U_0 = \exp(iH_0), \qquad \operatorname{Tr} H_0 = 0, \qquad \det U_0 = 1. \tag{2.135}$$

If $U = \exp(iH)$ is a matrix of $U(n)$, it can be associated with a matrix $U_0 = \exp(iH_0)$ of SU(n) as follows:

$$H = H_0 + \frac{\alpha}{n} I, \qquad U = \exp\left(\mathrm{i}\frac{\alpha}{n}\right)U_0 = U_0 \exp\left(i\frac{\alpha}{n}\right). \tag{2.136}$$

Since the phase factor belongs to U(1), we may represent these relations by the expression

$$\mathrm{U}(n) = \mathrm{SU}(n) \otimes \mathrm{U}(1), \tag{2.137}$$

meaning that a matrix of U(n) can be expressed as a *direct product* of a matrix in SU(n) and a matrix (the phase factor) in U(1).

Matrices of SU(n) are characterized by $n^2 - 1$ real parameters. For example, a general matrix of SU(2) has the form

$$U_0(\alpha,\beta,\gamma) = \begin{pmatrix} \cos\alpha\cos\beta\, e^{i\gamma} & \sin\beta + i\sin\alpha\cos\beta \\ -\sin\beta + \mathrm{i}\sin\alpha\cos\beta & \cos\alpha\cos\beta\, e^{-i\gamma} \end{pmatrix} \tag{2.138}$$

containing three real parameters α, β, and γ. It is easy to verify that U_0 is both unitary and unimodular.

2.11.3. Continuous groups and infinitesimal generators

A matrix group is called a *continuous group* if the matrices are continuous functions of the group parameters. Since the identity matrix I belongs to all the matrix groups, there are matrices of a continuous group G infinitesimally close to I. Suppose the matrices of G in a small neighborhood of I can be specified by a small number of matrices S of G. Then following the idea of integral calculus, we see that all matrices of G can be reached by making an infinite number of successive infinitesimal transformations of the type shown in Eq. (2.121), each involving only the generators S. Indeed, from what we know for rotation operators, we see that matrices of G are exponential functions of the generators and that the number of generators is equal to the number of real parameters characterizing the group matrices.

We have met such group generators before. The rotation matrix $\exp(i\boldsymbol{\theta}\cdot\mathbf{J})$ of Eq. (2.123) is expressible in terms of the three generators J_i of Eqs. (2.116). Other generators we have used are the frequency operator W of Eq. (2.89), which generates the time displacement operators $\exp(-i\tau W)$ of the time displacement group, and the propagation operator $\mathbf{K}$ of Eq. (2.97), which generates the space translation operator $\exp(i\boldsymbol{\rho}\cdot\mathbf{K})$ of the space translation group.

It is possible to obtain the generators of a matrix group when we are given the general form of the matrices of the group. For example, in the case of SU(2), one obtains directly from Eq. (2.138) the matrices for infinitesimal changes from I:

$$\begin{aligned} U_0(d\alpha,d\beta,d\gamma) &= \begin{pmatrix} 1 + i\,d\gamma & d\beta + i\,d\alpha \\ -d\beta + i\,d\alpha & 1 - i\,d\gamma \end{pmatrix} \\ &= I + i(\sigma_1\,d\alpha + \sigma_2\,d\beta + \sigma_3\,d\gamma), \end{aligned} \tag{2.139}$$

where

$$\sigma_1 = \begin{pmatrix} 0 & 1 \\ 1 & 0 \end{pmatrix}, \qquad \sigma_2 = \begin{pmatrix} 0 & -i \\ i & 0 \end{pmatrix}, \qquad \sigma_3 = \begin{pmatrix} 1 & 0 \\ 0 & -1 \end{pmatrix} \tag{2.140}$$

are the Pauli matrices of Eq. (2.12).

If

$$\begin{gathered} \boldsymbol{\phi} = \alpha\mathbf{i} + \beta\mathbf{j} + \gamma\mathbf{k}, \qquad \boldsymbol{\sigma} = \sigma_1\mathbf{i} + \sigma_2\mathbf{j} + \sigma_3\mathbf{k}, \\ U_0(d\boldsymbol{\phi}) = I + i\boldsymbol{\sigma} \cdot d\boldsymbol{\phi}. \end{gathered} \tag{2.141}$$

This matrix for an infinitesimal change from the identity matrix I can be integrated (in a way similar to the treatment of the rotation operator in Section 2.10) to give the general form of matrices in SU(2):

$$U_0(\boldsymbol{\phi}) = \exp(i\boldsymbol{\sigma} \cdot \boldsymbol{\phi}). \tag{2.142}$$

It is clear from the above derivation that this $U_0(\boldsymbol{\phi})$ is exactly equal to $U_0(\alpha,\beta,\gamma)$ of Eq. (2.138).

The generators $i\sigma_k$, $k = 1,2,3$, of SU(2) can also be obtained directly from Eq. (2.135) if one knows that all 2×2 Hermitian matrices (which are characterized by four parameters) can be expressed in terms of the Pauli matrices:

$$H = \boldsymbol{\sigma} \cdot \boldsymbol{\phi} + \frac{\alpha}{2} I, \qquad H_0 = \boldsymbol{\sigma} \cdot \boldsymbol{\phi}, \qquad \operatorname{Tr} H_0 = 0. \tag{2.143}$$

We further note the useful multiplicative relation described by Eq. (2.14a)

$$\sigma_i \sigma_j = \delta_{ij} + i \sum_k \varepsilon_{ijk} \sigma_k. \tag{2.144a}$$

This shows that 2×2 Hermitian matrices are closed under matrix multiplication.

2.11.4. Lie algebra

It was recognized by Lie, a Norwegian mathematician of the nineteenth century, that the important property of the generators of continuous groups is not so much that they can be identified with numerical matrices, as in Eq. (2.140) or (2.116). It is rather that they satisfy certain commutation relations. Thus for the three generators J_k of O(3), we have Eq. (2.120),

$$[J_i, J_j] = i \sum_k \varepsilon_{ijk} J_k,$$

while for the three Pauli spin matrices of SU(2) we have

$$[\sigma_i, \sigma_j] = 2i \sum_k \varepsilon_{ijk} \sigma_k \tag{2.144b}$$

or

$$[\tfrac{1}{2}\sigma_i, \tfrac{1}{2}\sigma_j] = i \sum_k \varepsilon_{ijk} \tfrac{1}{2}\sigma_k. \tag{2.144c}$$

Thus commutators of generators are linear combinations of the generators themselves. Such a mathematical structure is called a *Lie algebra.* The linear coefficients $i\varepsilon_{ijk}$ are called its *structure constants.*

What we have shown is that the three matrices $\frac{1}{2}\sigma_k$ (but not the σ_k themselves) constitute the same Lie algebra as the three generators J_k for rotations. Therefore, if $\exp(i\boldsymbol{\theta}\cdot\mathbf{J})$ is a rotation matrix for rotation of a vector angle $\boldsymbol{\theta}$, $\exp(i\boldsymbol{\phi}\cdot\frac{1}{2}\boldsymbol{\sigma})$ must be a rotation matrix for rotation of a vector angle $\boldsymbol{\phi}$.

But wait! $\exp(i\boldsymbol{\theta}\cdot\mathbf{J})$ is a 3×3 matrix which operates on three-dimensional position vectors, but $\exp(i\boldsymbol{\phi}\cdot\frac{1}{2}\boldsymbol{\sigma})$ is a 2×2 matrix. What does it operate on? The answer, obtained by Pauli from studies of atomic spectra, is that it operates on an internal attribute of atomic particles that can have one of only two possible values. Since we are concerned with rotation, the property is called an *intrinsic spin,* which can either point up or point down. (These directions may be chosen arbitrarily to be the $+z$ or $-z$ direction.) An intrinsic-spin state can thus be represented by a two-dimensional *state vector*

$$\mathbf{u} = \begin{pmatrix} a \\ b \end{pmatrix} \tag{2.145}$$

of unit length ($a^2 + b^2 = 1$) on which the rotation matrices act.

There is in addition an extra factor $\frac{1}{2}$ in the rotation matrix for intrinsic spins. As a result, a rotation of $\phi = 2\theta$ is necessary to produce the same effect in the spin vector $\mathbf{u}$ as a rotation of θ produces in the position vector $\mathbf{r}$. This is readily seen by considering successive rotations of angle π about the y axis. The relevant rotation matrices are

$$R_y(\theta_2) = e^{i\theta_2 J_2} = \begin{pmatrix} \cos\theta_2 & 0 & -\sin\theta_2 \\ 0 & 1 & 0 \\ \sin\theta_2 & 0 & \cos\theta_2 \end{pmatrix},$$

$$R_y(\pi) = \begin{pmatrix} -1 & 0 & 0 \\ 0 & 1 & 0 \\ 0 & 0 & -1 \end{pmatrix}; \tag{2.146}$$

$$U_y(\phi_2) = e^{i\phi_2(1/2)\sigma_2} = \begin{pmatrix} \cos\frac{1}{2}\phi_2 & \sin\frac{1}{2}\phi_2 \\ -\sin\frac{1}{2}\phi_2 & \cos\frac{1}{2}\phi_2 \end{pmatrix},$$

$$U_y(\pi) = \begin{pmatrix} 0 & 1 \\ -1 & 0 \end{pmatrix} = i\sigma_y. \tag{2.147}$$

Therefore

$$R_y(\pi)\begin{pmatrix} x \\ y \\ z \end{pmatrix} = \begin{pmatrix} -x \\ y \\ -z \end{pmatrix}, \qquad R_y(2\pi)\begin{pmatrix} x \\ y \\ z \end{pmatrix} = R_y(\pi)\begin{pmatrix} -x \\ y \\ -z \end{pmatrix} = \begin{pmatrix} x \\ y \\ z \end{pmatrix}; \tag{2.148}$$

$$U_y(\pi)\begin{pmatrix} a \\ b \end{pmatrix} = \begin{pmatrix} b \\ -a \end{pmatrix}, \qquad U_y(2\pi)\begin{pmatrix} a \\ b \end{pmatrix} = \begin{pmatrix} -a \\ -b \end{pmatrix},$$
$$U_y(3\pi)\begin{pmatrix} a \\ b \end{pmatrix} = \begin{pmatrix} -b \\ a \end{pmatrix}, \qquad U_y(4\pi)\begin{pmatrix} a \\ b \end{pmatrix} = \begin{pmatrix} a \\ b \end{pmatrix}. \tag{2.149}$$

We see that it takes two full turns, not just one, to restore the spin vector $\mathbf{u}$ to its original form.

Problems

2.11.1. Show that for square matrices λ of order n, the orthogonality relation $\lambda^T\lambda = I = \lambda\lambda^T$ gives $\frac{1}{2}n(n+1)$ conditions on its matrix elements.

2.11.2. Show that for square matrices U of order n, the unitarity relation $U^\dagger U = I = UU^\dagger$ gives n^2 conditions on its matrix elements.

2.11.3. Relate Eqs. (2.138) and (2.142).

Appendix 2. Tables of mathematical formulas

1. Coordinate transformations

$$\mathbf{x}_{\text{new}} = \lambda \mathbf{x}_{\text{old}}, \qquad \lambda_{ij} = \mathbf{e}_i^{\text{new}} \cdot \mathbf{e}_j^{\text{old}}$$
$$\lambda^T = \lambda^{-1}$$

$$R_z(\theta) = \begin{pmatrix} \cos\theta & \sin\theta & 0 \\ -\sin\theta & \cos\theta & 0 \\ 0 & 0 & 1 \end{pmatrix}, \qquad R_y(\theta) = \begin{pmatrix} \cos\theta & 0 & -\sin\theta \\ 0 & 1 & 0 \\ \sin\theta & 0 & \cos\theta \end{pmatrix},$$

$$R_x(\theta) = \begin{pmatrix} 1 & 0 & 0 \\ 0 & \cos\theta & \sin\theta \\ 0 & -\sin\theta & \cos\theta \end{pmatrix}$$

$R(\alpha, \beta, \gamma) = R_z(\gamma)R_x(\beta)R_z(\alpha)$ for the Euler angles α, β, γ.

2. Determinant

$$\det A = \sum_{ij\cdots l} \varepsilon_{ij\ldots l} A_{i1} A_{j2} \cdots A_{ln}$$
$$= \sum_k A_{k\kappa} C_{k\kappa} = \sum_k A_{k\kappa} (-1)^{k+\kappa} M_{k\kappa} \quad \text{(Laplace development of the } \kappa\text{th column)}$$
$$= \sum_\kappa A_{k\kappa} C_{k\kappa} \qquad \text{(Laplace development of the } k\text{th row)}$$

$$\delta_{ik} \det A = \sum_j A_{ij} C_{kj}$$
$$A^{-1} = C^T / \det A$$
$$\det A^T = \det A$$
$$\det(kA) = k^n \det A, \qquad \text{if } k \text{ is a scalar}$$
$$\det(AB) = (\det A)(\det B) = \det(BA).$$

3. Matrix equations

$Ax = c$ has solutions $x_i = \sum_j c_j C_{ji}/\det A$ (Cramer's rule); $Ax = 0$ has nontrivial solutions if $\det A = 0$. The solutions $\mathbf{x}$ are orthogonal to the row vectors $\mathbf{a}_i$ of A.

4. Matrix eigenvalue problem

$(K - \lambda M)\mathbf{q} = 0$ requires $\det(K - \lambda M) = 0$. If H in $H\mathbf{e}_i = \lambda_i \mathbf{e}_i$ is Hermitian, the eigenvalues λ_i are real and the eigenvectors $\hat{e}_i$ are orthogonal or can be orthogonalized. If the square matrix U contains the orthonormalized eigenvectors $\mathbf{e}_i$ columnwise, then

$$U^\dagger U = I, \qquad U^\dagger H U = D = (\lambda_i \delta_{ij}).$$

$\operatorname{Tr} A = \sum_i \lambda_i$, where λ_i are the eigenvalues of A.

5. Infinitesimal generators of transformations

$$f(t+\tau) = \exp\left(\tau \frac{d}{dt}\right) f(t)$$

$$\phi(\mathbf{r} + \boldsymbol{\rho}) = \exp(\boldsymbol{\rho} \cdot \boldsymbol{\nabla})\phi(\mathbf{r})$$

$$R(\theta) = \exp(i\boldsymbol{\theta} \cdot \mathbf{J})$$

$$J_1 = \begin{pmatrix} 0 & 0 & 0 \\ 0 & 0 & -i \\ 0 & i & 0 \end{pmatrix}, \qquad J_2 = \begin{pmatrix} 0 & 0 & i \\ 0 & 0 & 0 \\ -i & 0 & 0 \end{pmatrix},$$

$$J_3 = \begin{pmatrix} 0 & -i & 0 \\ i & 0 & 0 \\ 0 & 0 & 0 \end{pmatrix}$$

$$[J_i, J_j] = i \sum_k \varepsilon_{ijk} J_k$$

$$U_0(\boldsymbol{\phi}) = \exp(i\boldsymbol{\phi} \cdot \tfrac{1}{2}\boldsymbol{\sigma})$$

$$\sigma_1 = \begin{pmatrix} 0 & 1 \\ 1 & 0 \end{pmatrix}, \qquad \sigma_2 = \begin{pmatrix} 0 & -i \\ i & 0 \end{pmatrix}, \qquad \sigma_3 = \begin{pmatrix} 1 & 0 \\ 0 & -1 \end{pmatrix}$$

$$[\tfrac{1}{2}\sigma_i, \tfrac{1}{2}\sigma_j] = i \sum_k \varepsilon_{ijk} \tfrac{1}{2}\sigma_k$$

$$\sigma_i \sigma_j = \delta_{ij} + i \sum_k \varepsilon_{ijk} \sigma_k.$$

6. Differential eigenvalue equations

$K^2 u(\mathbf{r}) = k^2 u(\mathbf{r})$, $\quad \mathbf{K} = -\mathrm{i}\boldsymbol{\nabla}$ (Helmholtz equation)

then $u(\mathbf{r}) = e^{\pm i\mathbf{k} \cdot \mathbf{r}}$;

$(p_{op}^2 c^2 - H^2) u(\mathbf{r},t) = -m^2 c^4 u(\mathbf{r},t)$,

$H = i\hbar \dfrac{\partial}{\partial t}$ (Klein–Gordon equation)

then $(mc^2/\hbar)^2 = \omega^2 - k^2c^2$, $\quad u(\mathbf{r},t) = e^{\pm i\mathbf{k} \cdot \mathbf{r}} e^{\pm i\omega t}$.

7. Matrix relations

$\mathrm{Tr}(AB) = \mathrm{Tr}(BA)$.
$P^{-1}AP$ is an orthogonal (unitary) transformation of A if P is an orthogonal (unitary) matrix.
$O^{\mathrm{T}}O = I = OO^{\mathrm{T}}$ if O is orthogonal ($O^{\mathrm{T}} = O^{-1}$).
$U^{\dagger}U = I = UU^{\dagger}$ if U is unitary ($U^{\dagger} = U^{-1}$).
$(\boldsymbol{\sigma} \cdot \mathbf{A})(\boldsymbol{\sigma} \cdot \mathbf{B}) = \mathbf{A} \cdot \mathbf{B} + i\boldsymbol{\sigma} \cdot \mathbf{A} \times \mathbf{B}$, where $\mathbf{A}$ and $\mathbf{B}$ are vectors.
$\det(e^{A}) = e^{\mathrm{Tr}\,A}$ if A is a normal matrix, that is, one that can be diagonalized by a unitary transformation.

3

FOURIER SERIES AND FOURIER TRANSFORMS

3.1. Wave–particle duality: Quantum mechanics

In a famous lecture delivered on December 14, 1900, Planck proposed an explanation for the observed energy density per unit frequency of the radiation emitted by a black body in thermal equilibrium. His formula contained as special cases known theoretical expressions that worked well only for high frequencies (Wien's radiation formula) or for very low frequencies (Rayleigh–Jeans law). However, he had to introduce a totally new idea to achieve this—the assumption that the radiation energy is emitted or absorbed by mechanical oscillators only in discrete amounts ε proportional to its frequency ν

$$\varepsilon = h\nu. \tag{3.1}$$

The universal proportionality constant h, Planck's constant, can readily be deduced from the measured radiation density. It characterizes the discontinuous character of Planck's *quantum* oscillators, in marked contrast to the continuous values of the energies or amplitudes of classical oscillators. Planck's hypothesis of the energy quantum marked the beginning of quantum physics.

In technical terms, Planck's quantum hypothesis is concerned with the partition of energy in statistical mechanics. A similar observation on energy partition led Einstein in 1905 to the "heuristic point of view" that light may appear as particles. He pointed out that Wien's radiation formula could be obtained by imaging that the black-body radiation itself was a gas of particles, the light quanta, of energies $h\nu$. He immediately applied this theory to several physical phenomena, including the photoelectric effect (which describes the observation that the energies of electrons emitted from a metal irradiated by ultraviolet light depend only on its frequency and not on its intensity).

It was Niels Bohr who first saw the far-reaching implications of the quantum postulate in understanding the structure of atoms. Bohr was able to explain the distinctive colors of light emitted by excited atoms by making a number of revolutionary postulates. He supposed that an atom consisted of Z electrons bound electrostatically to a very small nucleus with a positive charge Ze, e being the magnitude of the electronic charge. He next supposed that the electron of an excited atomic system existed in a *stationary state* of discrete energy E_i, and that, when the electron made a transition to a final atomic state of energy E_f, the excess energy was emitted as light of frequency ν:

$$h\nu = E_i - E_f = \hbar\omega, \tag{3.2}$$

where $\omega = 2\pi\nu$, $\hbar = h/2\pi$. By assuming that the orbital angular momentum of the electron moving around the positively charged atomic nucleus could exist only in integral multiples of $\hbar$, he was able to derive expressions for the energies of these stationary states that agreed with values deduced from experiment.

Bohr's work stimulated the search for quantum conditions or *quantization rules*. In this, much progress was made by following Bohr's correspondence principle: This states that the quantum theory of matter should approach classical mechanics in the domain (the so-called limit of large quantum numbers) where classical mechanics is known to be valid.

Successes of this "old" quantum theory in accounting for the structure of simple atomic systems were followed by significant difficulties with more complicated systems. The quantization rules also appeared empirical and *ad hoc*. This unsatisfactory state of affairs led in 1925 to two different but equivalent formulations that gave quantum mechanics its present logical foundation—the matrix mechanics of Heisenberg, Born, and Jordan, and the wave mechanics of de Broglie and Schrödinger.

According to Heisenberg, the difficulties of the old quantum theory arose because of its reliance on physically unobservable concepts such as electronic orbits, that is, the position $x(t)$ of classical mechanics. The reason is that in a quantum system it is not possible to determine a position without introducing disturbances that radically change the state of motion of the system. One should use only physical observables instead.

The obvious question now is how these physical observables are to be constructed. Heisenberg's solution is as follows: The position of an atomic state n should be expanded in the Fourier series

$$x_n(t) = \sum_{m=-\infty}^{\infty} x(n,m) \exp[i\omega(n,m)t],$$

where $x(n,m)$ has the physical interpretation of a transition amplitude between the states n and m, and

$$\omega(n, m) = (E_m - E_n)/\hbar \tag{3.3}$$

is the quantum frequency of the transition as given by the Bohr–Einstein relation of Eq. (3.2). The physical observables are $x(n,m)$ and $\omega(n,m)$, not $x_n(t)$ itself.

The distinction between these two types of quantities might appear trivial, but the situation is quite subtle. The reason is that, in order to preserve consistency in the theoretical description, the amplitudes $x(n,m)$ should satisfy a certain recombination rule that Heisenberg worked out. This turned out to be the rule of matrix multiplication. The resulting theory was soon formalized by Born, Jordan, and Heisenberg into the matrix formulation of quantum mechanics.

In the same year (1925), de Broglie noted that the symmetrical treatment of energy and momentum in relativity implied that momentum should also have a quantum character, namely

$$\mathbf{p} = \hbar \mathbf{k}. \tag{3.4}$$

For light

$$p = E/c = \hbar\omega/c = \hbar k;$$

hence k is just the wave number of its wave motion. de Broglie then suggested that since the energy–momentum relation also held for material particles, they should also have wave properties. In this way, the concept of wave–particle duality was now applied to massive particles as well as to light waves.

This was a remarkable suggestion. The wave–particle nature of light had been a subject of speculation since ancient times. Newton himself preferred a corpuscular theory, while his respected contemporary Huygens advocated the wave theory. Newton's preference remained unquestioned for over a century until Young demonstrated the diffraction of light by two slits in 1803. This is a wave property that cannot be explained in the corpuscular picture. Gradually, with the mathematical analyses and experimental observations on interference phenomena by Fresnel and others, the wave theory gained the upper hand. We have seen how, a century later, Einstein finally rescued the particle theory of light from disrepute. However, it had never been seriously suspected before that massive particles had wave properties.

The wave nature of particles must be verified by observing interference phenomena for particles. This was provided a few years later by Davisson and Germer, and by G. P. Thomson and others who successfully observed the diffraction of electrons by crystals.

The significance of the concept of matter waves in quantum mechanics was realized by Schrödinger, who applied it in 1926 to the archetypical conservative mechanical system in nonrelativistic mechanics described by the expression

$$\frac{p^2}{2m} + V(\mathbf{r}) = E, \tag{3.5}$$

where $V(\mathbf{r})$ is the potential energy. We recall from Section 2.7 that the frequency $\omega = E/\hbar$ is the eigenvalue of the time displacement operator $i\,\partial/\partial t$. Similarly the wave vector $\mathbf{k}$ in Eq. (3.4) is the eigenvalue of the space translation operator $-i\boldsymbol{\nabla}$. Hence the terms in Eq. (3.5) may be interpreted instead as operators operating on what is now called the time-dependent Schrödinger wave function $\Psi(\mathbf{r}, t)$:

$$\left(\frac{1}{2m}(-i\hbar\nabla)^2 + V(\mathbf{r})\right)\Psi(\mathbf{r}, t) = i\hbar\frac{\partial}{\partial t}\Psi(\mathbf{r},t). \tag{3.6}$$

For a system in a stationary state of energy $\hbar\omega$, the time dependence in $\Psi(\mathbf{r},t)$ has to be an eigenfunction, say $\exp(-i\omega t)$, of the Hamiltonian operator $H = i\hbar\, \partial/\partial t$ belonging to the eigenvalue $E = \hbar\omega$. Hence

$$\Psi(\mathbf{r},t) = \psi(\mathbf{r}) \exp(-i\omega t). \tag{3.7}$$

This allows a differential equation for the time-independent wave function $\psi(\mathbf{r})$ to be separated from Eq. (3.6):

$$\left(-\frac{\hbar^2}{2m}\nabla^2 + V(\mathbf{r}) - E\right)\psi(\mathbf{r}) = 0. \tag{3.8}$$

The result is an eigenvalue problem in which the eigenvalue E and the eigenfunction $\psi(\mathbf{r})$ are to be determined. This can be done by requiring that $\psi(\mathbf{r})$ satisfies certain boundary conditions at $r = 0$ and ∞. Thus at one stroke, the mathematics of differential equations, which had been developed since the time of Newton and Leibniz, was brought to bear on the problem of quantum mechanics.

Schrödinger himself demonstrated the equivalence between his wave mechanics and the matrix mechanics of Heisenberg, Born, and Jordan. The matrix element F_{fi} in the latter formulation describing the transition amplitude from the initial stationary state i to the final state f caused by the operator F is just the integral

$$F_{fi} = \int \psi_f^*(\mathbf{r}) F(\mathbf{r}) \psi_i(\mathbf{r})\, d^3\mathbf{r} \tag{3.9}$$

involving the wave functions of these states. The operator F is here "represented" by an appropriate function or operator $F(\mathbf{r})$ of $\mathbf{r}$. This turns out to be part of the mathematics of Fourier analysis of operators and functions describing ordinary wave phenomena. This familiar language gives wave mechanics a very intuitive appeal. It enables the physical basis of quantum mechanics to be described in concrete physical terms. In contrast, matrix mechanics is much more compact and abstract, and hence more elegant.

In this chapter, we study the representation of functions by Fourier series and integrals. We also point out certain important wave properties of quantum mechanics, including the uncertainty principle (1927) of Heisenberg.

3.2. Fourier series

The world is full of vibrations. The sound we hear is an acoustic wave; the sight we see is an electromagnetic wave; surfers in Malibu, California, ride on the gravity waves of the ocean.

The simplest wave motion in one dimension is described by the one-dimensional wave equation

$$\left(\frac{\partial^2}{\partial x^2} - \frac{1}{v^2}\frac{\partial^2}{\partial t^2}\right)u(x,t) = 0. \tag{3.10}$$

which we have derived in Section 2.6. If the time dependence is $\exp(\pm i\omega t)$, the wave motion has a definite frequency ω. We then obtain the Helmholtz equation

$$\left(\frac{d^2}{dx^2}+k^2\right)X(x)=0, \qquad k=\frac{\omega}{v}. \tag{3.11}$$

It can be shown by direct substitution that a solution of this equation is the function

$$X(x)=a\cos kx+b\sin kx, \tag{3.12}$$

where a and b are constants. (The theory of differential equations is discussed in Chapter 4). In particular, the shape of a vibrating string of length π fixed at its ends is given by

$$X_n(x)=\sin nx, \tag{3.13}$$

where n is a positive integer. This shape is called the nth *normal mode* of vibration of the string.

Historically, the wave equation (3.10) was first studied around the middle of the eighteenth century. Around 1742, D. Bernoulli stated that vibrations of different modes could coexist in the same string. In 1753, stimulated by the work of D'Alembert and of Euler, he further proposed that all possible shapes of a vibrating string even when its ends were not fixed were representable as

$$f(x)=\sum_{n=1}^{\infty} b_n \sin nx. \tag{3.14}$$

His argument was that the infinity of coefficients could be used to reproduce any function at an infinity of points. The possibility of adding a cosine series to take care of even functions of x was considered in the controversy that followed this claim. However, all the other eighteenth-century mathematicians, while agreeing that the series in Eq. (3.14) solved Eq. (3.11), disputed the possibility that *all* solutions could be so represented.

In 1807, Fourier submitted a paper on heat conduction to the Academy of Sciences in Paris in which he claimed that *every* function in the closed interval $[-\pi,\pi]$ (i.e., $-\pi \leq x \leq \pi$) could be represented in the form

$$S=\tfrac{1}{2}a_0+\sum_{n=1}^{\infty}(a_n\cos nx+b_n\sin nx). \tag{3.15}$$

His integral formulas for the coefficients a_n, b_n were not new, as they had been obtained by Clairaut in 1757 and by Euler in 1777. However, Fourier broke new ground by pointing out that these integral formulas were well defined even for very arbitrary functions and that the resulting coefficients were identical for different functions that agreed inside the interval, but not outside it.

This paper of Fourier was rejected by Lagrange, Laplace, and Legendre on behalf of the Academy on the grounds that it lacked mathematical rigor. Although a revised version won the Academy's grand prize in 1812, it too was not published in the academy's *Mémoires* until 1824, when Fourier himself became the secretary of the Academy. This occurred 2 years after he published his book, *Theorique Analytique de la Chaleur* (the analytic theory of heat), in which arguments were advanced that an aribtrary function could be represented by the series in Eq. (3.15). This book had great impact on the development of mathematical physics in the nineteenth century and on the concept of functions in modern analysis.

Let us suppose that we are interested in those series (3.15) that converge uniformly to functions $f(x)$ in the interval $[-\pi, \pi]$. By *uniform convergence* in the specified interval, we mean convergence for any value of x in the interval. [Uniform convergence for a series of terms $f_k(x)$ is assured if $|f_k(x)| \leq M_k$ and the series for M_k is convergent.] A uniformly convergent infinite series is known to be integrable and differentiable term by term.

Given a uniformly convergent series $f(x)$ of the above type, the series

$$f(x)\cos mx = \tfrac{1}{2}a_0 \cos mx + \sum_{n=1}^{\infty} (a_n \cos nx \cos mx + b_n \sin nx \cos mx) \tag{3.16}$$

also converges uniformly, because $|\cos mx| \leq 1$. It can therefore be integrated term by term, with the result that

$$\int_{-\pi}^{\pi} f(x)\cos mx\, dx = \tfrac{1}{2}a_0 \int_{-\pi}^{\pi} \cos mx\, dx + \sum_{m=1}^{\infty} \left(a_n \int_{-\pi}^{\pi} \cos nx \cos mx\, dx + b_n \int_{-\pi}^{\pi} \sin nx \cos mx\, dx \right). \tag{3.17}$$

The right-hand side can be evaluated with the help of the formulas

$$\int_{-\pi}^{\pi} \sin mx\, dx = 0; \tag{3.18a}$$

$$\int_{-\pi}^{\pi} \cos mx\, dx = \begin{cases} 2\pi, & m = 0 \\ 0, & m \neq 0; \end{cases} \tag{3.18b}$$

$$\int_{-\pi}^{\pi} \sin nx \cos mx\, dx = 0, \qquad \text{for all } m,n; \tag{3.18c}$$

$$\int_{-\pi}^{\pi} \cos nx \cos mx\, dx = \begin{cases} 0, & m \neq n \\ \pi, & m = n; \end{cases} \tag{3.18d}$$

$$\int_{-\pi}^{\pi} \sin nx \sin mx\, dx = \begin{cases} 0, & m \neq n \\ \pi, & m = n, \end{cases} \tag{3.18e}$$

which can be derived by using trigonometric identities such as

$$\sin nx \sin mx = \tfrac{1}{2}\cos(n-m)x - \tfrac{1}{2}\cos(n+m)x. \tag{3.19}$$

One can then show that

$$\int_{-\pi}^{\pi} f(x)\cos mx\,dx = a_m\pi, \tag{3.20a}$$

$$\int_{-\pi}^{\pi} f(x)\sin mx\,dx = b_m\pi, \tag{3.20b}$$

$$\int_{-\pi}^{\pi} f(x)\,dx = a_0\pi. \tag{3.20c}$$

If the function $f(x)$ in Eq. (3.20) is replaced by an arbitrary function $F(x)$ in the interval $[-\pi,\pi]$ (meaning $-\pi \leqslant x \leqslant \pi$), the resulting coefficients a_0, a_n, and b_n are called its *Fourier coefficients,* and the trigonometric series (3.15), namely,

$$f(x) = \tfrac{1}{2}a_0 + \sum_{n=1}^{\infty}(a_n\cos nx + b_n\sin nx), \tag{3.21}$$

is called its *Fourier representation* or *Fourier series.* Note that two distinct steps are involved here. The Fourier coefficients are first calculated by using the original function $F(r)$. They are then used to reconstruct the original function with the help of the trigonometric series (3.21). In numerical calculations, the infinite series is truncated after a sufficiently large number of terms.

A number of questions immediately arise. Do these Fourier coefficients exist at all? Is the Fourier series convergent? Does it converge to the original function $F(x)$? These are among the questions that will be discussed in this chapter.

3.3. Fourier coefficients and Fourier-series representation

Whether the Fourier coefficients

$$\begin{aligned} a_n &= \frac{1}{\pi}\int_{-\pi}^{\pi} F(x)\cos nx\,dx, \qquad n \geqslant 0, \\ b_n &= \frac{1}{\pi}\int_{-\pi}^{\pi} F(x)\sin nx\,dx, \qquad n > 0, \end{aligned} \tag{3.22}$$

exist depends not only on the nature of the function $F(x)$, but also on the precise meaning of an integral.

The definite integral of a continuous function, defined in elementary calculus as the limit of a sum, is due to Cauchy. Riemann generalized the concept to certain bounded functions that may have an infinite number of discontinuities in the interval of integration. In 1902, Lebesque put forward a new theory of integrals in which more functions are integrable. As a result, more functions can be represented by Fourier series.

In this book, we restrict ourselves to ordinary, that is, Riemann, integrals. For these, it is known that a function is integrable in the finite closed interval $[a,b]$ if it is continuous, or bounded and continuous except for a finite number of points. If it is bounded, the function is integrable even when it has an infinite number of discontinuities, provided that they can be enclosed in a finite or infinite number of intervals with a total length that can be made as small as one pleases. In particular, this bounded function is integrable if its points of discontinuity are "countable" like the integers $1, 2, 3, \ldots$.

Besides the integrability of $F(x)$ itself, there are two other useful concepts of integrability that will be used in our discussion. A function $F(x)$ is *absolutely integrable* if its absolute value $|F(x)|$ is integrable. If $F(x)$ is absolutely integrable, it is certainly integrable, but an integrable function may not be absolutely integrable. A function is *square integrable* if both $F(x)$ and $F^2(x)$ are integrable. Every bounded integrable function is also square integrable, but an unbounded integrable function may not be square integrable. A simple example of the latter is $x^{-1/2}$ in the interval $[0,b]$ since the square integral $(\ln x)$ is infinite at $x=0$.

In elementary applications, we shall deal only with functions that are both absolutely and square integrable. The relevant integrals are often tabulated in mathematical books and tables (such as the *Handbook of Chemistry and Physics*) as finite sine and cosine transforms. Let us now concentrate on the construction and use of Fourier series.

3.3.1. Calculation of Fourier coefficients

Example 3.3.1. Obtain the Fourier coefficients for $F(x)=x,\ -\pi \leqslant x \leqslant \pi$.

Since $F(x)$ is odd, all a_n vanish. For b_n we have

$$\begin{aligned}
\int_{-\pi}^{\pi} x \sin nx \, dx &= 2\int_0^{\pi} x \sin nx \, dx \\
&= 2\left(-\frac{d}{dk}\int_0^{\pi} \cos kx \, dx\right) = 2\left(-\frac{d}{dk}\right)\left(\frac{1}{k}\sin k\pi\right) \\
&= 2\left(\frac{1}{k^2}\sin k\pi - \frac{\pi \cos k\pi}{k}\right)_{k=n} \\
&= \frac{2\pi}{n}(-1)^{n+1}
\end{aligned}$$

$$\therefore b_n = \frac{2}{n}(-1)^{n+1}.$$

Example 3.3.2.

$$G(x)=\begin{cases} 1, & 0 \leqslant x \leqslant \pi \\ -1, & -\pi \leqslant x \leqslant 0. \end{cases}$$

Again $a_n = 0$, while

$$b_n = \frac{2}{\pi}\int_0^{\pi} \sin nx\, dx = -\frac{2}{\pi n}\cos nx\Big|_0^{\pi}$$

$$= \begin{cases} 0, & n \text{ even} \\ \dfrac{4}{\pi}, & n \text{ odd.} \end{cases}$$

Three partial sums of the Fourier series for $G(x)$ are shown in Fig. 3.1.

Example 3.3.3. Obtain the Fourier-series representation for the function

$$F(x) = x^2, \qquad -\pi \leqslant x \leqslant \pi,$$

and use it to show that

(a) $$\pi^2 = 6\sum_{n=1}^{\infty}\frac{1}{n^2},$$

(b) $$\pi^2 = 12\sum_{n=1}^{\infty}\frac{(-1)^{n+1}}{n^2}.$$

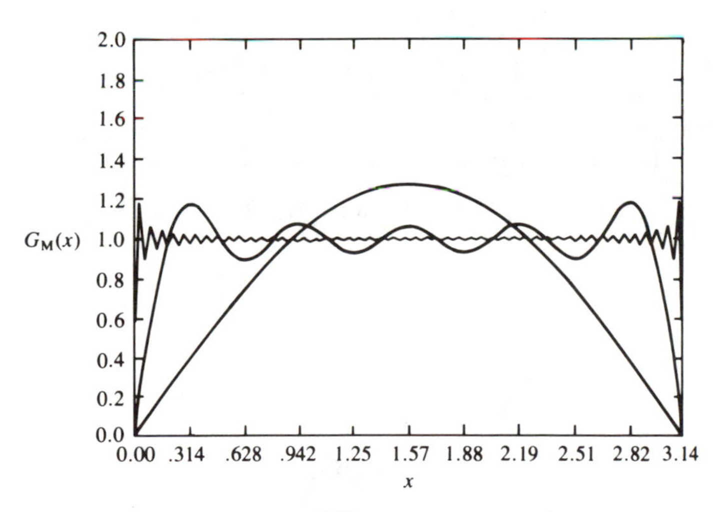

Fig. 3.1. The partial sums $G_M(x) = (4/\pi)\sum_{n=1}^{M} \sin(2n+1)x/(2n+1)$ for $M = 1, 10, 100$ and for positive x.

$$a_0 = \frac{1}{\pi}\int_{-\pi}^{\pi} x^2\, dx = \tfrac{2}{3}\pi^2,$$

$$\begin{aligned}
a_n &= \frac{1}{\pi}\int_{-\pi}^{\pi} x^2 \cos nx\, dx = \frac{2}{\pi}\left(-\frac{d^2}{dk^2}\int_0^{\pi} \cos kx\, dx\right)_{k=n} \\
&= -\frac{2}{\pi}\left[\frac{d}{dk}\left(-\frac{1}{k^2}\sin k\pi + \frac{\pi}{k}\cos k\pi\right)\right]_{k=n} \\
&= -\frac{2}{\pi}\left[-\frac{2\pi}{k^2}\cos k\pi + \sin k\pi\left(\frac{2}{k^3} - \frac{\pi^2}{k}\right)\right]_{k=n} \\
&= \frac{4}{n^2}(-1)^n,
\end{aligned}$$

$$b_n = 0.$$

$$\therefore x^2 = \frac{\pi^2}{3} + 4\sum_{n=1}^{\infty}(-1)^n\frac{\cos nx}{n^2}.$$

(a) At $x = \pi$:

$$\pi^2 = \frac{\pi^2}{3} + 4\sum_{n=1}^{\infty}\frac{1}{n^2}, \qquad \text{i.e.,} \qquad \pi^2 = 6\sum_{n=1}^{\infty}\frac{1}{n^2}.$$

(b) At $x = 0$:

$$\frac{\pi^2}{3} = -4\sum_{n=1}^{\infty}\frac{(-1)^n}{n^2}, \qquad \text{i.e.,} \qquad \pi^2 = 12\sum_{n=1}^{\infty}\frac{(-1)^{n+1}}{n^2}.$$

(Question: Which series would you rather use for a numerical calculation of π^2? Why?)

3.3.2. Modified Fourier series

We see in these examples that, when a given function is odd in x, only the sine terms will appear. Similarly, the Fourier series representation of a function even in x contains only cosine terms. More generally, any function $F(x)$ can be separated into an even and an odd part:

$$F(x) = F_{\text{even}}(x) + F_{\text{odd}}(x),$$

where

$$F_{\text{even}}(x) = \tfrac{1}{2}[F(x) + F(-x)] = G_1(x),$$
$$F_{\text{odd}}(x) = \tfrac{1}{2}[F(x) - F(-x)] = G_2(x).$$

Therefore it follows that

$$F_{\text{even}}(x) = \tfrac{1}{2}a_0 + \sum_n a_n \cos nx, \tag{3.23a}$$

$$F_{\text{odd}}(x) = \sum_n b_n \sin nx. \tag{3.23b}$$

This separation is illustrated in Figure 3.2.

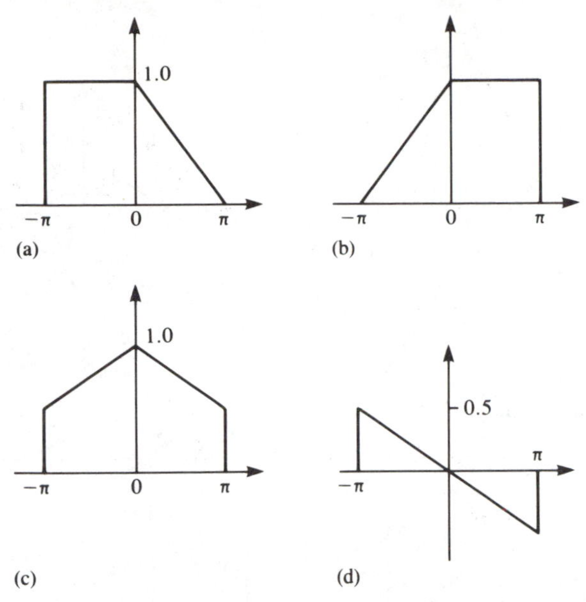

Fig. 3.2. The even and odd parts of a function. (a) $F(x)$; (b) $F(-x)$; (c) $F_{\text{even}}(x)$; (d) $F_{\text{odd}}(x)$.

If the function $G_1(x)$ is originally defined only in the positive half-interval $[0,\pi]$, we can expand it in a cosine series by first making an *even extension* to the negative half-interval $[-\pi,0]$. The extended function is just $F_{\text{even}}(x)$, and the cosine series in question is just Eq. (3.23a). In a similar way, a function $G_2(x)$ in the half-interval $[0,\pi]$ has an *odd extension* $F_{\text{odd}}(x)$ to $[-\pi,0]$, that is the sine series of Eq. (3.23b).

Other pieces of $F(x)$ can be isolated in a similar way. For example, if

$$F_{\text{even,even}}(x) = \text{cosine terms with even } n \text{ only,} \tag{3.24a}$$

it must also be even about the points $x = \pm\pi/2$ for each even half of $F_{\text{even}}(x)$, since this is the property of $\cos nx$ with even n. Similarly,

$$F_{\text{even,odd}}(x) = \text{cosine terms with odd } n \text{ only} \tag{3.24b}$$

has two even halves, each of which is odd about its own middle point. Figure 3.3 shows these Fourier series constructed from the functions of Fig. 3.2, together with the corresponding forms for the sine terms:

$$F_{\text{odd,even}}(x) = \text{sine terms with } \textit{odd } n \text{ only} \tag{3.24c}$$

$$F_{\text{odd,odd}}(x) = \text{sine terms with } \textit{even } n \text{ only.} \tag{3.24d}$$

Note that the subscripts on F refer to the symmetry of the function, not the evenness or oddness of n.

Interesting results can also be obtained by adding Fourier series

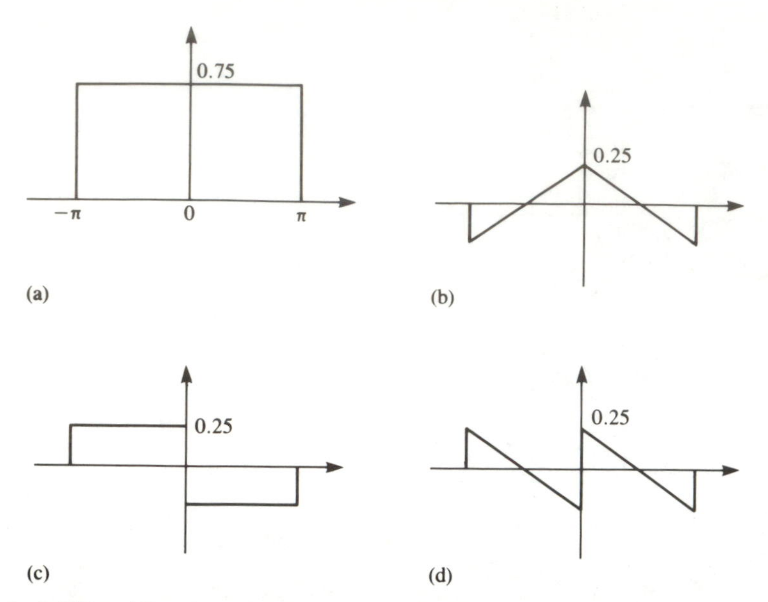

Fig. 3.3. Additional Fourier series that can be constructed from the function $F(x)$ of Fig. 3.2(a). (a) $F_{\text{even,even}}(x)$: cosine terms with even n only; (b) $F_{\text{even,odd}}(x)$: cosine terms with odd n only; (c) $F_{\text{odd,even}}(x)$: sine terms with odd n only; (d) $F_{\text{odd,odd}}(x)$; sine terms with even n only.

together. For example, the sum of

$$\sum_{n=1}^{\infty} \frac{\cos nx}{n} = -\ln\left(2 \sin\frac{x}{2}\right), \qquad (0,2\pi),$$

and

$$\sum_{n=1}^{\infty} (-1)^{n+1} \frac{\cos nx}{n} = \ln\left(2 \cos\frac{x}{2}\right), \qquad (-\pi,\pi)$$

is

$$\sum_{n=1}^{\infty} \frac{\cos(2n+1)x}{2n+1} = -\tfrac{1}{2} \ln\left(\tan\frac{x}{2}\right).$$

This sum is valid only in the open interval $(0,\pi)$ (i.e., $0 < x < \pi$) common to both the original intervals. [The original intervals are chosen differently in order to avoid the infinity caused by ln(0).]

3.3.3. Periodic extension of F(x) by its Fourier representation

While $F(x)$ is defined only in the interval $[-\pi,\pi]$, its Fourier representation $f(x)$, assuming that it does exist in this interval, is also defined for all other values of x. In particular,

$$\begin{aligned} f(x+2m\pi) &= \tfrac{1}{2}a_0 + \sum_{n=0}^{\infty} \{a_n \cos[n(x+2m\pi)] + b_n \sin[n(x+2m\pi)]\} \\ &= f(x), \qquad \text{if } m = \text{any integer}. \end{aligned}$$

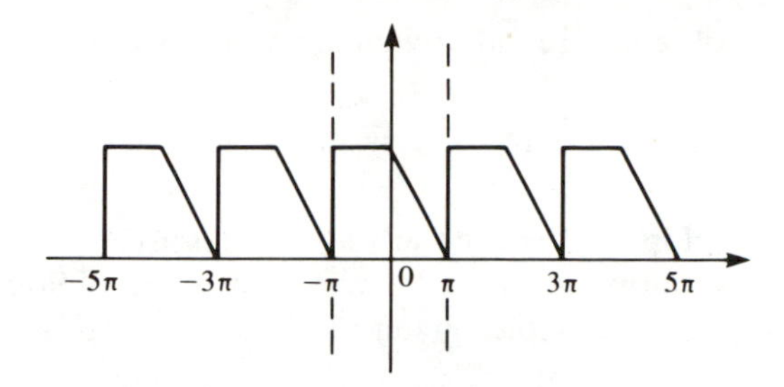

Fig. 3.4. Periodic extension of Fig. 3.2(a) outside the interval $[-\pi, \pi]$.

Therefore, $f(x)$ for $|x| > \pi$ is a periodic repetition of the basic function defined in the original interval, with a period of 2π, the length of the original interval. For example, Fig. 3.4 shows the periodic extension of Fig. 3.2(a).

3.3.4. Fourier series for an arbitrary interval

If $f(x)$ is defined for the interval $[-L, L]$ of length $2L$, a simple change of variables to

$$y = \frac{\pi}{L} x$$

will modify the interval to the standard form $[-\pi, \pi]$, now for the variable y. We then have

$$f(x) = \tfrac{1}{2}a_0 + \sum_{n=0}^{\infty} \left(a_n \cos n \frac{\pi x}{L} + b_n \sin n \frac{\pi x}{L} \right), \tag{3.25}$$

where

$$\begin{aligned} a_0 &= \frac{1}{\pi} \int_{-\pi}^{\pi} F\left(x = \frac{L}{\pi} y \right) dy \\ &= \frac{1}{L} \int_{-L}^{L} F(x)\, dx, \end{aligned}$$

and similarly

$$\begin{aligned} a_n &= \frac{1}{L} \int_{-L}^{L} F(x) \cos\left(n \frac{\pi x}{L} \right) dx, \\ b_n &= \frac{1}{L} \int_{-L}^{L} F(x) \sin\left(n \frac{\pi x}{L} \right) dx. \end{aligned} \tag{3.26}$$

For example, if $L = 1$, the trigonometric functions to be used are $\cos n\pi x$ and $\sin n\pi x$.

An asymmetric interval $A \leqslant x \leqslant B$ is sometimes used. The necessary formulas can be obtained readily by changing the variable to

$$y = \frac{\pi}{L}(x - x_0), \qquad \text{where} \qquad L = \tfrac{1}{2}(B - A), \qquad x_0 = \tfrac{1}{2}(B + A).$$

The resulting development will be left as an exercise.

A change of variable can also be made in a given Fourier series to generate another. For example, given the Fourier series

$$\sum_{n=0}^{\infty} \frac{\sin(2n+1)x}{2n+1} = \frac{\pi}{4}$$

for the open interval $(0, \pi)$ (or $0 < x < \pi$), a change of variables to

$$t = x - \tfrac{1}{2}\pi$$

yields

$$\sum_{n=0}^{\infty} (-1)^n \frac{\cos(2n+1)t}{2n+1} = \frac{\pi}{4}.$$

Integrations and usually differentiations of a given Fourier series will also generate others. Care should be taken with differentiation since the resulting infinite series is always less convergent than the original one.

Problems

3.3.1. Find the Fourier-series representations of the following functions in the interval $-\pi \leqslant x \leqslant \pi$:
  (a) $F(x) = 0$ if $x < 0$; $F(x) = x$, if $x > 0$;
  (b) $F(x) = 0$, if $x < 0$; $F(x) = 1$, if $x > 0$;
  (c) $F(x) = \sin ax$, where a is an arbitrary constant;
  (d) $F(x) = \cos ax$.

3.3.2. Find the Fourier-series representation of the following functions in the interval $-L \leqslant x \leqslant L$:
  (a) $F(x) = e^x$;
  (b) $F(x) = |x|$;
  (c) $F(x) = |x|$, if $|x| \leqslant L/2$; $F(x) = L - |x|$, if $|x| > L/2$;
  (d) $F(x) = -1$, if $x < -L/3$; $F(x) = 1$, if $x > L/3$; $F(x) = 0$, if $-L/3 < x < L/3$;
  (e) $F(x) = x$, if $|x| < L/2$, $F(x) = (L - |x|)x/|x|$, if $x > L/2$.

3.3.3. Find the Fourier *cosine* series in the interval $-L \leqslant x \leqslant L$, which will reproduce the functions of Problem 3.3.2 in the half-interval $0 \leqslant x \leqslant L$.

3.3.4. Find the Fourier *sine* series in the interval $-L \leqslant x \leqslant L$, which will reproduce the functions of Problem 3.3.2 in the half-interval $0 \leqslant x \leqslant L$.

3.3.5. Obtain the Fourier series for an asymmetric interval $A \leqslant x \leqslant B$.

3.3.6. Find the Fourier-series representations of the functions of Problem 3.3.2 in the interval $0 \leqslant x \leqslant L$ with period L.

3.3.7. Obtain the function in $[-\pi,\pi]$ whose Fourier-series representation is

$$f(x) = \cos x + \tfrac{1}{9}\cos 3x + \tfrac{1}{25}\cos 5x + \cdots.$$

(Hint: This is related to the Fourier series for x^2.)

3.3.8. Describe the functions

(a) $g_1(x) = \dfrac{1}{\pi}\sum_{n=1}^{\infty} \sin nx \int_{-\pi}^{\pi} F(t) \sin nt\, dt,$

(b) $g_2(x) = \dfrac{1}{\pi}\sum_{n=1}^{\infty} \sin(2n-1)x \int_{-\pi}^{\pi} F(t) \sin(2n-1)t\, dt,$

(c) $g_3(x) = \dfrac{2}{\pi}\sum_{n=1}^{\infty} \sin nx \int_{0}^{\pi} F(t) \sin nt\, dt,$

(d) $g_4(x) = \dfrac{2}{\pi}\sum_{n=1}^{\infty} \sin(2n-1)x \int_{0}^{\pi} F(t) \sin(2n-1)t\, dt$

for $-2\pi \leqslant x \leqslant 2\pi$ in terms of a given arbitrary function $F(t)$ defined for $-\pi \leqslant t \leqslant \pi$.

3.3.9. Show that

(a) $\tfrac{1}{2} - x = \dfrac{1}{\pi}\sum_{n=1}^{\infty} \dfrac{1}{n} \sin 2n\pi x = u(x), \qquad$ for $\qquad 0 < x < 1;$

(b) $\tfrac{1}{4} = \dfrac{1}{\pi}\sum_{n=1}^{\infty} \dfrac{\sin(2n-1)\pi x}{(2n-1)} = v(x), \qquad$ for $\qquad 0 < x < 1;$

(c) $n(x) = x + u(x) - 8v^2(x)$ is an integer and that it gives the largest integer in *any* positive number x.

(Hint: $u(x)$ is a Fourier representation of $\tfrac{1}{2} - x$ in [0,1] with a period of 1, while $v(x)$ is an odd extension of $\tfrac{1}{4}$ in [0,1] to the interval [−1,1] in a Fourier sine series with a period of 2.)

3.3.10. By using a Fourier cosine series given in Appendix 3B, show that

$$\sum_{n=0}^{\infty} (-1)^n \frac{\sin(2n+1)x}{(2n+1)^2} = \frac{\pi}{4} \times \begin{cases} \pi - x, & \pi/2 \leqslant x \leqslant \pi \\ c, & |x| \leqslant \pi/2 \\ -\pi - x, & -\pi \leqslant x \leqslant -\pi/2. \end{cases}$$

3.4. Complex Fourier series and the Dirac δ function

The Fourier representation [Eq. (3.25)] of a function $F(x)$ in the interval $[-L,L]$ can be written in another useful form involving complex coefficients:

$$\begin{aligned} f(x) &= \tfrac{1}{2}a_0 + \sum_{n=1}^{\infty} \left(a_n \frac{\exp[i(n\pi/L)x] + \exp(-i(n\pi/L)x]}{2} \right. \\ &\qquad \left. + b_n \frac{\exp[i(n\pi/L)x] - \exp[-i(n\pi/L)x]}{2i} \right) \\ &= \sum_{n=-\infty}^{\infty} c_n \exp[i(n\pi/L)x], \end{aligned} \tag{3.27}$$

if

$$c_0 = \tfrac{1}{2}a_0, \qquad c_{n>0} = \tfrac{1}{2}(a_n - ib_n), \qquad c_{n<0} = \tfrac{1}{2}(a_{|n|} + ib_{|n|}).$$

The $\exp[-i(n\pi/L)x]$ terms can be written as $\exp[i(n\pi/L)x]$ with negative integers n. Hence the sum over n should now be extended to all the negative integers. The coefficients can be expressed compactly in the form

$$c_n = \frac{1}{2L}\int_{-L}^{L} F(x)\exp[-i(n\pi/L)x]\,dx, \tag{3.28}$$

where use has been made of the formulas (3.26) for a_n, b_n.

Alternatively, we can look for the complex Fourier representation

$$F(x) \simeq \sum_{n=-\infty}^{\infty} c_n \exp[i(n\pi/L)x]$$

with coefficients c_n calculated according to the method of Section 3.2. The result for c_n agrees with Eq. (3.28), as we shall show explicitly in Section 3.9.

3.4.1. Dirac δ function

By substituting c_n back into the Fourier series of Eq. (3.27), we obtain

$$\begin{aligned} f(x) &= \sum_{n=-\infty}^{\infty} \exp[i(n\pi/L)x]\left(\frac{1}{2L}\int_{-L}^{L} \exp[-i(n\pi/L)x']F(x')\,dx'\right) \\ &= \int_{-L}^{L} F(x')\,\delta(x-x')\,dx', \end{aligned}$$

where

$$\delta(x-x') = \frac{1}{2L}\sum_{n=-\infty}^{\infty} \exp[i(n\pi/L)(x-x')] \tag{3.29}$$

is called a *Dirac δ function*. If we restrict ourselves to functions $F(x)$ that can be represented in some sense by the complex Fourier series of Eq. (3.27), then to the same extent $f(x)$ may be considered a "good enough" copy of $F(x)$, and also be denoted by the symbol $F(x)$. Under the circumstances we have

$$F(x) = \int_{-L}^{L} F(x')\,\delta(x-x')\,dx'. \tag{3.30}$$

Thus the δ function picks out one value $F(x)$ of the function $F(x')$ when integrated over x'. In particular, if $F(x) = 1$ over the entire interval, when we look up the function at the point x in $[-L, L]$, we get 1, too. Hence

$$\int_{-L}^{L} \delta(x-x')\,dx' = 1, \qquad \text{if} \qquad -L \leqslant x \leqslant L. \tag{3.31}$$

3.4.2. *Property of the δ function*

The Dirac δ function has some very unusual properties. At $x'=x$, each term in the infinite sum shown in Eq. (3.29) is 1. Hence $\delta(x-x')$ is infinite there. Second, Eq. (3.30) states that the integral is completely independent of $F(x')$ at $x'\neq x$. This is possible only if $\delta(x-x')$ vanishes whenever $x'\neq x$. According to Dirac, this function may be visualized as follows:

> To get a picture of $\delta(x)$, take a function of the real variable x which vanishes everywhere except inside a small domain, of length ε say, surrounding the origin $x=0$, and which is so large inside this domain that its integral over this domain is unity. The exact shape of the function inside this domain does not matter, provided that there are no unnecessarily wild variations (for example, provided the function is always of order ε^{-1}). Then in the limit $\varepsilon\to 0$ this function will go over into $\delta(x)$.
>
> Dirac, *The Principles of Quantum Mechanics*, Fourth Edition (Clarendon Press, Oxford, 1958), p. 58.

The simplest such object is a rectangular "spike" of width ε and height ε^{-1} centered at x:

$$D_\varepsilon(x-x')=\begin{cases}\varepsilon^{-1}, & |x-x'|\leq\frac{1}{2}\varepsilon\\ 0, & \text{otherwise.}\end{cases} \tag{3.32}$$

Since the area under $D_\varepsilon(x-x')$ is 1, Eq. (3.31) is satisfied. For Eq. (3.30), we get

$$\int_{-L}^{L}F(x')D_\varepsilon(x-x')\,dx'=\int_{x-\frac{1}{2}\varepsilon}^{x+\frac{1}{2}\varepsilon}F(x')\varepsilon^{-1}\,dx'$$
$$=F(x)+\frac{1}{24}\frac{d^2F(x)}{dx^2}\varepsilon^2+\cdots$$

in a Taylor expansion, where the higher-order terms not shown contain higher powers of ε^2. Thus Eq. (3.30) is exactly satisfied in the limit $\varepsilon\to 0$, or

$$\delta(x-x')=\lim_{\varepsilon\to 0}D_\varepsilon(x-x'). \tag{3.33}$$

Such an object is not a well-defined mathematical function, which must have a definite value at every point at which it is defined. Dirac called it an "improper function," which has the characteristic that "when it occurs as a factor in an integrand the integral has a well-defined value." Such generalized functions are now called *distributions*.

It is easy to see that the formal difficulty one gets into in visualizing $\delta(x)$ is a simple consequence of performing the sum n in Eq. (3.29) before the integral over x', that is, in the wrong order. If we always integrate first, we will never get into trouble, assuming of course that the Fourier series itself converges and that the functions of interest are all square integrable.

The point is that it is notationally simpler to show only the integration part of the problem. A number of simple and useful relations can then be deduced about δ functions. They should be taken to mean that each of the two sides of an equation gives the same result when used in an integral in which the other factors should in general be first expanded in Fourier series. The nice thing about these relations is that they often are fairly obvious results of simple manipulations, and may be understood or even "derived" without referring to the underlying Fourier series. Examples are

$$\delta(-x) = \delta(x), \tag{3.34a}$$

$$x\,\delta(x) = 0, \tag{3.34b}$$

$$\delta(ax) = \delta(x)/|a|. \tag{3.34c}$$

$$\delta(x^2 - b^2) = [\delta(x-b) + \delta(x+b)]/2\,|b|. \tag{3.34d}$$

These properties can be derived as follows. Equation (3.34a) comes from the fact that $D_\varepsilon(x)$ is an even function of x. Equation (3.34b) results because

$$\int_{-L}^{L} f(x) x\,\delta(x)\,dx$$

consists simply of looking up the function $xf(x)$ at $x = 0$ where the argument of the δ function is zero. If $f(0)$ is finite, the result is zero. Equation (3.34c) comes from a change of varibles $ax = y$

$$\int_{-L}^{L} f(x)\,\delta(ax)\,dx = \int_{-aL}^{aL} f\left(\frac{y}{a}\right)\delta(y)\frac{dy}{a} = \frac{f(0)}{a},$$

if $a > 0$. This is just what is obtained if $\delta(ax) = \delta(x)/a$. If $a < 0$, the integration over y goes from a positive value $-aL$ to a negative value aL. To bring the direction of integration back to the normal order of lower limit to upper limit, we must write

$$\int_{-aL}^{aL} \frac{dy}{a} = -\int_{aL}^{-aL} \frac{dy}{a} = \int_{aL}^{-aL} \frac{dy}{|a|}.$$

The additional factor $|a|^{-1}$ can be interpreted as the change in area under the rectangular function $D_\varepsilon(ax)$ when it is plotted as a function of x rather than as a function of ax.

The demonstration of Eq. (3.34d) requires a number of steps. An integration over a δ function represents a function lookup at the zero of the argument, here $x^2 - b^2$. As a function of the integration variable x, there are actually two zeros: $x = b$ and $-b$. Since $x^2 - b^2 = (x-b)(x+b)$, it behaves like $(x-b)2b$ near $x = b$, and like $-2b(x+b)$ near $x = -b$. Hence

$$\delta(x^2 - b^2) = \delta(2b(x-b)) + \delta(-2b(x+b)),$$

where we have used the fact that an integration requires an addition of contributions. Each of the δ functions can be evaluated with the help of Eq. (3.34c) to give the stated result.

An even more complicated relation is the identity

$$\int_{-\infty}^{\infty} f(x)\,\delta(g(x))\,dx = \frac{f(x_0)}{|g'(x_0)|}, \tag{3.35a}$$

if $g(x)$ has a single root or zero at x_0. Here the slope

$$g'(x) = \frac{d}{dx}g(x)$$

at x_0 must be nonzero. Thus near x_0 the function $g(x)$ behaves like $(x - x_0)g'(x_0)$. Such zeros are called *simple zeros*. In a similar way, if $g(x)$ has N "simple" zeros at $x = x_i$ where $g(x) \simeq g'(x_i)(x - x_i)$ with nonzero slopes, then

$$\delta(g(x)) = \sum_{i=1}^{N} \frac{\delta(x - x_i)}{|g'(x_i)|}. \tag{3.35b}$$

The demonstration of these relations will be left as exercises. (See Problem 3.4.5.)

Example 3.4.1.

$$\int_{-\infty}^{\infty} e^{-x}\,\delta(x^2 - a^2)\,dx = \int_{-\infty}^{\infty} e^{-x}\,\frac{\delta(x - a) + \delta(x + a)}{2a}\,dx$$
$$= (e^{-a} + e^{a})/2a, \qquad \text{if} \qquad a > 0.$$

Example 3.4.2.

$$\int_{-\infty}^{\infty} e^{-x^2}\,\delta(\sin x)\,dx = \int_{-\infty}^{\infty} e^{-x^2}\left(\sum_{n=-\infty}^{\infty} \frac{\delta(x - n\pi)}{|\cos n\pi|}\right)dx$$
$$= \sum_{n=-\infty}^{\infty} \int_{-\infty}^{\infty} e^{-x^2}\,\delta(x - n\pi)\,dx$$
$$= \sum_{n=-\infty}^{\infty} e^{-(n\pi)^2}.$$

Example 3.4.3.

$$\int_0^5 \sin x\,\delta((x-2)(x-4)(x-6))\,dx = \frac{\sin 2}{|g'(2)|} + \frac{\sin 4}{|g'(4)|}$$

on applying Eq. (3.35b). There is no contribution for the δ function at $x = 6$ because it is outside the limits of integration. To evaluate the derivatives, we note that

$$g'(x) = (x-4)(x-6) + (x-2)(x-6) + (x-2)(x-4).$$

Hence $g'(2) = 8$, $g'(4) = -4$, and the integral is $\frac{1}{8}\sin 2 + \frac{1}{4}\sin 4$.

Even derivatives of the δ function can be defined and used. For example, if

$$\delta'(x-x') \equiv \frac{d}{dx'}\delta(x-x'), \tag{3.36}$$

then an integral involving it can be calculated by an integration by parts:

$$\int_{-L}^{L} F(x')\,\delta'(x-x')\,dx' = F(L)\,\delta(x-L) - F(-L)\,\delta(x+L) - \int_{-L}^{L} F'(x')\,\delta(x-x')\,dx'. \tag{3.37}$$

This is well defined if $F(L)=F(-L)=0$ because the boundary terms then vanish.

An advanced but noteworthy feature involving the δ function is connected with Eq. (3.34b). This states that $x\,\delta(x)$ is identically zero. As a result, we may add any finite multiple of this zero to one side of an equation

$$A(x) = B(x) = B(x) + cx\,\delta(x),$$

where c is an arbitrary finite constant. However, if we should divide both sides by x, the addition $c\,\delta(x)$ is not longer zero at $x=0$. Then

$$A(x)/x = B(x)/x + c\,\delta(x) \tag{3.38}$$

is not necessarily true for arbitrary values of c.

For example,

$$x\frac{d}{dx}\ln x = 1 = 1 + cx\,\delta(x)$$

is valid for any finite constant c. However,

$$\frac{\mathrm{d}}{\mathrm{d}x}\ln x = \frac{1}{x} + c\,\delta(x) \tag{3.39}$$

is true only for special values of c. To determine these values, we must examine the behavior of these functions in the neighborhood of $x=0$ by integrating both sides of the expression from $-\varepsilon$ to ε:

$$I = \int_{-\varepsilon}^{\varepsilon} \frac{d}{dx}\ln x\,dx = \ln\varepsilon - \ln(-\varepsilon) = \ln(-1),$$

while on the right-hand side

$$I = \int_{-\varepsilon}^{\varepsilon} \frac{dx}{x} + c = c,$$

because $1/x$ is an odd function of x. Thus

$$c = \ln(-1) = \ln(e^{i(2n+1)\pi}) = i(2n+1)\pi, \tag{3.40}$$

where n is any integer. The need for the additive constant c in Eq. (3.39) will become very clear in Chapter 6, where we shall find that $d \ln x/dx$ is a multivalued function, while x^{-1} is a single-valued function. The different values of c are thus needed to match the former function's many values.

The δ-function term in Eq. (3.39) is not just a mathematical curiosity. It plays an important role in quantum mechanics.

Problems

3.4.1. show that the Dirac δ function for the real Fourier series in the interval $[-L,L]$ is

$$\delta(x-x') = \frac{1}{2L} + \frac{1}{L}\sum_{n=1}^{\infty}\left[\cos\left(\frac{n\pi x}{L}\right)\cos\left(\frac{n\pi x'}{L}\right) + \sin\left(\frac{n\pi x}{L}\right)\sin\left(\frac{n\pi x'}{L}\right)\right].$$

3.4.2. Obtain the Dirac δ function $\delta(x-x')$ for

(a) The Fourier cosine series; and

(b) the Fourier sine series, in the half interval $[0,L]$. Why are these two answers not the same?

3.4.3. By integrating the Dirac δ function in Problem 3.4.1 or 3.4.2, obtain the Fourier series for the function

$$G(x) = \begin{cases} 1, & x > 0 \\ -1, & x < 0. \end{cases}$$

3.4.4. Given a function

$$D_\varepsilon(x) = \begin{cases} \varepsilon^{-1}, & |x| < \tfrac{1}{2}\varepsilon \\ 0, & |x| > \tfrac{1}{2}\varepsilon \end{cases}$$

in the interval $[-L,L)$.

(a) Obtain its Fourier-series representation.

(b) Show that in the limit $\varepsilon \to 0$ this Fourier-series representation agrees with that for the Dirac δ function obtained in Problem 3.4.1 or 3.4.2.

3.4.5. Derive Eqs. (3.35a) and (3.35b).

3.4.6. If $F(x)$ is discontinuous at $x=0$, show that

(a) $$\int_0^{\infty} F(x)\,\delta(x)\,dx = \tfrac{1}{2}\lim_{\varepsilon\to 0} F(\varepsilon),$$

(b) $$\int_{-\infty}^{0} F(x)\,\delta(x)\,dx = \tfrac{1}{2}\lim_{\varepsilon\to 0} F(-\varepsilon).$$

3.5. Fourier transform

The complex Fourier series has an important limiting form when $L \to \infty$. Suppose that as $L \to \infty$

1. $n\pi/L \equiv k$ remains large (ranging in fact from $-\infty$ to ∞), and
2. $c_n \to 0$ (because it is proportional to L^{-1}), but

$$g(k) \equiv \lim_{\substack{L\to\infty \\ c_n\to 0}} \frac{L}{\pi} c_n = \frac{1}{2\pi} \int_{-\infty}^{\infty} f(x) e^{-ikx}\, dx = \text{finite}. \tag{3.41}$$

Then

$$\begin{aligned} f(x) &= \sum_{n=-\infty}^{\infty} c_n e^{ikx}, \qquad k = \frac{n\pi}{L} \\ &= \lim_{\substack{L\to\infty \\ c_n\to 0}} \sum_{n=-\infty}^{\infty} \frac{\pi}{L} g(k) e^{ikx}. \end{aligned}$$

The sum over n in steps of $\Delta n = 1$ can be written as a sum over k, which is proportional to n. Hence

$$\sum_n \frac{\pi}{L} = \left(\sum_{\Delta k} \frac{L}{\pi} \right) \frac{\pi}{L} = \sum_{\Delta k}.$$

However, $\Delta k = (\pi/L)\,\Delta n$ becomes infinitesimally small when L becomes large. The sum over Δk then becomes an integral

$$\lim_{\substack{L\to\infty \\ c_n\to 0}} \sum_n \frac{\pi}{L} = \int dk, \tag{3.42}$$

and

$$f(x) = \int_{-\infty}^{\infty} dk\, g(k) e^{ikx}. \tag{3.43}$$

We call $g(k)$ of Eq. (3.41) the *Fourier transform of* $f(x)$, and Eq. (3.43) the *Fourier inversion formula.*

To obtain the Dirac δ function for this transformation, we substitute Eq. (3.41) into Eq. (3.43) to get

$$\begin{aligned} f(x) &= \int_{-\infty}^{\infty} dk\, e^{ikx} \frac{1}{2\pi} \int_{-\infty}^{\infty} dx'\, e^{-ikx'} f(x') \\ &= \int_{-\infty}^{\infty} dx'\, f(x')\, \delta(x - x'). \end{aligned}$$

Hence

$$\delta(x - x') = \frac{1}{2\pi} \int_{-\infty}^{\infty} dk\, e^{ik(x-x')}. \tag{3.44}$$

3.5.1. *Conjugate symmetry*

The Fourier transform and its inverse are similar in structure. This symmetry can be made explicit by using the symmetrical definitions:

$$g(k)=\frac{1}{\sqrt{2\pi}}\int_{-\infty}^{\infty} e^{-ikx}f(x)\,dx, \tag{3.45a}$$

$$f(x)=\frac{1}{\sqrt{2\pi}}\int_{-\infty}^{\infty} e^{ikx}g(k)\,dk. \tag{3.45b}$$

Thus the roles of $f(x)$ and $g(k)$ can be interchanged under the substitutions $x \leftrightarrow k$ and $i \to -i$. We call this substitution a *conjugate transformation*. The variables x and k are said to form a *conjugate pair* of variables.

The conjugate pair x and k are joined together by the fact that *either* $f(x)$ *or* $g(k)$ contains all the information about the function. This is an important feature of Fourier transforms and of wave functions to which we shall return in a later section.

3.5.2. *Properties and applications of Fourier transforms*

Fourier transforms involve complex integrations, which are usually studied as part of the theory of functions of complex variables. This theory is described in Chapter 6. For the time being, we must be content with the transforms of very simple functions, or with the use of tables of Fourier transforms in mathematical handbooks. A short table is appended to this chapter for the reader's convenience.

In referring to Fourier transforms, it is convenient to use the notation

$$\mathscr{F}\{f(x)\}=\frac{1}{\sqrt{2\pi}}\int_{-\infty}^{\infty} e^{-ikx}f(x)\,dx=g(k). \tag{3.46}$$

Example 3.5.1. The Fourier transform of the box function

$$b(x)=\begin{cases}1, & |x|\leqslant\alpha\\ 0, & |x|\geqslant\alpha\end{cases} \tag{3.47}$$

is

$$\begin{aligned}\mathscr{F}\{b(x)\}&=\frac{1}{\sqrt{2\pi}}\int_{-\alpha}^{\alpha} e^{-ikx}\,dx=\frac{1}{\sqrt{2\pi}}\frac{e^{-ik\alpha}}{-ik}\bigg|_{-\alpha}^{\alpha}\\ &=\frac{1}{\sqrt{2\pi}}\frac{2}{k}\sin k\alpha.\end{aligned} \tag{3.48}$$

Example 3.5.2.

$$\begin{aligned}\mathscr{F}\left\{\frac{d}{dx}f(x)\right\}&=\frac{1}{\sqrt{2\pi}}\int_{-\infty}^{\infty} e^{-ikx}\left(\frac{d}{dx}f(x)\right)dx\\ &=\frac{1}{\sqrt{2\pi}}f(x)e^{-ikx}\bigg|_{-\infty}^{\infty}-\frac{(-ik)}{\sqrt{2\pi}}\int_{-\infty}^{\infty} e^{-ikx}f(x)\,dx\end{aligned}$$

Table 3.1. Properties of Fourier transforms

Property	If $f(x)$ is	Then its Fourier transform
Complex conjugation	Real	$g^*(k) = g(-k)$
	Real and even	$g^*(k) = g(-k) = g(k)$
	Real and odd	$g^*(k) = g(-k) = -g(k)$
Translation	$f(x-a)$	$e^{-ika}g(k)$
Attenuation	$f(x)e^{ax}$	$g(k+ai)$
Derivatives	$\frac{d}{dx}f(x)$	$ikg(k)$
	$\frac{d^n}{dx^n}f(x) = f^{(n)}(x)$	$(ik)^n g(k)$

after an integration by parts. If $f(x) \to 0$ as $|x| \to \infty$, only the second term survives. Therefore

$$\mathscr{F}\left\{\frac{d}{dx}f(x)\right\} = ik\mathscr{F}\{f(x)\}.$$

This and other simple properties of the Fourier series are summarized in Table 3.1. Their derivations are left as exercises.

3.5.3. Calculation of Fourier transforms

The formulas for derivatives are particularly useful because they reduce differential expressions to algebraic expressions, as the following example shows.

Example 3.5.3. Solve the inhomogeneous differential equation

$$\left(\frac{d^2}{dx^2} + p\frac{d}{dx} + q\right)f(x) = R(x), \qquad -\infty \leqslant x \leqslant \infty, \tag{3.49}$$

where p and q are constants.

We transform both sides

$$\begin{aligned}\mathscr{F}\left\{\frac{d^2f}{dx^2} + p\frac{df}{dx} + qf\right\} &= [(ik)^2 + p(ik) + q]\tilde{f}(k) \\ &= \mathscr{F}\{R(x)\} = \tilde{R}(k),\end{aligned} \tag{3.50}$$

where a tilde is used to denote a Fourier transform. Hence

$$\begin{aligned}f(x) &= \frac{1}{\sqrt{2\pi}}\int_{-\infty}^{\infty} e^{ikx}g(k)\,dk \\ &= \frac{1}{\sqrt{2\pi}}\int_{-\infty}^{\infty} e^{ikx}\frac{\tilde{R}(k)}{-k^2 + ipk + q}\,dk.\end{aligned} \tag{3.51}$$

The formal solution is called an *integral representation* of the solution. Of course, it will not do us any good if we do not know how to evaluate this complex integral. Fortunately, this is a simple problem in the theory of functions of complex variables, which will be discussed in Chapter 6. Hence we have gained by obtaining the integral representation.

Example 3.5.4. Use the translation property of Table 3.1 to obtain the Fourier transform of the translated box function $b(x-\beta)$ of Eq. (3.47).

$$\mathscr{F}\{b(x-\beta)\} = e^{-ik\beta}\mathscr{F}\{b(x)\} = e^{-ik\beta}\frac{1}{\sqrt{2\pi}}\frac{2}{k}\sin k\alpha,$$

where $\mathscr{F}\{b(x)\}$ is from Eq. (3.48).

Problems

3.5.1. Obtain the Fourier transform of

$$f(x) = \begin{cases} 0, & x<0 \\ e^{-ax}\sin bx, & x>0. \end{cases}$$

3.5.2. Find the three-dimensional Fourier transform of the wave function of a 1s electron in the hydrogen atom:

$$\psi_{1s}(\mathbf{r}) = \frac{1}{(\pi a_0^2)^{1/2}}\exp(-r/a_0),$$

where a_0 is the radius of the orbit.

3.5.3. Use the Fourier transform

$$\mathscr{F}\left\{\frac{\sinh ax}{\sinh \pi x}\right\} = \frac{1}{\sqrt{2\pi}}\frac{\cos a}{\cosh k + \cos a}, \qquad |a|<\pi,$$

to find

$$\mathscr{F}\left\{\frac{1-\exp(-bx)}{1-\exp(-2\pi x)}e^{-cx}\right\},$$

where $b>0$ and $c>0$. Should the value of b be restricted?

3.5.4. If $g(k) = \mathscr{F}\{f(x)\}$, show that

(a) $\mathscr{F}\{f(ax)e^{ibx}\} = \dfrac{1}{a}g\left(\dfrac{k-b}{a}\right), \qquad a>0;$

(b) $\mathscr{F}\{f(ax)\cos bx\} = \dfrac{1}{2a}\left[g\left(\dfrac{k-b}{a}\right)+g\left(\dfrac{k+b}{a}\right)\right], \qquad a>0;$

(c) $\mathscr{F}\{f(ax)\sin bx\} = \dfrac{1}{2ai}\left[g\left(\dfrac{k-b}{a}\right)-g\left(\dfrac{k+b}{a}\right)\right], \qquad a>0.$

What are the results if $a<0$?

3.5.5. Use Eq. (3.44) to show that

$$\int_{-\infty}^{\infty} |f(x)|^2\,dx = \int_{-\infty}^{\infty} f^*(x)\,dx \int_{-\infty}^{\infty} \delta(x-x')f(x')\,dx'$$
$$= \int_{-\infty}^{\infty} |g(k)|^2\,dk.$$

3.5.6. Use Entry 2 of the table of Fourier transforms given in Appendix 3C to obtain the following results:

(a) $$\mathcal{F}\left\{\frac{1}{x-b+ia}\right\} = -i\sqrt{2\pi}\,e^{-ka-ikb}\Theta(k),$$
$$\mathcal{F}\left\{\frac{1}{x-b-ia}\right\} = i\sqrt{2\pi}\,e^{ka-ikb}\Theta(-k), \qquad \text{if} \qquad a>0.$$

Here $\Theta(x)$ is 0 if $x<0$, and 1 if $x>0$.

(b) $$\mathcal{F}\left\{\frac{1}{x^2+1}\right\} = \sqrt{\pi/2}\exp(-|k|).$$

3.6. Green function and convolution

If the inhomogeneity function in Eq. (3.49) is

$$R(x) = \delta(x),$$

its Fourier transform $\tilde{R}(k)$ is just $(2\pi)^{-1/2}$. Equation (3.51) then states that the solution of the differential equation

$$\left(\frac{d^2}{dx^2}+p\frac{d}{dx}+q\right)G(x) = \delta(x), \tag{3.52}$$

which is now denoted by $G(x)$, has a Fourier transform that is $(2\pi)^{-1/2}$ times the reciprocal of the Fourier transform of the differential operator. Being essentially the inverse of a differential operator, the function $G(x)$, called a *Green function*, is potentially a very useful idea.

To appreciate its power, let us suppose that the solution of *any* inhomogeneous differential equation

$$\mathcal{D}(x)f(x) = R(x) \tag{3.53}$$

involving an arbitrary differential operator $\mathcal{D}(x)$ can be written in the integral form

$$f(x) = \int_{-\infty}^{\infty} G(x-x')R(x')\,dx'. \tag{3.54}$$

Substitution of Eq. (3.54) into Eq. (3.53) gives the result

$$\int_a^b [\mathcal{D}(x)G(x-x')]R(x')\,dx' = R(x).$$

Thus the original differential equation is satisfied if

$$\mathscr{D}(x)G(x-x') = \delta(x-x'), \tag{3.55}$$

that is, if $G(x-x')$ is the Green function for the differential operator $\mathscr{D}(x)$.

An integral of the form (3.54) is called a *convolution*, and may be denoted by the symbol

$$(G*R)_x \equiv \frac{1}{\sqrt{2\pi}} \int_{-\infty}^{\infty} G(x-x')R(x')\,dx'. \tag{3.56}$$

Its Fourier transform turns out to be always the product $\tilde{G}(k)\tilde{R}(k)$ of transforms. This *convolution theorem* can be derived directly as follows:

$$\begin{aligned}
\mathscr{F}\{(G*R)_x\} &\equiv \frac{1}{\sqrt{2\pi}} \int e^{-ikx}\,dx\left(\frac{1}{\sqrt{2\pi}} \int G(x-x')R(x')\,dx'\right) \\
&= \left(\frac{1}{\sqrt{2\pi}} \int e^{-ik(x-x')}G(x-x')\,d(x-x')\right) \\
&\quad \times \left(\frac{1}{\sqrt{2\pi}} \int d^{-ikx'} R(x')\,dx'\right) \\
&= \mathscr{F}\{G\}\mathscr{F}\{R\}.
\end{aligned} \tag{3.57}$$

This shows, perhaps more clearly than Eq. (3.54), that the Green function may be interpreted as the inverse of the differential operator $\mathscr{D}(x)$. If $\mathscr{D}(x)$ involves functions of x rather than just constant coefficients, it is not so easy to calculate $\tilde{G}(k)$, but that is another story. (See Chapter 4 for a general method for calculating Green functions.)

In engineering, the inhomogeneity function $R(x)$ in Eq. (3.53) is called an *input* to, and the solution $f(x)$ is called an *output* from, the system, while the Green function in Eq. (3.54) is called a *response function,* since it describes how the system responds to the input. Newton's equation of motion in the presence of a driving force

$$m\frac{d^2}{dt^2}\mathbf{r}(t) = \mathbf{F}(t)$$

is another example of Eq. (3.53). A δ-function force proportional to $\delta(t)$ is called an *impulsive* force. Thus the Green function also describes the response of a mechanical system to an impulsive driving force.

Example 3.6.1. Obtain a solution to the equation of a driven harmonic oscillator

$$\ddot{x}(t) + 2\beta\dot{x}(t) + \omega_0^2 x(t) = R(t), \tag{3.58}$$

where β and ω_0 are positive real constants.

According to Eq. (3.54), $x(t)$ has the form

$$x(t)=\int_{-\infty}^{\infty} G(t-t')R(t')\,dt', \tag{3.59}$$

where the Green function $G(t-t')$ is the solution of the differential equation with an impulsive driving force:

$$\left(\frac{d^2}{dt^2}+2\beta\frac{d}{dt}+\omega_0^2\right)G(t-t')=\delta(t-t').$$

Let

$$\mathscr{F}\{G(t-t')\}=\tilde{G}(\omega)\equiv\frac{1}{\sqrt{2\pi}}\int_{-\infty}^{\infty} e^{-i\omega(t-t')}G(t-t')\,d(t-t').$$

Proceeding in the same way as in Eq. (3.50), we obtain

$$\tilde{G}(\omega)=\frac{1}{\sqrt{2\pi}}\frac{1}{(\omega_0^2-\omega^2)+i2\beta\omega}.$$

The Green function

$$G(t)=\frac{1}{\sqrt{2\pi}}\int_{-\infty}^{\infty} e^{i\omega t}\tilde{G}(\omega)\,d\omega$$

can now be computed with the help of Entry 2 of the table of Fourier transforms given in Appendix 3C. The result (to be worked out in Problem 3.6.2) is

$$G(t)=\frac{1}{\omega_1}e^{-\beta t}\sin\omega_1 t\,\Theta(t), \tag{3.60}$$

where

$$\Theta(t)=\begin{cases}0 & \text{if} \quad t<0\\ 1 & \quad\quad\; t>0\end{cases}$$

is the unit *step function* and

$$\omega_1=\sqrt{\omega_0^2-\beta^2}.$$

The use of Eq. (3.60) in Eq. (3.59) gives the explicit solution

$$x(t)=\frac{1}{\omega_1}\int_{-\infty}^{t} e^{-\beta(t-t')}\sin\omega_1(t-t')\,R(t')\,dt'. \tag{3.61a}$$

We should note the interesting feature that the integral over t' takes into account the effects of all driving forces occurring in the past ($t'<t$). It contains no effect due to driving forces in the future ($t'>t$), because these forces have not yet occurred. Hence the result is explicitly consistent with the physical requirement of causality.

To check that the Green function in Eq. (3.60) is correct, we can apply it to obtain the response to the simple driving force

$$R(t') = e^{i\Omega t}.$$

Equation (3.61a) then gives (to be worked out in Problem 3.6.2)

$$x(t) = \frac{e^{-i\Omega t}}{(\omega_0^2 - \Omega^2) + i2\beta\Omega}. \tag{3.61b}$$

It can be verified by direct substitution that this is the solution of Eq. (3.58) for the given driving force.

Problems

3.6.1. Suppose the input (as a function of time t) to a system

$$R(t) = \begin{cases} 0, & t<0 \\ e^{-\alpha t}, & t>0 \end{cases}$$

gives rise to the output

$$f(t) = \begin{cases} 0, & t<0 \\ (1-e^{-\beta t})e^{-\alpha t}, & t>0, \end{cases}$$

where α and β are positive constants.

(a) Find the Fourier transform

$$\tilde{G}(\omega) = \frac{1}{\sqrt{2\pi}} \int_{-\infty}^{\infty} e^{-i\omega t} G(t)\, dt$$

of the response function $G(t)$. [$\tilde{G}(\omega)$ is also called a response function.]

(b) Obtain the response of the system to the input $R(t) = A\,\delta(t)$.

3.6.2. (a) Use Entry 2 of the table of Fourier transforms given in Appendix 3C to verify Eq. (3.60).

(b) Verify Eq. (3.61b).

3.7. Heisenberg's uncertainty principle

The Fourier transform of the Gaussian function

$$f(x) = N \exp(-\tfrac{1}{2}cx^2) \tag{3.62}$$

is of considerable interest:

$$g(k) = \frac{1}{\sqrt{2\pi}} \int_{-\infty}^{\infty} e^{-ikx} N e^{-(1/2)cx^2}\, dx.$$

It can be calculated by completing the square in the exponent

$$-\tfrac{1}{2}cx^2 - ikx = -\frac{c}{2}\left(x + \frac{ik}{c}\right)^2 - \frac{k^2}{2c} = -\tfrac{1}{2}cy^2 - k^2/2c\,,$$

so that

$$g(k) = \frac{N}{\sqrt{c\pi}}e^{-k^2/2c}\left(\int_{-\infty}^{\infty} e^{-(1/2)cy^2}\sqrt{c/2}\,dy\right) = \frac{N}{\sqrt{c}}e^{-k^2/2c}. \tag{3.63}$$

We see that a Gaussian transforms into a Gaussian but with an inverted falloff constant c^{-1} that is inversely proportional to the old falloff constant c in $f(x)$. Thus a narrow $f(x)$ gives rise to a broad $g(k)$, and vice versa. This is illustrated in Fig. 3.5.

The reciprocal width relations shown in Fig. 3.5 turn out to be a general property of Fourier transforms. It can be characterized more precisely as follows: Let us denote integrals by a "bracket" symbol called a *scalar*, or *inner*, *product*.

$$(f_1, f_2) \equiv \int_{-\infty}^{\infty} f_1^*(x) f_2(x)\,dx. \tag{3.64}$$

If g_i is the Fourier transform of $f_i(x)$, then

$$\begin{aligned}
(f_1, f_2) &= \int_{-\infty}^{\infty} f_1^*(x) f_2(x)\,dx \\
&= \int_{-\infty}^{\infty}\left(\frac{1}{\sqrt{2\pi}}\int_{-\infty}^{\infty} e^{ikx} g_1(k)\,dk\right)^*\left(\frac{1}{\sqrt{2\pi}}\int_{-\infty}^{\infty} e^{ik'x} g_2(k')\,dk'\right)dx \\
&= \iint_{-\infty}^{\infty} g_1^*(k) g_2(k')\left(\frac{1}{2\pi}\int_{-\infty}^{\infty} dx\, e^{-i(k-k')x}\right)dk\,dk' \\
&= \int_{-\infty}^{\infty} g_1^*(k) g_2(k)\,dk = (g_1, g_2).
\end{aligned} \tag{3.65a}$$

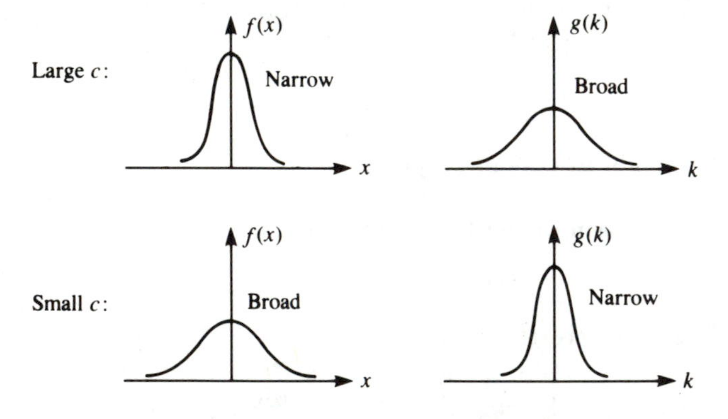

Fig. 3.5. Reciprocal width relations in Fourier transforms.

This shows that the inner product can be calculated by using either f or g. This is to be expected since g contains the same information as f; they describe the same mathematical system, which may be denoted by the symbol ϕ such that $\phi = f(x)$ or $g(k)$ depending on whether the integration variable is x or k. We call $f(x)$ or $g(k)$ the x, or k, *representation* of ϕ. In this notation, we may write Eq. (3.65a) as

$$(\phi_1, \phi_2) = (f_1, f_2) = (g_1, g_2). \tag{3.65b}$$

More general integrals can also be represented readily in this new notation. For example,

$$\begin{aligned}\int_{-\infty}^{\infty} f_1^*(x) A(x) f_2(x)\, dx &= (f_1, A(x) f_2)\\ &= (\phi_1, A(x)\phi_2),\end{aligned} \tag{3.66}$$

while

$$\begin{aligned}\int_{-\infty}^{\infty} g_1^*(k) B(k) g_2(k)\, dx &= (g_1, B(k) g_2)\\ &= (\phi_1, B(k)\phi_2).\end{aligned} \tag{3.67}$$

Here we refer explicitly to the x dependence of $A(x)$ or the k dependence of $B(k)$ to remind ourselves in which representation we know these functions.

To illustrate these expressions, let us calculate a few inner products for the Gaussian function (3.62):

$$(\phi, \phi) = (f, f) = N^2 \int_{-\infty}^{\infty} e^{-cx^2}\, dx = N^2\sqrt{\pi/c},$$

$$(\phi, x\phi) = N^2 \int_{-\infty}^{\infty} e^{-cx^2} x\, dx = 0,$$

$$(\phi, x^2\phi) = N^2 \int_{-\infty}^{\infty} e^{-cx^2} x^2\, dx = \frac{N^2}{2c}\sqrt{\pi/c}.$$

The average value of x in the system ϕ is called its *expectation value*

$$\langle x \rangle \equiv (\phi, x\phi)/(\phi, \phi) = \bar{x}, \tag{3.68}$$

while its *uncertainty* or *dispersion* Δx is defined by

$$\begin{aligned}(\Delta x)^2 &= (\phi, (x - \bar{x})^2\phi)/(\phi, \phi)\\ &= (\phi, x^2\phi)/(\phi, \phi) - \bar{x}^2.\end{aligned} \tag{3.69}$$

Hence the Gaussian function (3.62) has

$$\bar{x} = 0, \qquad (\Delta x)^2 = \frac{1}{2c}. \tag{3.70}$$

Similarly, it is easily to see that for the same Gaussian function

$$\bar{k} = 0, \qquad (\Delta k)^2 = \frac{c}{2}, \tag{3.71}$$

since the Gaussian falloff constant in $g(k)$ is c^{-1}.

Thus for the Gaussian function (3.62), the product of uncertainties has a unique value

$$\Delta x \, \Delta k = \tfrac{1}{2}, \tag{3.72}$$

independent of c.

We have thus found that the conjugate variables x and k in ϕ cannot simultaneously be known with infinite precision: If one is known better, knowledge of the other must unavoidably be reduced proportionally. A complete knowledge of one, say k, is possible only when there is complete ignorance of the other. To see in physical terms why this should be so, let us recall that, if x is the position of a wave, then k is its wave number. A wave with a unique value of k is infinitely long, for otherwise it will have a beginning and an end where the wave does not oscillate in the same manner as over its length. An infinitely long wave like this does not have a definite position, since x can be anywhere along its length. Hence the position uncertainty Δx must be infinite in order that Δk be zero.

For Gaussian functions, the uncertainty product $\Delta x \, \Delta k$ is a constant. It turns out that for other functions the uncertainty product can be larger, but never smaller. That is,

$$\Delta x \, \Delta k \geqslant \tfrac{1}{2} \tag{3.73}$$

is generally true, as we shall prove in the next section. For matter waves, $\hbar k$ is the particle momentum according to Eq. (3.4). Hence the uncertainty relation can be written in the form

$$\Delta x \, \Delta p \geqslant \hbar/2, \tag{3.74}$$

called the *Heisenberg uncertainty principle* (1927). This states that the position and the momentum of a massive particle cannot be known simultaneously with infinite precision, in dramatic contrast to the situation assumed in Newtonian mechanics. The uncertainty principle describes an important feature of quantum-mechanical systems. It has interesting epistemological implications as well, since it imposes a limit on the precision of our knowledge of physical systems.

Problems

3.7.1. Obtain the Fourier transform of the sequence of functions

$$\delta_n(x) = \frac{n}{\sqrt{\pi}} \exp(-n^2 x^2)$$

and show that $\lim(n \to \infty) \, \delta_n(x) = \delta(x)$.

3.7.2. For the Gaussian functions (3.62) and (3.63) show directly that

(a) $(\phi,k^2\phi) = -(\phi,(d/dx)^2\phi)$

(b) $(\phi,e^{ika}\phi) = (\phi,e^{a(d/dx)}\phi)$.

(Hint: Use the displacement property discussed in Section 2.7.)

3.8. Conjugate variables and operators in wave mechanics

Given mathematical systems ϕ_i that have x representations $f_i(x)$ and k representations $g_i(k)$, it is clear that the integrals $(\phi_i,A(x)\phi_j)$ involving a given function $A(x)$ is conveniently calculated by an x integration, while the integral $(\phi_i,B(k)\phi_j)$ is readily calculated by a k integration.

It is also possible to calculate $(f_i,A(x)f_j)$ in the k representation, however. To do this, we note that, according to Eq. (3.65b), an inner product is unchanged under Fourier transformation

$$(\phi_i,\phi_j) = (f_i,f_j) = (\mathscr{F}\{f_1(x)\},\mathscr{F}\{f_j(x)\}) = (g_i,g_j). \tag{3.65c}$$

Hence

$$(\phi_i,A(x)\phi_j) = (f_i,A(x)f_j) = (\mathscr{F}\{f_i(x)\},\mathscr{F}\{A(x)f_j(x)\}).$$

$\mathscr{F}\{f_i(x)\}$ is just $g_i(k)$, while $\mathscr{F}\{A(x)f_j(x)\}$ can be evaluated readily with the help of the convolution theorem

$$\begin{aligned}\mathscr{F}\{A(x)f_j(x)\} &= \frac{1}{\sqrt{2\pi}}\int_{-\infty}^{\infty} dx\, e^{-ikx}\left(\frac{1}{\sqrt{2\pi}}\int_{-\infty}^{\infty} dk'\, e^{ik'x}\tilde{A}(k')\right)\\ &\quad\times\left(\frac{1}{\sqrt{2\pi}}\int_{-\infty}^{\infty} dk''\, e^{ik''x}g_j(k'')\right)\\ &= \iint \delta(k'+k''-k)\,dk'\,dk''\frac{1}{\sqrt{2\pi}}\tilde{A}(k')g_j(k'')\\ &= \frac{1}{\sqrt{2\pi}}\int dk''\,\tilde{A}(k-k'')g_j(k'').\end{aligned}$$

According to Eq. (3.56), the final expression is a *convolution* and may be denoted $(\tilde{A}*g_j)_k$, or simply $\tilde{A}*g_j$. Thus

$$\mathscr{F}\{A(x)f_j(x)\} = \tilde{A}*g_j, \tag{3.75}$$

a result that actually can be read off directly from Eq. (3.57). Hence

$$(\phi_1,A(x)\phi_j) = (f_i,A(x)f_j) = (g_i,\tilde{A}*g_j) \tag{3.76}$$

describes how the integral can be calculated in the k representation.

There is another way of representing this change of representation that is even more suggestive and useful. According to Table 3.1,

$$\mathscr{F}\left\{\frac{d}{dx}f(x)\right\} = ik\mathscr{F}\{f(x)\} = ikg(k),$$

and

$$\mathscr{F}\left\{\left(\frac{d}{dx}\right)^n f(x)\right\} = (ik)^n g(k), \tag{3.77a}$$

when certain boundary terms vanish, as shown explicitly in Example 3.5.2. Similarly,

$$\mathscr{F}\left\{\left(\frac{d}{dk}\right)^n g(k)\right\} = (-ix)^n f(x), \tag{3.77b}$$

where we have used the conjugate transformation $x \leftrightarrow k$, $i \leftrightarrow -i$. Hence a change of integration variable can be presented symbolically as follows:

$$(f_i, A(x) f_j) = (\phi_i, A(x)\phi_j) = \left(g_i, A\left(i\frac{d}{dk}\right) g_j\right), \tag{3.78a}$$

$$(g_i, B(k) g_j) = (\phi_i, B(k)\phi_j) = \left(f_i, B\left(\frac{1}{i}\frac{d}{dx}\right) f_j\right). \tag{3.78b}$$

Indeed, this is the more flexible procedure, because it can accommodate an expression appearing between the ϕs in which both conjugate variables appear simultaneously:

$$\begin{aligned}(\phi_i, C(x,k)\phi_j) &= \left(f_i, C\left(x, \frac{1}{i}\frac{d}{dx}\right) f_j\right)\\ &= \left(g_i, C\left(i\frac{d}{dk}, k\right) g_j\right).\end{aligned} \tag{3.79}$$

When this occurs, we must remember that one variable is expressible as a differential operator of the other, so that they do not commute:

$$\begin{aligned}[k,x] = kx - xk &= \left(-i\frac{d}{dx}\right)x - x\left(-i\frac{d}{dx}\right) = -i\\ &= k\left(i\frac{d}{dk}\right) - \left(i\frac{d}{dk}\right)k = -i.\end{aligned} \tag{3.80}$$

Thus when both conjugate variables x and k appear in C, they are no longer simple variables but noncommuting *operators* whose orderings in the expression must be carefully preserved.

We are now in a position to prove Heisenberg's uncertainty principle, which states that

$$(\Delta x)(\Delta k) \geqslant \tfrac{1}{2},$$

where Δu is the uncertainty of the operator u as defined by Eq. (3.69).

This inequality turns out to be related to the so-called *Schwarz inequality* in vector algebra, which states that for two real vectors $\mathbf{A}$ and $\mathbf{B}$

$$A^2 B^2 \geqslant (\mathbf{A}\cdot\mathbf{B})^2 = (AB\cos\theta)^2, \tag{3.81}$$

where $A^2 = \mathbf{A} \cdot \mathbf{A}$. If we now take $\mathbf{A} = (x - \bar{x})\phi$, $\mathbf{B} = (k - \bar{k})\phi$, then

$$A^2 = ((x - \bar{x})\phi, (x - \bar{x})\phi) = (\phi, (x - \bar{x})^2\phi) = (\Delta x)^2, \tag{3.82}$$

where we have used the fact that the expression appearing to the left of the comma is transposed.

Therefore,

$$\begin{aligned} A^2B^2 = (\Delta x)^2(\Delta k)^2 &\geqslant (\mathbf{A} \cdot \mathbf{B})^2 \\ &= |(\phi, (x - \bar{x})(k - \bar{k})\phi)|^2 \\ &= |\langle (x - \bar{x})(k - \bar{k}) \rangle|^2 \\ &= |\langle \alpha\beta \rangle|^2, \end{aligned}$$

where (Problem 3.8.1)

$$\alpha = x - \bar{x}, \qquad \beta = k - \bar{k} \tag{3.83}$$

are Hermitian operators. If we work in terms of symmetric and antisymmetric products, we can express the last expression as

$$\begin{aligned} |\langle \alpha\beta \rangle|^2 &= |\tfrac{1}{2}\langle \alpha\beta - \beta\alpha \rangle + \tfrac{1}{2}\langle \alpha\beta + \beta\alpha \rangle|^2 \\ &= \tfrac{1}{4}\,|\langle \alpha\beta - \beta\alpha \rangle|^2 + \tfrac{1}{4}\,|\langle \alpha\beta + \beta\alpha \rangle|^2 + \tfrac{1}{2}\,\mathrm{Re}\langle \alpha\beta - \beta\alpha \rangle \langle \alpha\beta + \beta\alpha \rangle. \end{aligned}$$

The third term on the right vanishes because $\langle \alpha\beta - \beta\alpha \rangle \langle \alpha\beta + \beta\alpha \rangle$ is purely imaginary. (This is because $\langle \alpha\beta + \beta\alpha \rangle$ is real and $\langle \alpha\beta - \beta\alpha \rangle$ is purely imaginary. See Problem 3.8.1.) The second term $\frac{1}{4}\,|\langle \alpha\beta + \beta\alpha \rangle|^2$ is non-negative. Hence

$$(\Delta x)^2(\Delta k)^2 \geqslant |\langle \alpha\beta \rangle|^2 \geqslant \tfrac{1}{4}\,|\langle \alpha\beta - \beta\alpha \rangle|^2. \tag{3.84}$$

The remaining term involves the commutator

$$[\alpha, \beta] = [x - \bar{x}, k - \bar{k}] = [x, k] = i,$$

where use has been made of Eq. (3.80). Hence $|\langle \alpha\beta - \beta\alpha \rangle|^2 = 1$ and

$$(\Delta x)(\Delta k) \geqslant \tfrac{1}{2}.$$

Problems

3.8.1. An operator H is said to be Hermitian if

$$H_{ij} \equiv (\phi_i, H\phi_j) = H_{ji}^* = (\phi_j, H\phi_i)^*.$$

(a) Show that the position operator x is Hermitian if it has only real values in the x representation.

(b) Show that if $x = i\dfrac{d}{dk}$ is a Hermitian operator in the k representation, that is

$$\left\{\int_{-\infty}^{\infty} g_j^*(k)\left[i\frac{d}{dk}g_i(k)\right]dk\right\}^*$$

$$=\left\{-i\int_{-\infty}^{\infty}\left[\frac{d}{dk}g_j^*(k)\right]g_i(k)\,dk+\text{boundary terms}\right\}^*$$

$$=\int_{-\infty}^{\infty} g_i^*(k)\left[i\frac{d}{dk}g_j(k)\right]dk,$$

only if $g_j(k)$ and $g_i(k)$ satisfy suitable boundary conditions at $k=\pm\infty$.

3.8.2. If x,k are Hermitian operators and $\alpha=x-\bar{x}$, $\beta=k-\bar{k}$, show that $\langle\alpha\beta+\beta\alpha\rangle$ is real, while $\langle\alpha\beta-\beta\alpha\rangle$ is purely imaginary.

3.8.3. Show that an eigenvector of a matrix is simultaneously an eigenvector of all other matrices that commute with it.

3.9. Generalized Fourier series and Legendre polynomials

The developments of the preceding sections suggest that there is considerable similarity between the expansion of functions by Fourier series and transforms and the decomposition of simple vectors into components. It is now useful to develop this similarity in more detail in order to gain additional understanding into the nature of the Fourier-series expansion, or equivalently into the mathematics of wave motion.

Let us start by reminding ourselves that a vector $\mathbf{A}$ in three-dimensional space has three Cartesian components

$$\mathbf{A}=A_1\mathbf{e}_1+A_2\mathbf{e}_2+A_3\mathbf{e}_3=\sum_i A_i\mathbf{e}_i. \tag{3.85}$$

Given $\mathbf{A}$, its component A_i along $\mathbf{e}_i$ can be calculated by means of the scalar product

$$\mathbf{e}_i\cdot\mathbf{A}=\mathbf{e}_i\cdot(A_1\mathbf{e}_1+A_2\mathbf{e}_2+A_3\mathbf{e}_3)=A_i. \tag{3.86}$$

The Fourier series in Eq. (3.27) [or Eq. (3.21)] has a similar structure. The Fourier series $f(x)$, now called $F(x)$ if we believe it to be as good as the original function $F(x)$,

$$F(x)=\sum_{n=-\infty}^{\infty} c_n\psi_n(x) \tag{3.87}$$

is also a sum of terms, each of which is made up of a Fourier coefficient c_n (the analog of a vector component) and a unique function

$$\psi_n(x)=\exp\left(i\frac{n\pi}{L}x\right) \tag{3.88}$$

which plays the role of $\mathbf{e}_i$. The functions $\psi_n(x)$ satisfy the integral relation

$$\int_{-L}^{L} \psi_m^*(x)\psi_n(x)\,dx = \int_{-L}^{L} \exp\left(i\frac{\pi x}{L}(m-n)\right) dx$$

$$= \begin{cases} 2L, & m = n \\ \dfrac{2L}{(m-n)\pi}\sin(m-n)\pi, & m \neq n \end{cases}$$

$$= 2L\delta_{mn}. \tag{3.89}$$

If we left multiply Eq. (3.87) by $\psi_m^*(x)$ and integrate over x, we get

$$\int_{-L}^{L} \psi_m^*(x)F(x)\,dx = \sum_n c_n \int_{-L}^{L} \psi_m^*(x)\psi_n(x)\,dx$$

$$= \sum_n c_n 2L\delta_{mn} = 2Lc_m. \tag{3.90}$$

In this way, the formula (3.28) for Fourier coefficients can be derived directly.

In the language of vector algebra, the extraction of a Fourier coefficient has involved a scalar product, like the situation in Eq. (3.86). Hence the integrals over x in both Eq. (3.89) and Eq. (3.90) may also be called scalar or inner products.

Scalar products of vectors describe their orthonormalization properties. In a similar way, inner products of functions describe the orthonormalization of functions. In particular, Eq. (3.89) states that the functions $\psi_m(x)$ are orthogonal to each other and that they can be normalized into the *unit functions*

$$e_n(x) = \frac{1}{\sqrt{2L}}\exp(in\pi x/L) \tag{3.91}$$

satisfying the orthonormality relations

$$(e_m, e_n) \equiv \int_{-L}^{L} e_m^*(x)e_n(x)\,dx = \delta_{mn}, \tag{3.92}$$

where we have used the inner-product notation of Section 3.7. If we now write the Fourier series in the vectorial form

$$F(x) = \sum_{n=-\infty}^{\infty} F_n e_n(x), \tag{3.93}$$

reminiscent of Eq. (3.85), the components F_n of $F(x)$ must necessarily be given by the inner product

$$F_n = (e_n, F) \equiv \int_{-L}^{L} e_n^*(x)F(x)\,dx. \tag{3.94}$$

Even the Dirac δ function (3.29) has the simple form

$$\delta(x-x') = \sum_{n=-\infty}^{\infty} e_n(x)e_n^*(x') \tag{3.95}$$

in this notation.

The number of distinct unit functions $e_n(x)$ appearing in these expressions is infinite. Hence the space involved is said to be *infinite dimensional.* A particularly important feature of this space is that inner products are defined, so that the concepts of length and orthogonality become meaningful. In recognition of this fact, the space is called an *inner-product space.*

It is well known that any choice of three perpendicular axes can be used in the expansion (3.85) of vectors in space. In the case of functions, a different choice of coordinate axes means the use of different unit functions $e_n(x)$.

To illustrate this idea, let us consider the complex Fourier series (3.87) for the interval $[-1,1]$, that is, for $L=1$. The original functions in Eq. (3.88) can be expanded into a convergent infinite series in powers of x:

$$\psi_n(x) = \exp(in\pi x) = 1 + in\pi x + \frac{1}{2!}(in\pi x)^2 + \cdots .$$

Since powers of x are even simpler than exponential functions, we ask if we cannot use these powers directly in the expansion instead of $\psi_n(x)$.

To study this question, let us take the first two basis functions to be

$$P_0(x) = 1, \qquad P_1(x) = x. \tag{3.96a}$$

These satisfy the orthogonality relations

$$(P_0,P_0) = \int_{-1}^{1} dx = 2, \tag{3.97a}$$

$$(P_0,P_1) = (P_1,P_0) = \int_{-1}^{1} x\,dx = 0, \tag{3.97b}$$

$$(P_1,P_1) = \int_{-1}^{1} x^2\,dx = \tfrac{2}{3}. \tag{3.97c}$$

The next power x^2 is orthogonal to $P_1(x)$

$$(P_1,x^2) = (x^2,P_1) = \int_{-1}^{1} x^3\,dx = 0,$$

but not to $P_0(x)$

$$(P_0,x^2) = (x^2,P_0) = \int_{-1}^{1} x^2\,dx = \tfrac{2}{3}. \tag{3.98}$$

Hence it cannot be one of a set of mutually orthogonal functions that can be used for expanding functions in what might be called generalized Fourier series.

If x^2 is not orthogonal to P_0, part of it must be parallel to it. Indeed, Eq. (3.97a) shows that the part parallel to P_0 in x^2 must be $\frac{1}{3}P_0$, since it gives the same inner product $\frac{2}{3}$ with P_0, according to Eqs. (3.97a) and (3.98). In other words, $x^2 - \frac{1}{3}P_0$ must be orthogonal to P_0. This can be checked explicitly

$$(P_0, x^2 - \tfrac{1}{3}P_0) = (P_0, x^2) - \tfrac{1}{3}(P_0, P_0) = \tfrac{2}{3} - \tfrac{1}{3}(2) = 0.$$

It has become customary to use not $x^2 - \frac{1}{3}P_0$ but the function

$$P_2(x) = \tfrac{3}{2}(x^2 - \tfrac{1}{3}) = (3x^2 - 1)/2 \tag{3.96b}$$

proportional to it chosen such that $P_2(x = 1) = 1$.

The next power is x^3, which is orthogonal to all even powers of x, but not to the odd powers. The combination

$$P_3(x) = (5x^3 - 3x)/2 \tag{3.96c}$$

can easily be shown to be orthogonal to $P_1(x)$ and also satisfies the (arbitrary) convention $P_3(x = 1) = 1$. Similarly,

$$P_4(x) = (35x^4 - 30x^2 + 3)/8, \tag{3.96d}$$

$$P_5(x) = (63x^5 - 70x^3 + 15x)/8 \tag{3.96e}$$

are orthogonal to polynomials of degree $m < 4$ and 5, respectively.

Proceeding in this manner, we can construct the polynomials $P_n(x)$, $n = 0, 1, 2, \ldots, \infty$, such that $P_n(x = 1) = 1$ and $P_n(x)$ is orthogonal to all $P_m(x)$ with $m < n$. These polynomials are called *Legendre polynomials.* They can be shown to satisfy the orthogonality relation

$$\int_{-1}^{1} P_m(x)P_n(x)\,dx = \frac{2}{2n+1}\,\delta_{mn}. \tag{3.99}$$

a result we shall derive in Chapter 5.

Given a function $F(x)$ defined in $[-1,1]$, we can expand it in the *Legendre series*

$$F(x) = \sum_{n=0}^{\infty} c_n P_n(x), \tag{3.100}$$

where the *Legendre coefficient* c_n can be obtained by multiplying Eq. (3.100) by $P_m(x)$ and then integrating over x:

$$\begin{aligned}\int_{-1}^{1} P_m(x)F(x)\,dx &= \sum_n c_n \int_{-1}^{1} P_m(x)P_n(x)\,dx \\ &= c_m \frac{2}{2m+1}.\end{aligned}$$

That is,

$$c_n = \frac{2n+1}{2}\int_{-1}^{1} P_n(x)F(x)\,dx. \tag{3.101}$$

The Legendre polynomials are sketched in Fig. 3.6. The expansion of functions in Legendre series is illustrated by two examples.

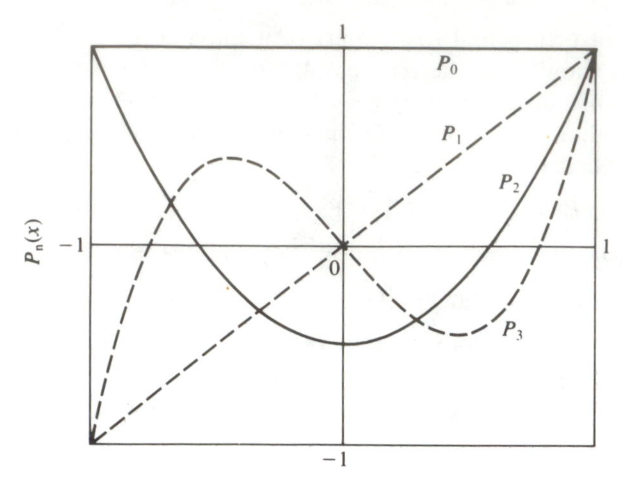

Fig. 3.6. $P_n(x)$ for $n = 0$, 1, 2, and 3.

Example 3.9.1. Expand the function

$$G(x) = \begin{cases} 1, & 0 \leqslant x \leqslant 1 \\ -1, & -1 \leqslant x \leqslant 0, \end{cases} \tag{3.102}$$

in a Legendre series.

Since $G(x)$ is odd in x, the Legendre coefficients of even n vanish. The odd-n coefficients are

$$c_1 = \frac{3}{2}\int_{-1}^{1} G(x)x\,dx = 3\int_0^1 x\,dx = \tfrac{3}{2},$$

$$c_3 = \frac{7}{2}\int_{-1}^{1} G(x)\frac{5x^3 - 3x}{2}\,dx$$

$$= \frac{7}{2}\int_0^1 (5x^2 - 3x)\,dx = -\tfrac{7}{8},$$

$$c_5 = \frac{11}{2}\int_0^1 \frac{63x^5 - 70x^3 + 15x}{4}\,dx = \tfrac{11}{16}, \qquad \text{etc.}$$

Hence

$$G(x) = \tfrac{3}{2}P_1(x) - \tfrac{7}{8}P_3(x) + \tfrac{11}{16}P_5(x) - \cdots. \tag{3.103}$$

The first three partial sums of this Legendre series are shown in Fig. 3.7. They should be compared with those for the Fourier expansion of the same function given in Example 3.3.2.

Example 3.9.2. Expand $F(x) = \cos \pi x$ in $[-1,1]$ in a Legendre series.

Since $F(x)$ is even, only the coefficients with even n can be nonzero.

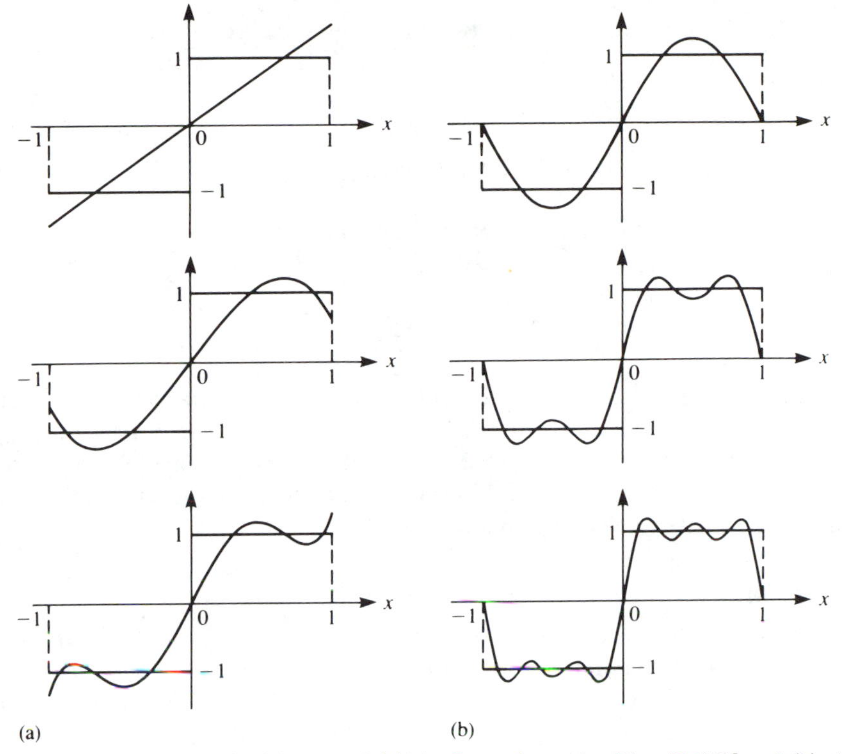

Fig. 3.7. The first few partial sums of (a) the Legendre series [Eq. (3.103)] and (b) the Fourier series (Example 3.3.2) for the same function (3.102).

To calculate these we need the integrals

$$\int_{-1}^{1} x^2 \cos \pi x \, dx = 2\left(-\frac{d^2}{dk^2}\right) \int_0^1 \cos kx \, dx \bigg|_{k=\pi}$$

$$= -2 \frac{d^2}{dk^2}\left(\frac{\sin k}{k}\right)\bigg|_{k=\pi}$$

$$= -\frac{4}{\pi^2},$$

$$\int_{-1}^{1} x^4 \cos \pi x \, dx = 2 \frac{d^4}{dk^4}\left(\frac{\sin k}{k}\right)\bigg|_{k=\pi} = \frac{8}{\pi^2}\left(1 - \frac{6}{\pi^2}\right), \qquad \text{etc.}$$

Hence the Legendre coefficients are

$$c_0 = \frac{1}{2}\int_{-1}^{1} \cos \pi x \, dx = 0,$$

$$c_2 = \frac{5}{2}\int_{-1}^{1} \tfrac{3}{2}(x^2 - \tfrac{1}{3}) \cos \pi x \, dx = \frac{15}{4}\int_{-1}^{1} x^2 \cos \pi x \, dx$$
$$= -\frac{15}{\pi^2},$$
$$c_4 = \frac{9}{2}\frac{1}{8}\int_{-1}^{1} (35x^4 - 30x^2 - 3) \cos \pi x \, dx$$
$$= \frac{9}{16}\left[35\frac{8}{\pi^2}\left(1 - \frac{6}{\pi^2}\right) - 30\left(-\frac{15}{\pi^2}\right)\right]$$
$$= \frac{9}{16}\left(\frac{730}{\pi^2} - \frac{1680}{\pi^4}\right), \qquad \text{etc.}$$

If there are different sets of functions in terms of which a given function in a given interval can be expanded, it is often necessary to decide which set is the best to use. The choice is not always obvious, and may depend on the structure of a problem as well as the efficacy of the expansion for the specific problem.

Problems

3.9.1. Show that the Legendre polynomials $P_3(x)$ and $P_4(x)$ are orthogonal to polynomials of lower degrees.

3.9.2. Obtain the first three coefficients of the Legendre expansions of the following functions in the interval $[-1,1]$:
(a) e^{ikx};
(b) $e^{-\gamma x}$;
(c) $x^{-1/n}$, $\quad n > 1$.

3.9.3. Give the orthonormal version [of the form of Eq. (3.93)] of the Legendre expansion. Obtain an explicit expression for the Dirac δ function for the Legendre expansion.

3.10. Orthogonal functions and orthogonal polynomials

Other functions of interest to mathematical physics can be introduced in a rather natural way by generalizing the discussion of the previous section on Legendre polynomials.

To do this, let us consider the expansion of a function $F(x)$ in an interval $[a,b]$ in terms of a suitable set of functions ψ_n in $[a,b]$:

$$F(x) = \sum_n c_n \psi_n(x),$$

as we have done for the Fourier series (3.21) or (3.27). The calculation of the expansion coefficient c_n becomes simple if these $\psi_n(x)$ are *orthogonal functions* satisfying an orthogonality relation of the general form

$$\int_a^b \psi_m^*(x)\psi_n(x)w(x)\, dx = h_n\delta_{mn}. \tag{3.104}$$

It can then be extracted by a simple integration:

$$c_n = \frac{1}{h_n}\int_a^b \psi_n^*(x)F(x)w(x)\,dx. \tag{3.105}$$

Equation (3.104) differs from the orthogonality relations used earlier in this chapter in the appearance of an additional *weight function* $w(x)$.

The separation of $w(x)$ from $\psi_n(x)$ is just a matter of convenience, but it might also be desirable on physical grounds. For example, it might give the load on a loaded string (related to the mass matrix M of Sections 2.6 and 2.7) which we might want to distinguish from the shape of its vibration. In particular, Eq. (3.104) has the same structure as the corresponding expression [Eq. (2.80)] for matrices.

The simplest functions one can use for $\psi_n(x)$ are the powers of x. However, the integral

$$\int_a^b x^m x^n\,dx = \frac{1}{m+n+1}(b^{m+n+1} - a^{m+n+1}) \tag{3.106}$$

are not orthogonal; they may not even be finite if $|a|$ or $|b|$ becomes infinite. The weight function $w(x)$ now comes to the rescue: It can be chosen to fall off sufficiently rapidly with increasing $|x|$ so as to give finite integrals.

The chosen powers must next be orthogonalized. One systematic way of doing this is called a *Schmidt orthogonalization,* which we have already used in the previous section. To formalize this procedure, let us note that, given a set of n linearly independent but nonorthogonal vectors $\boldsymbol{\phi}_i$ in an n-dimensional linear vector space, we can construct from them a set of orthogonal vectors $\boldsymbol{\psi}_i$ in the following way: We take

$$\boldsymbol{\psi}_1 = \boldsymbol{\phi}_1 = \psi_1\mathbf{e}_1, \qquad \mathbf{e}_1 \equiv \mathbf{e}(\boldsymbol{\psi}_1). \tag{3.107}$$

For $\boldsymbol{\psi}_2$ we can take that part of $\boldsymbol{\phi}_2$ that is perpendicular to $\boldsymbol{\psi}_1$ or $\mathbf{e}_1$. That is,

$$\boldsymbol{\psi}_2 = \boldsymbol{\phi}_2 - \mathbf{e}_1(\mathbf{e}_1 \cdot \boldsymbol{\phi}_2), \tag{3.108}$$

since $\mathbf{e}_1 \cdot \boldsymbol{\phi}_2$ is the component along $\mathbf{e}_1$. The operator

$$\mathbf{e}_1\mathbf{e}_1\,\cdot = \mathscr{P}_1 = \cdot\,\mathbf{e}_1\mathbf{e}_1 \tag{3.109}$$

is called the *projection operator* on $\mathbf{e}_1$. It can be used to simplify Eq. (3.108) into

$$\boldsymbol{\psi}_2 = (1 - \mathscr{P}_1)\boldsymbol{\phi}_2 = \psi_2\mathbf{e}_2. \tag{3.110}$$

We now define $\boldsymbol{\psi}_3$ to be that part of $\boldsymbol{\phi}_3$ perpendicular to both $\boldsymbol{\psi}_1$ and $\boldsymbol{\psi}_2$:

$$\boldsymbol{\psi}_3 = (1 - \mathscr{P}_1 - \mathscr{P}_2)\boldsymbol{\phi}_3 = \psi_3\mathbf{e}_3, \tag{3.111}$$

and more generally

$$\boldsymbol{\psi}_j = \left(1 - \sum_{i=1}^{k-1} \mathscr{P}_i\right)\boldsymbol{\phi}_j = \psi_j\mathbf{e}_j. \tag{3.112}$$

The linear independence of the original ϕ_i ensures that none of the ψ_k will be zero.

This is exactly what we have done in the previous section in constructing the Legendre polynomials from the powers x^n, $n = 0, 1, 2, \ldots, \infty$. We shall further illustrate the procedure by constructing orthogonal polynomials in the interval $[a,b] = [-\infty,\infty]$.

For this infinite interval, the integrals (3.106) are infinite. Hence a weight function $w(x)$ is needed to ensure convergence. The interval is symmetric about $x = 0$; hence we look for an even function of x. One possibility is the Gaussian function $\exp(-x^2)$ for which

$$\int_{-\infty}^{\infty} e^{-x^2}\, dx = f(1),$$

where

$$f(a) = \int_{-\infty}^{\infty} e^{-ax^2}\, dx = \sqrt{\pi/a},$$

while

$$\int_{-\infty}^{\infty} x^m e^{-x^2}\, dx = 0, \qquad \text{if } m \text{ is odd,} \tag{3.113a}$$

and

$$\int_{-\infty}^{\infty} x^{2m} e^{-x^2}\, dx = \left(-\frac{d}{da}\right)^m f(a)\bigg|_{a=1} \tag{3.113b}$$

are all finite.

We begin by noting that even and odd powers are orthogonal to each other by virtue of Eq. (3.113a). That is,

$$(x^m, x^n) = \int_{-\infty}^{\infty} e^{-x^2} x^{m+n}\, dx = 0, \qquad \text{if} \qquad m \neq n.$$

Hence we may use for the first two polynomials

$$H_0(x) = 1, \qquad \text{and} \qquad H_1(x) = 2x. \tag{3.114}$$

Here we have used the convention that the term in $H_n(x)$ with the highest power of x is $(2x)^n$. The functions (3.114) may next be normalized by computing their lengths

$$(H_0, H_0) = \int_{-\infty}^{\infty} e^{-x^2}\, dx = \sqrt{\pi},$$

$$(H_1, H_1) = 4\int_{-\infty}^{\infty} e^{-x^2} x^2\, dx = 4\left(-\frac{d}{da} f(a)\right)_{a=1} = 2\sqrt{\pi}.$$

Hence the orthonormal polynomials are

$$e_0(x) = \pi^{-1/4}, \qquad e_1(x) = 2x(4\pi)^{-1/4}.$$

The next orthogonal polynomial $H_2(x)$ is just that part of $(2x)^2$

orthogonal to $e_0(x)$:

$$\begin{aligned} H_2(x) &= (1-\mathscr{P}_0)4x^2 \\ &= 4x^2 - e_0(x)(e_0, 4x^2) \\ &= 4x^2 - \pi^{-1/4}\left(\pi^{-1/4}\int_{-\infty}^{\infty} e^{-x^2}4x^2\,dx\right) \\ &= 4x^2 - 2. \end{aligned} \tag{3.115}$$

Similarly, $H_3(x)$ is just that part of $(2x)^3$ orthogonal to $e_1(x)$:

$$\begin{aligned} H_3(x) &= (1-\mathscr{P}_1)8x^3 \\ &= 8x^3 - e_1(x)(e_1, 8x^3) \\ &= 8x^3 - \frac{2x}{(4\pi)^{1/4}}\int_{-\infty}^{\infty} e^{-x^2}\frac{2x}{(4\pi)^{1/4}}8x^3\,dx. \end{aligned}$$

Since the integral involved is [from Eq. (3.113b)]

$$\int_{-\infty}^{\infty} e^{-x^2}x^4\,dx = \left[\left(-\frac{d}{da}\right)^2\sqrt{\pi/a}\right]_{a=1} = \tfrac{3}{4}\sqrt{\pi},$$

we get

$$H_3(x) = 8x^3 - 12x. \tag{3.116}$$

This can be continued until we get the general formula

$$\begin{aligned} H_n(x) &= \left(1 - \sum_{i=0}^{n-1}\mathscr{P}_i\right)(2x)^n \\ &= (2x)^n - \sum_{i=0}^{n-1} e_i(x)(e_i, 2^n x^n). \end{aligned} \tag{3.117}$$

These orthogonalized polynomials are called *Hermite polynomials*.

A number of such orthogonal polynomial systems appear frequently in physics and engineering. They may be specified in terms of the interval $[a,b]$, the weight function $w(x)$, and the convention commonly agreed on for the normalization constant h_n in Eq. (3.104). A few common examples are defined in Table 3.2. The first few polynomials themselves are given in Table 3.3.

Table 3.2. Orthogonality relations of certain orthogonal polynomials

$\psi_n(x)$	Name	a	b	$w(x)$	h_n
$P_n(z)$	Legendre	-1	1	1	$2/(2n+1)$
$L_n(x)$	Laguerre	0	∞	e^{-x}	1
$H_n(x)$	Hermite	$-\infty$	∞	e^{-x^2}	$\sqrt{\pi}\,2^n n!$
$T_n(x)$	Chebyshev of the first kind	-1	1	$(1-x^2)^{-1/2}$	$\begin{cases}\pi/2, & n\neq 0\\ \pi, & n=0\end{cases}$

Table 3.3. Special cases of certain orthogonal polynomials

$\psi_n(x)$	$\psi_0(x)$	$\psi_1(x)$	$\psi_2(x)$	$\psi_3(x)$
$P_n(x)$	1	x	$(3x^2-1)/2$	$(5x^3-3x)/2$
$L_n(x)$	1	$-x+1$	$(x^2-4x+2)/2!$	$(-x^3+9x^2-18x+6)/3!$
$H_n(x)$	1	$2x$	$4x^2-2$	$8x^3-12x$
$T_n(x)$	1	x	$2x^2-1$	$4x^3-3x$

We note that the *Chebyshev polynomials* $T_n(x)$ are defined over the same interval as the Legendre polynomials, but involve a different weight function, namely $(1-x^2)^{-1/2}$ rather than 1. This awkward-looking weight function becomes easily recognizable when $x = \cos\theta$, for then

$$\frac{dx}{(1-x^2)^{1/2}} = -d\theta.$$

When the weight function is changed, the meaning of orthogonality changes, too. The polynomials

$$P_0(x) = T_0(x) = 1$$

and

$$P_1(x) = T_1(x) = x$$

are naturally orthogonal because of their different parities. However, the part $[\frac{2}{3}P_2(x)]$ of x^2 orthogonal to $P_0(x) = 1$ is not the same as the part $[\frac{1}{2}T_2(x)]$ of the same x^2 orthogonal to the same $T_0(x) = 1$. Here everything is the same except for the meaning of orthogonality, that is, the choice of the weight function.

One family of orthogonal polynomials defined over the same interval with a different weight function might be better than another in the expansion of a certain class of functions. Indeed, in the expansion of functions, the weight function itself can often be chosen to improve the quality of the expansion.

Problems

3.10.1. Use Schmidt orthogonalization to obtain the four polynomials shown in Table 3.3 for the Laguerre and Chebyshev polynomials.

3.11. Mean-square error and mean-square convergence

Now that we have seen a variety of functions expanded in terms of different sets of orthogonal functions, it is time to ask if these expansions are good representations of the original functions.

To be more precise, let a function $F(x)$ in the interval $[a,b]$ be expanded in terms of a set of orthonormal functions $e_n(x)$, $n =$

$-\infty, \ldots, \infty$, for which

$$\int_a^b e_m^*(x) e_n(x)\, dx = \delta_{mn}. \tag{3.118}$$

In terms of the functions of Eq. (3.104), these orthonormal functions are

$$e_n(x) = [w(x)/h_n]^{1/2} \psi_n(x).$$

Suppose the generalized Fourier coefficients

$$f_n = \int_a^b e_n^*(x) F(x)\, dx \tag{3.119}$$

exist and the generalized Fourier series

$$f(x) = \sum_{n=-\infty}^{\infty} f_n e_n(x) \tag{3.120}$$

is uniformly convergent. Is $f(x)$ exactly identical to the original function $F(x)$?

The answer is "not necessarily so." The reason is that it is very difficult to reproduce a function exactly over the uncountably infinite number of points in the interval $[a,b]$ when there are only a countably infinite number of expansion coefficients to adjust. However, it is possible to analyze the situation a little further in an illuminating way.

To begin, we must quantify the discrepancy between an infinite series, say

$$g(x) = \sum_n g_n e_n(x) \tag{3.121}$$

and the given function. One simple (but not unique) way of doing this is through the *mean square deviation* or *error*

$$D \equiv \int_a^b |F(x) - g(x)|^2\, dx, \tag{3.122}$$

which has the virtue of being a single number, thus giving an average measure of the discrepancy.

The integrand of D is non-negative; hence D itself is non-negative. It is obvious that there is no upper limit to how large this error can be. The interesting question is how small one can reduce the error. This question is answered by the following theorem: The Fourier series $f(x)$ of Eq. (3.120) gives the best infinite-series representation of $F(x)$ in the sense that the error D is minimized when $g(x) = f(x)$. Any other choice of $g(x)$ will result in a larger error.

This theorem can be easily proved in the following way. For simplicity of notation, let us assume that $F(x)$ and $e_n(x)$ are real. Then using Eq.

(3.121), we have

$$\begin{aligned}
D &= \int_a^b [F^2(x) - 2g(x)F(x) + g^2(x)]\,dx \\
&= \int_a^b F^2(x)\,dx - 2\sum_{n=-\infty}^{\infty} g_n \int_a^b e_n(x)F(x)\,dx \\
&\quad + \sum_{n,n'} g_n g_{n'} \int_a^b e_n(x)e_{n'}(x)\,dx \\
&= \int_a^b F^2(x)\,dx - 2\sum_n g_n f_n + \sum_n g_n^2.
\end{aligned}$$

This shows that for any n

$$\frac{\partial D}{\partial g_n} = -2f_n + 2g_n. \tag{3.123}$$

Hence the choice $g_n = f_n$ gives (3.124)

$$\frac{\partial D}{\partial g_n} = 0, \qquad \frac{\partial^2 D}{\partial g_n^2} = 2 > 0, \tag{3.125}$$

so that the Fourier series (3.120) minimizes the mean-square error D.

The minimum error is

$$\begin{aligned}
D_{\min} &= \int_a^b |F(x) - f(x)|^2\,dx \\
&= \int_a^b F^2(x)\,dx - \sum_n f_n^2 \geqslant 0.
\end{aligned} \tag{3.126}$$

This is non-negative because the integrand $|F(x) - f(x)|^2$ is non-negative. Thus we have derived the *Bessel inequality*

$$\int_a^b F^2(x)\,dx \geqslant \sum_{n=-\infty}^{\infty} f_n^2. \tag{3.127}$$

3.11.1. Mean-square convergence

The Fourier series $f(x)$ of the function $F(x)$ is said to converge *in the mean* to $F(x)$ if $D_{\min} = 0$.

When this happens, the Bessel inequality becomes the *Parseval equation*

$$\int_a^b F^2(x)\,dx = \sum_{n=-\infty}^{\infty} f_n^2. \tag{3.128}$$

The set of functions $\{e_n(x)\}$ is said to be *complete* with respect to all functions $F(X)$ satisfying Parseval's equation. Parseval's equation itself is often called a *completeness relation*.

What type of functions satisfy Parseval's equation for the trigonometric functions of the Fourier series (3.21)? This question is answered by *Parseval's Theorem* (1893):

> The set of functions $\{e_n(x),\ n=-\infty \text{ to } \infty\}$ is complete with respect to piecewise continuous functions in the interval $[-\pi,\pi]$.

A function is piecewise continuous if it is continuous in the interval except at a finite number of points. For a proof of this theorem, see, for example, Carslaw, *An Introduction to the Theory of Fourier Series and Integrals* (Dover, New York, 1950), p. 284.

The following example illustrates the usefulness of Parseval's equation, which applies also to inner products of two different functions:

Example 3.11.1. Obtain the result $\pi^2 = 8\sum_{\text{odd}}^{\infty} 1/n^2$ with the help of the following two Fourier series in $[-\pi,\pi]$

$$F(x) = x \simeq f(x) = \sum_{n=1}^{\infty} \frac{2}{n}(-1)^{n+1} \sin nx, \tag{3.129}$$

$$G(x) = x/|x| \simeq g(x) = \sum_{\text{odd } n} \frac{4}{n\pi} \sin nx. \tag{3.130}$$

(These Fourier series have been obtained in Examples 3.3.1 and 3.3.2.)

According to Parseval's equation (3.128)

$$\begin{aligned} (F,G) &= 2\int_0^\pi x\,dx = \pi^2 \\ &= \sum_{m=1}^{\infty} \sum_{\text{odd } n} 2\frac{2}{m}(-1)^{m+1} \frac{4}{n\pi} \int_0^\pi \sin mx \sin nx\,dx \\ &= 8 \sum_{\text{odd } n} \frac{1}{n^2}. \end{aligned}$$

In technical terms, we have been considering an infinite-dimensional inner product space. If this space is complete (roughly in the sense of the Parseval equation), it is called a *Hilbert space*. Sometimes it is also required that the dimension is a countable infinity, as in the case in Eq. (3.120). The space is then said to be *separable*.

To further illustrate the idea of completeness for the interested reader, we note that the space of all continuous functions $F(x)$ in the interval $[-\pi,\pi]$ is not considered to be complete because there are smooth functions such as

$$F_n(x) = \begin{cases} 0, & x < 0 \\ x^{1/n}, & x > 0, \end{cases} \tag{3.131}$$

which becomes discontinuous in the limit $n \to \infty$. Thus discontinuous functions must also be included in order to have a Hilbert space. The space of wave functions of the simple wave equation (2.83) contains discontinuous functions, and is a Hilbert space.

Finally we note that the converse of Parseval's theorem is also true:

> A series of the form (3.120) for which the sum $\sum_n f_n^2$ converges is the Fourier series of a function that is square integrable.

This result is known as the *Fischer–Riesz theorem.*

Problems

3.11.1. Use Parseval's equations and the Fourier series of Appendix 3A or 3B to evaluate the following infinite sums:

(a) $\displaystyle\sum_{n=1}^{\infty} \frac{1}{n^4}$,

(b) $\displaystyle\sum_{n=1}^{\infty} \frac{(-1)^{n+1}}{n^6}$.

3.11.2. Use Parseval's equation and the Fourier series of Appendix 3A and 3B to evaluate the integral

$$\int_0^{\pi} \left[\ln\left(2\cos\frac{x}{2}\right)\right]^2 dx.$$

3.12. Convergence of Fourier series

Fourier's claim for his series representation of functions has given rise to many important questions. Their study has led to a deeper understanding of the nature of functions. However, our study of Fourier series has been motivated by our interest in quantum-mechanical applications, rather than by a desire to understand its subtle structure. To compensate for this neglect, we offer in this section a few qualitative remarks of a more mathematical nature.

A basic question is whether a Fourier series like Eq. (3.21) converges at all. This convergence problem does not appear to have been solved. Certain properties are known to be necessary but not sufficient; other properties are known to be sufficient but not necessary. The property that is both necessary and sufficient is unknown.

In this connection, it is useful to recall that a sufficient condition describes a special example of a general property, while a necessary condition describes a general property characteristic of this example, and possibly of other examples as well. The statement "if p, then q" or "p implies q" can be read as "p is a *sufficient condition* for q," or as "q is a *necessary condition* for p." For example, "chicken implies bird" means "being a chicken is a sufficient condition for being a bird," or "a chicken is an example of a bird." It also means "being a bird is a necessary condition for being a chicken," that is, "a chicken is necessarily a bird." However, to be a chicken it is not sufficient to be a bird. Thus a necessary condition may not be sufficiently restrictive. In a similar way, to be a bird it is not necessary to be a chicken. That is, a sufficient condition may not be necessary when there are other possibilities.

A condition that is both *sufficient* and *necessary* satisfies also the converse statement "q implies p." In this case, p and q are said to be equivalent since one implies the other. The statement is equivalent to the statement that "p is true if, and only if, q is true," or "q is true, if and only if, p is true." For example, the statement "equilateral triangles are equivalent to equiangular triangles" means that each is a sufficient and necessary condition for the other.

It is clear from this discussion that we have not yet isolated the exact property of a function that makes its Fourier series convergent. Several sufficient conditions for the convergence of Fourier series are known. For example, if $F(x)$ is an absolutely integrable function in $[-\pi,\pi]$, then its Fourier series $f(x)$ converges to the value $F(x)$ at every continuity point (i.e., a point where the function is continuous) where the left-hand and right-hand derivatives exist. This includes the special case where $F(x)$ has a unique derivative. At a point of discontinuity where $F(x)$ has left and right derivatives, the Fourier series converges to the arithmetical mean value:

$$f(x) = \tfrac{1}{2}[F(x+0) + F(x-0)]. \tag{3.132}$$

The same result also holds for a piecewise smooth function, that is, a function that is smooth except for a finite number of *jump discontinuities* [where $F(x)$ is discontinuous] or *kink discontinuities* (where its derivative is discontinuous).

Since the Fourier series is made up of continuous trigonometric functions $\cos nx$ and $\sin nx$, it might be supposed that a *convergent* Fourier series must itself be continuous, at least in the basic interval. In 1826 Abel pointed out that the Fourier representation $f(x)$ of x in $[-\pi,\pi]$ (Example 3.2.1) is actually discontinuous at the end points $\pm\pi$ and other similar points $x = (2n+1)\pi$, n being any integer, when $f(x)$ is considered a periodic function of x.

Furthermore, when a Fourier series $f(x)$ tries to represent a finite discontinuity at $x = a$ of a function $F(x)$ that is piecewise continuous, it overshoots by a calculable amount on both sides of a. This overshoot can already be seen in Fig. 3.1. Indeed, one can show that

$$f(a-0) = F(a-0) - \tfrac{1}{2}pD, \qquad f(a+0) = F(a+0) + \tfrac{1}{2}pD,$$

where

$$D = F(a+0) - F(a-0)$$

is the size of the discontinuity, and

$$p = -\frac{2}{\pi}\int_{\pi}^{\infty} \frac{\sin x}{x}\, dx = 0.179,$$

so that

$$d = f(a+0) - f(a-0) = (1+p)D, \tag{3.133}$$

rather than the original discontinuity D. This overshoot is known as the *Gibbs phenomenon*.

A Fourier representation of a function is useful even when its convergence is not known. If the function $F(x)$ is absolutely integrable in $[-\pi, \pi]$, its integral in this interval can be found by term-by-term integrations:

$$\int_a^b F(x)\,dx = \frac{a_0}{2}(b-a) + \sum_{n=1}^{\infty} [a_n(\sin nb - \sin na)/n - b_n(\cos nb - \cos na)/n]. \tag{3.134}$$

Conversely, if $F(x)$ has an absolutely integrable derivative, a term-by-term differentiation of $F(x)$ gives the Fourier series for $F'(x)$. That is,

$$F'(x) \sim f'(x) = \sum_{n=1}^{\infty} n(-a_n \sin nx + b_n \cos nx). \tag{3.135}$$

Because of the additional factor n, the differentiated series may not be convergent even when $f(x)$ is convergent.

Another interesting problem concerns the reconstruction, or approximation, of the function $F(x)$ whose Fourier series is known to be convergent. One systematic procedure is to expand the Fourier coefficients in inverse powers of n and to replace each Fourier series in inverse powers so obtained by the appropriate function from tables of Fourier series such as those shown in Appendix 3B and 3C. This procedure is illustrated by an example.

Example 3.12.1. Approximate the function giving rise to the Fourier series

$$f(x) = \sum_{n=1}^{\infty} \frac{\cos nx}{n+s}.$$

Since

$$\frac{1}{n+s} = \frac{1}{n} - \frac{s}{n^2} + \frac{s^2}{n^3} + \cdots,$$

we find

$$f(x) = \sum_{n=1}^{\infty} \cos nx \left(\frac{1}{n} - \frac{s}{n^2} + \cdots\right)$$
$$= -\ln\left(2\sin\frac{x}{2}\right) - s\,\frac{3x^2 - 6\pi x + 2\pi^2}{12} + \cdots.$$

Note that each succeeding term in this expansion has a more rapid convergence than those preceding it.

It might appear that the operation of finding the original function $F(x)$, referred to technically as the *summation* of the given Fourier series $f(x)$, makes sense only if $f(x)$ is convergent. This is really a trivial case, since we already know that a sum does exist. The interesting case turns out to be the summation of a divergent series!

The heart of the matter is a definition for the sum of a divergent series. Many different definitions are possible that all yield the same answer for a convergent series, as they should. However, the summability of a divergent series may vary with the definition. If summable, the sum may also have different values in different definitions!

To give the reader a taste of this fascinating subject, let us discuss the summation by the method of *arithmetic means* (AM). The AM σ_n of the first n partial sums

$$\begin{aligned} &s_0 = f_0, \qquad s_1 = f_0 + f_1, \qquad \ldots, \\ &s_n = f_0 + f_1 + \cdots + f_n \end{aligned} \tag{3.136}$$

of an infinite series of terms f_k is defined to be

$$\sigma_n = (s_0 + s_1 + \cdots + s_n)/n. \tag{3.137}$$

The sum by AM is then defined as the limit

$$\sigma = \lim_{n\to\infty} \sigma_n. \tag{3.138}$$

For example, the infinite series of terms $1, -1, 1, -1, \ldots$ has the alternating partial sums

$$\begin{aligned} &s_0 = 1, \qquad s_1 = 0, \qquad s_2 = 1, \\ &s_3 = 0, \qquad \ldots. \end{aligned}$$

Hence the series is divergent, but its AM

$$\sigma_n = \tfrac{1}{2} + \begin{cases} 0, & n \text{ even} \\ 1/2n, & n \text{ odd} \end{cases}$$

has the limit $\sigma = \frac{1}{2}$. Hence the series is summable by the method of AM to a value of $\frac{1}{2}$.

The relevance of this discussion is that the Fourier series of a continuous function is summable by this method of arithmetic means.

Problems

3.12.1. From the Fourier series for x^2, obtain the Fourier series for x.

3.12.2. From the Fourier series for x, obtain the Fourier series for x^2.

3.12.3. Derive the following formula for the Fourier series of the integral

$$\int_0^x F(x)\,dx = \sum_{n=1}^{\infty} (1 - \cos nx)\frac{b_n}{n} + \sum_{n=1}^{\infty} [a_n + (-1)^{n+1}a_0]\frac{\sin nx}{n}$$

expressed in terms of the Fourier coefficients for $F(x)$ itself.

3.12.4. Obtain approximate functions for the Fourier series

(a) $\displaystyle\sum_{n=1}^{\infty} \frac{n^2}{n^3+1} \sin nx;$

(b) $\displaystyle\sum_{n=1}^{\infty} \frac{n}{n^3+2} \cos nx.$

Appendix 3A. A short table of Fourier cosine series

$$F(x) = \sum_{n=1}^{\infty} a_n \cos nx$$

a_n	$F(x)$	Interval
$\frac{1}{n}$	$-\ln\left(2 \sin \frac{x}{2}\right)$	$(0, 2\pi)$
$\frac{1}{n^2}$	$(3x^2 - 6\pi x + 2\pi^2)/12$	$[0, 2\pi]$
$(-1)^{n+1}/n$	$\ln\left(2 \cos \frac{x}{2}\right)$	$(-\pi, \pi)$
$(-1)^{n+1}/n^2$	$(\pi^2 - 3x^2)/12$	$[-\pi, \pi]$
$\frac{1}{n}$, odd n	$-\frac{1}{2} \ln \tan \frac{\lvert x\rvert}{2}$	$(-\pi, 0)$ and $(0, \pi)$
$\frac{1}{n^2}$, odd n	$(\pi^2 - 2\pi \lvert x\rvert)/8$	$[-\pi, \pi]$

Reference: G. P. Tolstov, *Fourier Series* (Dover, New York, 1976), p. 148.

Appendix 3B. A short table of Fourier sine series

$$F(x) = \sum_{n=1}^{\infty} b_n \sin nx$$

b_n	$F(X)$	Interval
n^{-1}	$(\pi - x)/2$	$(0, 2\pi)$
n^{-3}	$(x^3 - 3\pi x^2 + 2\pi^2 x)/12$	$[0, 2\pi]$
$(-1)^{n+1}/n$	$x/2$	$(-\pi, \pi)$
$(-1)^{n+1}/n^3$	$(\pi^2 x - x^3)/12$	$[-\pi, \pi]$
$\frac{1}{n}$, odd n	$(\pi/4)x/\lvert x\rvert$	$(-\pi, 0)$ and $(0, \pi)$
$\frac{1}{n^3}$, odd n	$x(\pi^2 - \pi \lvert x\rvert)/8$	$[-\pi, \pi]$

Reference: G. P. Tolstov, *Fourier Series* (Dover, New York, 1976), p. 148.

Appendix 3C. A short table of Fourier transforms

$f(x)$	$g(k)=\frac{1}{\sqrt{2\pi}}\int_{-\infty}^{\infty} e^{-ikx}f(x)\,dx$	Conditions
1. $\delta(x)$	$\frac{1}{\sqrt{2\pi}}$	
2. $\begin{cases}0, & x<0\\ e^{-ax}, & x>0\end{cases}$	$\frac{1}{\sqrt{2\pi}}\frac{1}{a+ik}$	$\mathrm{Re}\,a>0$
3. $\exp(-\frac{1}{2}cx^2)$	$\frac{1}{\sqrt{c}}\exp\left(-\frac{k^2}{2c}\right)$	
4. $\frac{1}{1+x^2}$	$\sqrt{\pi/2}\exp(-\lvert k\rvert)$	
5. $[(a^2+x^2)(b+ix)^p]^{-1}$	$\sqrt{\pi/2}\frac{e^{-ak}}{a(a+b)^p}$	$\mathrm{Re}\,p>-1$, $\mathrm{Re}\,a>0$, $\mathrm{Re}\,b>0$
6. $[(a^2+x^2)(b-ix)^p]^{-1}$	$\sqrt{\pi/2}\,(b-a)^p e^{ak}/a$	$\mathrm{Re}\,p>-1$, $\mathrm{Re}\,a>0$, $\mathrm{Re}\,b>0$, $a\neq b$
7. $\begin{cases}P_n(x), & \lvert x\rvert<1\\ 0, & \lvert x\rvert>1\end{cases}$	$(-i)^n J_{n+1/2}(k)/\sqrt{k}$	
8. $\frac{e^{-\lambda x}}{a+e^{-x}}$	$\sqrt{\pi/2}\,a^{\lambda-1+ik}\csc(\pi\lambda+i\pi k)$	$0<\mathrm{Re}\,\lambda<1$, $-\pi<\arg a<\pi$

Reference: A Erdélyi *et al.*, *Table of Integral Transforms* (Bateman Manuscript Project) (McGraw-Hill, New York, 1954), Vol. I, Chap. III.

Appendix 3D. Tables of mathematical formulas

1. Fourier series

$$F(x)\simeq f(x)=\tfrac{1}{2}a_0+\sum_{n=1}^{\infty}(a_n\cos nx+b_n\sin nx),\qquad -\pi\leqslant x\leqslant\pi$$

$$a_n=\frac{1}{\pi}\int_{-\pi}^{\pi}F(x)\cos nx\,dx$$

$$b_n=\frac{1}{\pi}\int_{-\pi}^{\pi}F(x)\sin nx\,dx$$

$$F(x)\simeq f(x)=\sum_i f_i e_i(x),\qquad a\leqslant x\leqslant b$$

$$\int_a^b e_i^*(x)e_j(x)\,dx=\delta_{ij}$$

$$f_i=\int_a^b e_i^*(x)F(x)\,dx$$

$$\int_a^b |F(x)|^2\,dx\geqslant\sum_i |f_i|^2,\qquad \text{Bessel inequality.}$$

2. Dirac δ function

$$F(x) = \int_a^b F(x')\,\delta(x - x')\,dx'$$

$$\delta(-x) = \delta(x)$$

$$\delta(ax) = \delta(x)/|a|$$

$$\delta(x^2 - a^2) = [\delta(x - a) + \delta(x + a)]/2\,|a|$$

$$\delta(g(x)) = \sum_i \delta(x - x_i)/|g'(x_i)|, \qquad \text{where} \qquad g(x_i) = 0.$$

$$\int_a^b F(x')\frac{d}{dx'}\delta(x - x')\,dx' = -F'(x) \qquad \text{under certain conditions}$$

$$\delta(x - x') = \sum_n e_n(x)e_n^*(x').$$

3. Fourier transform

$$g(k) = \frac{1}{\sqrt{2\pi}}\int_{-\infty}^{\infty} e^{-ikx} f(x)\,dx = \mathscr{F}\{f(x)\}$$

$$f(x) = \frac{1}{\sqrt{2\pi}}\int_{-\infty}^{\infty} e^{ikx} g(x)\,dx.$$

4. Conjugate variables and operators

$$(\phi_i, \phi_j) = (f_i, f_j) = (g_i, g_j), \qquad g_i = \mathscr{F}\{f_i(x)\}$$

$$\bar{x} = (\phi, x\phi)/(\phi, \phi)$$

$$(\Delta x)^2 = (\phi, (x - \bar{x})^2\phi)/(\phi, \phi)$$

$$(\Delta x)(\Delta k) \geqslant \tfrac{1}{2}, \qquad \text{Heisenberg's uncertainty principle}$$

$$(\phi_i, x\phi_j) = i\left(\phi_i, \frac{d}{dk}\phi_j\right)$$

$$(\phi_i, k\phi_j) = -i\left(\phi_i, \frac{d}{dx}\phi_j\right)$$

$$[k, x] = kx - xk = -i$$

$$(\phi_i, C(x,k)\phi_j) = \left(f_i, C\left(x, \frac{1}{i}\frac{d}{dx}\right)f_j\right)$$

$$= \left(g_i, C\left(i\frac{d}{dk}, k\right)g_j\right).$$

4

DIFFERENTIAL EQUATIONS IN PHYSICS

4.1. Introduction

We are familiar with the fact that, given a smooth function $\mathbf{p}(t)$, its derivative can be calculated:

$$\frac{d}{dt}\mathbf{p}(t) = \mathbf{f}(t). \tag{4.1}$$

If on the other hand the derivative $\mathbf{f}(t)$ is given, but $\mathbf{p}(t)$ is as yet unknown, the equation is called a *differential equation* (DE). A DE is solved when a function $\mathbf{p}(t)$ is obtained that satisfies the stated functional relationship. The most familiar of such DE in physics is probably the Newton's force law shown in Eq. (4.1). This states that an external force $\mathbf{f}(t)$ causes a time rate of change of the unknown momentum $\mathbf{p}(t)$ of a particle.

Several other DEs of interest in physics can be derived from Newton's force law. For example, we have seen in Section 2.8 that its application to the one-dimensional motion of a system of identical masses connected by identical springs yields the one-dimensional *wave equation*

$$\left(\frac{1}{v^2}\frac{\partial^2}{\partial t^2} - \frac{\partial^2}{\partial x^2}\right)u(x,t) = 0 \tag{4.2}$$

in the continuum limit. In this equation, the curvature term involving the spatial derivative gives the net force acting on the mass located at x by virtue of its displacement from equilibrium. The mass responds by accelerating, so that its kinetic energy is changed. When it returns to an equilibrium position, it does not immediately come to a stop, but overshoots that position because of its "inertia." The springs connected to it eventually bring it to rest. This happens when its kinetic energy is completely converted into the potential energy stored in the springs. The springs next push or pull masses around so that the potential energy reverts to the kinetic form. This completes one of the countless cycles of the transmutation of energies.

The mathematical solution of the wave equation tells us that the motion of each mass does not necessarily repeat itself in exactly the same way. Rather, the wave disturbance, where the energy is located, tends to move away from the source (with velocities $\pm v$) to distances beyond the range of motion of the original mass. It is the wave disturbance that is of primary importance here, like the story of the Norman Conquest depicted on the Bayeux tapestry. The motion of the masses themselves, like the makeup of that linen roll, is only of secondary interest.

The wave equation can be modified to describe a number of other interesting physical situations. We may ask for the equilibrium (i.e., time-independent) configuration of masses when they are subject to a given external force. The answer is given by a solution of the one-dimensional *Poisson equation*

$$-\frac{\partial^2}{\partial x^2}\phi(x)=f(x). \tag{4.3}$$

We may even ask for force-free configurations consistent with boundary constraints. These are provided by solutions to the *Laplace equation*

$$\frac{\partial^2}{\partial x^2}\phi(x)=0. \tag{4.4}$$

Returning to the time-dependent wave equation, we might add a frictional term for a resistive medium. If friction is proportional to the instantaneous velocity, we find

$$\left(\frac{1}{v^2}\frac{\partial^2}{\partial t^2}+\frac{1}{\kappa}\frac{\partial}{\partial t}-\frac{\partial^2}{\partial x^2}\right)u(x,t)=0. \tag{4.5}$$

The added frictional term will cause a damping of the wave function $u(x,t)$ as the wave propagates in space-time.

Other physical attributes might behave like the mechanical system described by Eq. (4.5). The special case in which their effective mass vanishes is of unusual interest. In this situation, the first, or inertial, term in Eq. (4.5) is absent, leading to the equation

$$\left(\frac{1}{\kappa}\frac{\partial}{\partial t}-\frac{\partial^2}{\partial x^2}\right)u(x,t)=0. \tag{4.6}$$

In the absence of inertia, the displacement from equilibrium cannot overshoot the equilibrium position. Instead it disperses in a way rather similar to the relaxation of a heavily damped mechanical system. The rate of dispersal is controlled by the frictional term, where the constant κ is called a *diffusion constant* in the diffusion of gas molecules or a *thermal conductivity* in the transmission of heat. Equation (4.5) itself is called a *diffusion equation*.

In addition to differential equations, we also use *integral equations* in physics. These have the form

$$\int K(t,t')y(t')\,dt'+M(t)y(t)=R(t), \tag{4.7}$$

where $K(t,t')$, $M(t)$, and $R(t)$ are given functions, while $y(t)$ is the unknown function to be determined. Integral equations are far less common in physics than differential equations.

There are good reasons for the popularity of differential equations. They have simpler mathematical structures, so that they are easier to solve and to use. Certain important invariance principles (such as the invariance under space translations and time displacements) can be included simply and explicitly in differential equations. These are important technical advantages, but the basic superiority of differential equations is probably this: They correctly reflect the fact that our experimental knowledge of physical events and physical systems tends to be rather limited both in details and in space-time extensions. As a result, the coherence and regularities that connect physical properties, and the disturbances and responses that describe sequences of physical events, tend to be the properties of our immediate neighborhood. This limitation and localization of our knowledge favor a differential approach.

It is also a historical fact that in science we have been successful in relating not only properties at essentially the same location but also only a few properties at a time. Thus a complete knowledge of many physical attributes all over the universe is not needed before objective predictions of the outcomes of certain events can be made. This is indeed fortunate, for otherwise science might not have emerged.

The possibility of doing science in relative ignorance of the actual nature of the universe has its obvious limitations. The regularities that we can discover and quantify are usually among the simplest possible relations. The physical properties so related tend to exist in space-time in a reasonably smooth manner. Their relations as a rule do not involve derivatives of very high orders, or very complicated functions of their low-ordered derivatives. As a result, known physical laws are often simple differential equations.

In this chapter, we shall concentrate on linear differential equations. Nonlinear equations will be discussed briefly in Sections 4.3 and 4.14.

4.2. Linear differential equations

The *order* and *degree* of a differential equation (DE) refers to those of the derivative of the highest order after the DE has been rationalized. Thus the DE

$$\frac{d^4y(x)}{dx^4} + \left(\frac{dy(x)}{dx}\right)^{1/2} + x^2y(x) = R(x) \tag{4.8}$$

is of fourth order and second degree, since after rationalization it contains $[d^4y(x)/dx^4]^2$. The DE is said to be inhomogeneous if there is a term $R(x)$ that is independent of the unknown $y(x)$. The DE is an *ordinary differential equation* if there is only one variable, as is the case of Eq. (4.8). If there are two or more variables, it is a *partial differential equation*.

A DE

$$\mathscr{L}[y(x)] = R(x) \tag{4.9}$$

is *linear* if the differential operator $\mathscr{L}$ is a linear operator satisfying the *linearity property*

$$\mathscr{L}[a_1 y_1(x) + a_2 y_2(x)] = a_1 \mathscr{L}[y_1(x)] + a_1 \mathscr{L}(y_2(x)], \tag{4.10}$$

where the a_i are constants. For example,

$$\mathscr{D}[y(x)] \equiv \left(\frac{d}{dx} y(x)\right)^{1/2}$$

is *nonlinear* because

$$\mathscr{D}[ay(x)] = \left(\frac{d}{dx} ay(x)\right)^{1/2} = \sqrt{a}\left(\frac{d}{dx} y(x)\right)^{1/2} \neq a\left(\frac{d}{dx} y(x)\right)^{1/2}.$$

In a similar way, the second-order DE

$$\frac{d^2}{dt^2} \theta(t) + \frac{g}{l} \sin \theta(t) = 0 \tag{4.11}$$

describing the large-amplitude motion of a pendulum is also nonlinear. This is because

$$\sin(\lambda\theta_1 + \mu\theta_2) \neq \lambda \sin \theta_1 + \mu \sin \theta_2.$$

However, if the angular displacement is so small that $\sin \theta \simeq \theta$ is a good approximation, Eq. (4.11) can be *linearized* to

$$\frac{d^2}{dt^2} \theta(t) + \frac{g}{l} \theta(t) = 0. \tag{4.12}$$

From this we can see that physical systems not far from states of equilibrium are basically linear systems.

Linear differential equations (LDE) are simpler than nonlinear DE. Their relative simplicity is a consequence of the following two *superposition principles* that their solutions satisfy:

1. If $y_1(x)$ and $y_2(x)$ are any two solutions of a *homogeneous* LDE,

   $$\mathscr{L}y(x) = 0,$$

   then

   $$y_\text{h}(x) = c_1 y_1(x) + c_2 y_2(x), \qquad c_i = \text{constant}, \tag{4.13}$$

   is also a solution. This is because

   $$\mathscr{L}(c_1 y_1 + c_2 y_2) = c_1 \mathscr{L} y_1 + c_2 \mathscr{L} y_2 = 0.$$

2. If $y_\text{h}(x)$ is a (i.e., any) solution of the homogeneous LDE $\mathscr{L}y = 0$, and $y_\text{p}(x)$ is a *particular* (i.e., any) solution of the *inhomogeneous* LDE

   $$\mathscr{L}[y_\text{p}(x)] = R(x),$$

then the linear combination

$$y(x) = ay_{\rm p}(x) + by_{\rm h}(x), \qquad a,b = \text{constant}, \tag{4.14}$$

is a solution of the inhomogeneous LDE

$$\mathscr{L}[y(x)] = aR(x).$$

This is because

$$\mathscr{L}(ay_{\rm p} + by_{\rm h}) = a\mathscr{L}_{\rm p} + b\mathscr{L}y_{\rm h} = aR + 0 = aR.$$

Examples of linear differential equations are the simple, or hyperbolic, wave equation discussed earlier

$$\phi_{tt} - v^2\phi_{xx} = 0, \qquad \text{where } \phi_{tt} = \frac{\partial^2}{\partial t^2}\phi, \text{ etc.},$$

and the linearized Korteweg–deVries equation

$$\phi_t + c_0\phi_x + \nu\phi_{xxx} = 0,$$

where c_0 and ν are constants. Examples of nonlinear differential equations are the Korteweg–deVries equation (describing long waves in shallow water)

$$\phi_t + (c_0 + c_1\phi)\phi_x + \nu\phi_{xxx} = 0,$$

and the sine–Gordon equation

$$\phi_{tt} - \phi_{xx} + \sin\phi = 0$$

[describing certain persistent (called solitary) waves in many physical situations].

Problems

4.2.1. Identify the linear differential equations in the following

(a) $x^2y' + xyy' = 0, \qquad y = y(x), \qquad y' = \dfrac{d}{dx}y(x);$

(b) $x^2y' + y^2 = 0;$

(c) $xy'^2 + x^2 = 0;$

(d) $y' = \dfrac{x(1+y^2)^{1/2}}{y(1+x^2)^{1/2}};$

(e) $y'' = e^x;$

(f) $y'' = e^y;$

(g) $a^2y''^2 = (1+y'^2)^3.$

4.3. First-order differential equations

First-order DEs can be solved by direct integration if the equation is *separable*:

$$\frac{dy}{dx} = \frac{P(x)}{Q(y)}.$$

Then

$$\int Q(y)\,dy = \int P(x)\,dx.$$

Example 4.3.1.

$$\frac{dy}{dx} - x\sqrt{1-y^2} = 0,$$

$$\frac{dy}{\sqrt{1-y^2}} = x\,dx.$$

$$\therefore \int \frac{dy}{\sqrt{1-y^2}} = \sin^{-1} y = \tfrac{1}{2}x^2 + c,$$

or

$$y = \sin(\tfrac{1}{2}x^2 + c).$$

This works for both linear and nonlinear DEs, but it is not always easy to separate the variables x and y. A change of variables will sometimes do it, as illustrated by the following example.

Example 4.3.2. Solve

$$\frac{dy}{dx} = (x - y + 3)^2.$$

Let us first change variables to $z = x - y + 3$. Then

$$\frac{dz}{dx} = 1 - \frac{dy}{dx} = 1 - z^2,$$

$$\int \frac{dz}{1-z^2} = \int dx,$$

$$\tanh^{-1} z = x + c,$$

$$z = \tanh(x + c).$$

Therefore

$$y = x + 3 - \tanh(x + c).$$

The next best thing one can do for a nonlinear DE is to change it into a linear DE by a change of variables. This is because LDEs can always be solved in closed form, as we shall see. The following example shows how a nonlinear DE can be changed into a linear DE by a change of variables.

Example 4.3.3. The nonlinear DE

$$\frac{dy}{dx}+f(x)y=g(x)y^n, \qquad n\neq 1,$$

can be written as

$$\frac{1}{y^n}\frac{dy}{dx}+f(x)y^{1-n}=g(x),$$

or

$$\frac{dv}{dx}+(1-n)f(x)v=(1-n)g(x),$$

where

$$v(x)=y^{1-n}(x).$$

We finally turn to linear DEs. They have the general form

$$\left(\frac{d}{dx}+p(x)\right)y(x)=R(x). \tag{4.15}$$

The solution to the homogeneous equations, with $R(x)=0$, can be obtained readily by separating out the dependence on x and y:

$$\frac{1}{y(x)}\,dy(x)=-p(x)\,dx.$$

Hence

$$\int_a^x d\ln y(x')=-\int_a^x p(x')\,dx'.$$

That is,

$$y(x)=y(a)\exp\left(-\int_a^x p(x')\,dx'\right), \tag{4.16}$$

where the multiplicative constant has been so chosen that $y(x=a)$ is equal to a predetermined value $y(a)$. The number $y(a)$ is called the *boundary condition* at the point $x=a$.

Example 4.3.4.

$$\left(\frac{d}{dx}+x\right)y(x)=0$$

has the solution

$$y(x)=y(0)\exp\left(-\int_0^x x'\,dx'\right)=y(0)\exp(-\tfrac{1}{2}x^2).$$

If the boundary condition is $y(0)=1$, then

$$y(x)=\exp(-\tfrac{1}{2}x^2).$$

Given a solution $y_h(x)$ of the homogeneous equation, a particular solution of the *inhomogeneous* equation can be obtained by first writing

$$y_{\mathrm{p}}(x) = c(x)y_{\mathrm{h}}(x). \tag{4.17}$$

Direct substitution into Eq. (4.15) yields

$$\begin{aligned} R(x) = \mathscr{L}y_{\mathrm{p}}(x) &= \left(\frac{d}{dx} + p(x)\right)c(x)y_{\mathrm{h}}(x) \\ &= y_{\mathrm{h}}(x)\frac{d}{dx}c(x) + c(x)\left(\frac{d}{dx} + p(x)\right)y_{\mathrm{h}}(x). \end{aligned}$$

Since the last term vanishes, we have

$$\frac{d}{dx}c(x) = \frac{R(x)}{y_{\mathrm{h}}(x)},$$

so that

$$c(x) = c(a) + \int_a^x \left(\frac{R(x')}{y_{\mathrm{h}}(x')}\right) dx'.$$

Thus

$$y_{\mathrm{p}}(x) = \left[\int_a^x \left(\frac{R(x')}{y_{\mathrm{h}}(x')}\right) dx'\right] y_{\mathrm{h}}(x), \tag{4.18}$$

where we have dropped the term involving $c(a)$, since according to Eq. (4.14), it can be made to appear in the second (or homogeneous) term of the general solution $y(x) = y_{\mathrm{p}}(x) + by_{\mathrm{h}}(x)$. The solution satisfying the boundary condition $y(x = a) = y(a)$ is obtained by choosing the factor b such that

$$y(a) = y_{\mathrm{p}}(a) + by_{\mathrm{h}}(a).$$

That is,

$$y(x) = y_{\mathrm{p}}(x) + y_{\mathrm{h}}(x)[y(a) - y_{\mathrm{p}}(a)]/y_{\mathrm{h}}(a). \tag{4.19}$$

This method is called a *variation of constants,* since $y_{\mathrm{p}}(x)$ is "proportional" to $y_{\mathrm{h}}(x)$, but the proportionality "constant" is itself a function of x.

Example 4.3.5. Solve the first-order LDE

$$\left(\frac{d}{dx} + s\right)y(x) = e^{-tx}.$$

a. Get a homogeneous solution: $y_{\mathrm{h}}(x) = e^{-sx}$.
b. Get the proportional function

$$c(x) = \int_0^x \frac{e^{-tx'}}{e^{-sx'}} dx' = -\frac{1}{t-s}(e^{-(t-s)x} - 1).$$

c. Thus

$$y_p(x) = -\frac{1}{t-s}(e^{-(t-s)x})e^{-sx}$$
$$= \frac{1}{s-t}e^{-tx},$$

where we have ignored a constant term in $c(x)$.

d. Check:

$$\left(\frac{d}{dx}+s\right)\left(\frac{1}{s-t}e^{-tx}\right) = \left(\frac{-t+s}{s-t}\right)e^{-tx} = e^{-tx}.$$

4.3.1. Uniqueness of the solution

The solution $y(x)$ has been obtained, for both homogeneous and inhomogeneous equations, by specifying one boundary condition $y(a)$ at $x = a$. We now show that this is enough to determine the solution uniquely for both linear and nonlinear first-order differential equations.

Consider the Taylor expansion of $y(x)$ about the boundary point $x = a$:

$$y(x) = y(a) + (x-a)y'(a) + \frac{1}{2!}(x-a)^2 y''(a) + \cdots. \qquad (4.20)$$

Given $y(a)$, the general DE of first order

$$y' = f(x, y(x))$$

gives

$$y'(a) = f(a, y(a)),$$

$$y''(a) = \frac{d}{dx}f(x, y(x))\Big|_{x=a},$$

and all higher derivatives by direct differentiation. Hence, if the Taylor series converges, the specification of $y(a)$ alone is enough to determine the solution $y(x)$ uniquely.

Problems

4.3.1. Suppose R radioactive atoms are produced per second in a sample by neutron irradiation. If each atom so produced decays with the probability λ per second, show that the number $N(T)$ of radioactive atoms in the sample satisfies the differential equation

$$\frac{dN(t)}{dt} = R - \lambda N(t)$$

while under neutron irradiation. Calculate $N(t)$ if $N(0) = 0$.

4.3.2. A raindrop initially at rest falls down from a height h. If air resistance gives rise to a retarding force $-\lambda\mathbf{v}(t)$ proportional to the instantaneous velocity $\mathbf{v}(t)$, calculate the speed of the raindrop
(a) as a function of time; and
(b) when it hits the ground.

4.4. Second-order linear differential equations

Many DEs in physics are second order in the differential operator with the general form

$$\left(\frac{d^2}{dx^2}+P(x)\frac{d}{dx}+Q(x)\right)y(x)=R(x). \qquad (4.21)$$

Examples are

1. Newton's force law: $\dfrac{d^2}{dt^2}\mathbf{r}(t)=\dfrac{\mathbf{F}(t)}{m}$.
2. The simple harmonic oscillator: $\left(\dfrac{d^2}{dt^2}+\omega^2\right)x(t)=0.$
3. The one-dimensional Helmholtz equation: $\left(\dfrac{d^2}{dx^2}+k^2\right)\psi(x)=0.$

4.4.1. Boundary conditions

Consider the Taylor expansion (4.20) about the boundary $x=a$:

$$y(x)=y(a)+(x-a)y'(a)+\frac{1}{2!}(x-a)^2y''(a)+\cdots.$$

For the second-order DE, the numbers $y(a)$ and $y'(a)$ are freely adjustable, but not the higher derivatives $y''(a)$, $y'''(a)$, etc. This is because

$$y''(a)=-[P(a)y'(a)+Q(a)y(a)]+R(a),$$
$$y'''(a)=-(P'y'+Py''+Q'y+Qy')_{x=a}+R'(a), \qquad \text{etc.},$$

are uniquely determined once $y(a)$ and $y'(a)$ are chosen. Thus these two numbers control the solution $y(x)$ uniquely, assuming again that the Taylor series converges.

The discussion can be generalized readily to an nth-order LDE. Its solution is uniquely determined by n boundary conditions that may be taken to be $y(a), y'(a), \ldots, y^{(n-1)}(a)$ at any point $x=a$.

4.4.2. *The homogeneous differential equation*

Suppose $y_1(x)$ and $y_2(x)$ are two solutions of the homogeneous equation satisfying the boundary conditions

$$y_1(a) = 1, \qquad y_1'(a) = 0; \tag{4.22a}$$

$$y_2(a) = 0, \qquad y_2'(a) = 1. \tag{4.22b}$$

It is clear that $y_1(x)$ and $y_2(x)$ can never be proportional to each other, since they already differ at the boundary a. Thus they are linearly independent. They are also the only linearly independent solutions because second-order DEs have only two independent boundary conditions, such as $y(a)$ and $y'(a)$. Indeed, a general solution satisfying the boundary conditions $y(a) = c_1$ and $y'(a) = c_2$ is just the linear combination

$$y(x) = c_1 y_1(x) + c_2 y_2(x).$$

4.4.3. *Test of linear independence of solutions: Wronskian*

The converse of the above situation occurs when we have to determine the linear independence of two given homogeneous solutions $y_3(x)$ and $y_4(x)$. This can be done by inspection since two functions are linearly independent if they are not simply proportional to each other.

This simple procedure cannot be used for n given homogeneous solutions of an nth-order LDE when $n > 2$. Hence a more general procedure, applicable to a DE of any order, is desirable. We now describe this general procedure, but, for the sake of notational simplicity, we shall specialize to the case of the second-order DE. It turns out that the result is useful even for second-order equations.

If the functions $y_1(x)$ and $y_2(x)$ are linearly independent, any solution $y(x)$ of a second-order LDE and its slope $y'(x)$ can be expressed as

$$\begin{aligned} c_1 y_1(x) + c_2 y_2(x) &= y(x) \\ c_1 y_1'(x) + c_2 y_2'(x) &= y'(x), \end{aligned} \tag{4.23a}$$

with *unique* linear coefficients c_1 and c_2. This is because, if the functions were linearly dependent, with say $y_2(x) = a y_1(x)$, then (1) only the combination $c_1 + a c_2$ can be determined, and (2) $y(x)$ is also proportional to $y_1(x)$ and therefore cannot be a general solution of a second-order LDE.

If Eq. (4.23a) is written in the matrix form

$$\begin{pmatrix} y_1(x) & y_2(x) \\ y_1'(x) & y_2'(x) \end{pmatrix} \begin{pmatrix} c_1 \\ c_2 \end{pmatrix} = \begin{pmatrix} y(x) \\ y'(x) \end{pmatrix}, \tag{4.23b}$$

we see that a unique solution requires that the determinant

$$W(x) = \begin{vmatrix} y_1(x) & y_2(x) \\ y_1'(x) & y_2'(x) \end{vmatrix} \tag{4.24}$$

does not vanish. $W(x)$ is called a *Wronskian,* named after the nineteenth century mathematician Wronski. Suppose $W(x=a)\neq 0$. Then the coefficients c_1 and c_2 can be determined at $x=a$. Since Eqs. (4.23) are satisfied for other values of x, it follows that $W(x)\neq 0$ is guaranteed everywhere. A more rigorous demonstration will be given below.

The Wronskian for n functions contains these functions and their first $n-1$ derivatives. It can be used to test for the linear independence of n solutions of an LDE of order n.

The Wronskian satisfies a number of interesting properties. For a second-order homogeneous LDE, $W(x)$ satisfies a first-order LDE:

$$\begin{aligned}\frac{d}{dx}W(x) &= \frac{d}{dx}(y_1y_2' - y_2y_1') = y_1'y_2' + y_1y_2'' - y_2'y_1' - y_2y_1'' \\ &= -y_1(Py_2' + Qy_2) + y_2(Py_1' + Qy_1) \\ &= -P(x)W(x). \end{aligned} \tag{4.25}$$

Therefore $W(x)$ has the explicit form

$$W(x) = W(a)\exp\left(-\int_a^x P(x')\,dx'\right). \tag{4.26}$$

This solution shows clearly that (1) if $W(a)\neq 0$, then $W(x)\neq 0$ everywhere, and (2) if $W(a)=0$, then $W(x)=0$ everywhere. Thus it is not necessary to test for the linear independence of two solutions at more than one point. Finally, Eq. (4.26) shows that the Wronskian can be calculated before the LDE is solved.

Example 4.4.1. Two solutions of the DE

$$\left(\frac{d^2}{dx^2} + k^2\right)y(x) = 0 \tag{4.27}$$

are easily seen to be $\cos kx$ and $(1/k)\sin kx$. Thus

$$W(x) = \begin{vmatrix} \cos kx & \dfrac{1}{k}\sin kx \\ -k\sin kx & \cos kx \end{vmatrix} = 1.$$

If we had written the second solution as $\sin kx$, the Wronskian would be k, which vanishes as $k\to 0$. In this limit, $\sin kx = 0$ is indeed not linearly independent of $\cos kx = 1$, while

$$\lim_{k\to 0}\frac{1}{k}\sin kx = x$$

is linear independent of 1.

Problems

4.4.1. Determine if the functions in each of the following groups are linearly independent:

(a) e^{ikx}, e^{-ikx}, $\sin kx$;
(b) e^{ikx}, e^{-ikx}, $\tan kx$;
(c) $x^2 - 1$, $x^2 + 1$, x.

State any additional condition under which your answer is valid.

4.4.2. Show that the solution of a third-order, homogeneous linear differential equation is determined uniquely by specifying three boundary conditions.

4.4.3. Calculate the displacement $x(t)$ of a damped linear oscillator that satisfies the equation

$$m\ddot{x} + b\dot{x} + kx = 0$$

with

(a) $x(t=0) = 1$, $\dot{x}(t=0) = 0$;
(b) $x(t=0) = 0$, $\dot{x}(t=0) = 1$.

4.4.4. Calculate the Wronskian $W(x)$ of the differential equation in Problem 4.4.3.

4.4.5. Calculate the Wronskian of the following differential equations

(a) Associated Legendre equation:

$$(1 - x^2)y'' - 2xy' + \left(l(l+1) - \frac{m^2}{1 - x^2}\right)y = 0;$$

(b) Bessel equation:

$$x^2y'' + xy' + (x^2 - m^2)y = 0;$$

(c) Spherical Bessel equation:

$$x^2y'' + 2xy' + [x^2 - l(l+1)]y = 0;$$

(d) Mathieu equation:

$$y'' + [c - \tfrac{1}{2}d^2\cos(2x)]y = 0;$$

(e) Confluent hypergeometric equation:

$$xy'' + (c - x)y' - ay = 0.$$

4.4.6. Show that the specification of $y(a)$ and $y'(a)$ at the boundary $x = a$ will also define a unique solution for a nonlinear differential equation of second order.

4.5. The second homogeneous solution and an inhomogeneous solution

If one solution $y_1(x)$ of a second-order *homogeneous* LDE is known, a second solution $y_2(x)$, linearly independent of the first, can be obtained with the help of the Wronskian:

$$\frac{d}{dx}\left(\frac{y_2}{y_1}\right)=\frac{y_1y_2'-y_2y_1'}{y_1^2}=\frac{W(x)}{y_1^2(x)}.$$

A simple integration gives

$$\frac{y_2(x)}{y_1(x)}-\frac{y_2(b)}{y_1(b)}=\int_b^x\frac{W(t)\,dt}{y_1^2(t)}=g(x).$$

Hence

$$y_2(x)=g(x)y_1(x),\qquad g(x)=\int_b^x\left[\exp\left(-\int_b^t P(t')\,dt'\right)\Big/y_1^2(t)\right]dt,\quad (4.28)$$

where we have dropped a term proportional to $y_1(x)$.

Example 4.5.1. If one solution of

$$\frac{d^2}{dx^2}\psi(x)=0\qquad (4.29)$$

is 1, show that the linearly independent second solution is x.

$$\begin{aligned}y_1(x)&=1\\ y_2(x)&=g(x)y_1(x)=g(x)\\ &=\int_b^x dt=x-b=x,\qquad \text{if}\qquad b=0,\end{aligned}$$

where we have used the result $P(x)=0$. Alternatively, given $y_2(x)$, then $y_1(x)=cg(x)x$, where

$$g(x)=\int_b^x\frac{1}{y_1^2(t)}\,dt=\int_b^x\frac{1}{t^2}\,dt=\frac{1}{b}-\frac{1}{x}.$$

We may take $b=\infty$, $c=-1$ to yield $y_1(x)=1$.

4.5.1. *Inhomogeneous solutions*

The solution of the inhomogeneous second-order LDE (4.21)

$$\mathscr{L}y(x)=R(x)$$

has the general form

$$y(x)=y_p(x)+[c_1y_1(x)+c_2y_2(x)],\qquad (4.30)$$

where $y_p(x)$ is a solution (i.e., any solution) of the inhomogeneous equation and $y_1(x)$, $y_2(x)$ are the two linearly independent homogeneous solutions. We call $y_p(x)$ a *particular solution* (or a *particular integral*) and $[c_1y_1(x)+c_2y_2(x)]$ a *complementary function.*

Why do we bother with $y(x)$ if we already have a $y_p(x)$? The reason is that $y_p(x)$ satisfies only the boundary conditions $y_p(a)$, $y_p'(a)$ at $x=a$.

Suppose we want instead a solution satisfying the boundary conditions $y(a) = \alpha$, $y'(a) = \beta$. We do not want to look for the particular solution with just the right boundary conditions. This is because particular solutions are hard to obtain; we are lucky if we manage to get one, with or without the correct boundary conditions.

The complementary function now comes to the rescue, because it can change boundary conditions without contributing anything to the inhomogeneity of the DE. The reader can easily verify that the required complementary function is $c_1 y_1(x) + c_2 y_2(x)$ with the coefficients chosen to satisfy the correct boundary conditions at $x = a$:

$$\begin{aligned} c_1 y_1(a) + c_2 y_2(a) &= \alpha - y_p(a) \\ c_1 y_1'(a) + c_2 y_2'(a) &= \beta - y_p'(a). \end{aligned} \tag{4.31}$$

Furthermore, the existence of the linear coefficients

$$\begin{pmatrix} c_1 \\ c_2 \end{pmatrix} = \begin{pmatrix} y_1(a) & y_2(a) \\ y_1'(a) & y_2'(a) \end{pmatrix}^{-1} \begin{pmatrix} \alpha - y_p(a) \\ \beta - y_p'(a) \end{pmatrix} \tag{4.32}$$

is guaranteed by the linear independence of the homogeneous solutions y_1 and y_2, since then the Wronskian $W(a) \neq 0$ and therefore the inverse matrix in Eq. (4.32) exists.

4.5.2. A particular solution: Method of variation of constants.

We still need one (any one) particular solution $y_p(x)$ of the inhomogeneous DE.

To obtain this, we first observe that the function $y_p(x)$ contains two degrees of freedom in the sense that at a point $x = x_1$ its value $y_p(x_1)$ and its slope $y_p'(x_1)$ can be chosen arbitrarily. These two arbitrary numbers may be expressed in terms of the values and slopes of the two linearly independent *homogeneous* solutions $y_i(x)$, $i = 1,2$:

$$y_p(x) = v_1 y_1(x) + v_2 y_2(x) \tag{4.33a}$$

$$y_p'(x) = v_1 y_1'(x) + v_2 y_2'(x), \tag{4.33b}$$

because the right-hand sides also describe a system with two degrees of freedom, as represented by the two linear coefficients v_1 and v_2. Indeed, these linear coefficients are uniquely determined by Eq. (4.33) at $x = x_1$ because the Wronskian $W(x_1)$ does not vanish if the $y_i(x)$ are linearly independent.

However, these linear coefficients v_1 and v_2 cannot be constants independent of x, for then $y_p(x)$ solves the homogeneous LDE, not the inhomogeneous equation. We therefore conclude that

$$v_i = v_i(x), \qquad i = 1,2,$$

are functions of x. It is to this necessary dependence on x that the name "variation of constants" refers.

Direct differentiation of Eq. (4.33a) does not yield Eq. (4.33b) unless

$$v_1'(x)y_1(x) + v_2'(x)y_2(x) = 0. \tag{4.34a}$$

This single requirement is insufficient to determine the two unknowns v_i'. We need another relation. This can be obtained from the original DE (4.21) with the help of Eqs. (4.33):

$$v_1'(x)y_1'(x) + v_2'(x)y_2'(x) = R(x). \tag{4.34b}$$

The two Eqs. (4.34) can now be used to solve for v_1' and v_2'. The results are

$$\begin{aligned} v_1'(x) &= -y_2(x)R(x)/W(x), \\ v_2'(x) &= y_1(x)R(x)/W(x), \qquad W(x) \neq 0. \end{aligned} \tag{4.35}$$

[The verification of Eqs. (4.34b, 4.35) is left as an exercise.] These are first-order DEs whose solutions are

$$v_1(x) = -\int_a^x \frac{y_2(t)R(t)}{W(t)}\,dt, \qquad v_2(x) = \int_a^x \frac{y_1(t)R(t)}{W(t)}\,dt, \tag{4.36}$$

where the integration constants have been chosen arbitrarily to give $v_1(a) = v_2(a) = 0$. With this choice, the particular solution satisfies the boundary conditions $y_p(a) = 0$ and $y_p'(a) = 0$, according to Eqs. (4.33).

Example 4.5.2. Obtain a particular solution of the DE

$$\left(\frac{d^2}{dx^2} + k^2\right)y(x) = A \sin qx. \tag{4.37}$$

From the earlier discussion on the homogeneous equation (4.27), we know that its two linearly independent solutions are

$$y_1(x) = \cos kx, \qquad y_2(x) = \frac{1}{k}\sin kx,$$

and that the Wronskian is $W(x) = 1$. Hence Eq. (4.36) gives

$$\begin{aligned} v_1(x) &= -\int_{-L}^x \frac{A}{k}\sin kx \sin qx\,dx \\ &= -\frac{A}{k}\left(\frac{\sin(k-q)x}{2(k-q)} - \frac{\sin(k+q)x}{2(k+q)}\right) + C_1, \end{aligned}$$

where the constant C_1 comes from the lower limit of integration and may be dropped. Similarly

$$\begin{aligned} v_2(x) &= A\int_{-L}^x \cos kx \sin qx\,dx \\ &= \begin{cases} A\left(\dfrac{\cos(k-q)x}{2(k-q)} - \dfrac{\cos(k+q)x}{2(k+q)}\right) + C_2, & k \neq q \\ \dfrac{A}{2k}\sin^2 kx, & k = q. \end{cases} \end{aligned}$$

Thus a particular solution is, for $q \neq k$,

$$\begin{aligned} y_p(x) &= -\frac{A}{2k(k-q)}[\sin(k-q)x \cos kx - \cos(k-q)x \sin kx] \\ &\quad + \frac{A}{2k(k+q)}[\sin(k+q)x \cos kx - \cos(k+q)x \sin kx] \\ &= A \sin qx/(k^2 - q^2). \end{aligned} \tag{4.38}$$

For $q = k$, the answer is

$$\begin{aligned} y_p(x) &= \frac{A}{k}\left(\frac{\sin 2kx}{4k} - \tfrac{1}{2}x\right)\cos kx + \frac{A}{2k^2}\sin^3 kx \\ &= -\frac{A}{2k}x \cos kx + \frac{A}{2k^2}\sin kx, \end{aligned} \tag{4.39}$$

where we may drop the $\sin kx$ term.

Equation (4.37) appears in classical mechanics as the equation of motion (if $x = t$, the time variable) of a driven harmonic oscillator with no damping. The possible zero in the denominator $k^2 - q^2$ of the solution (4.38) suggests the possible appearance of an amplitude resonance where the amplitude $y(t)$ of the oscillation is maximal. In the special case considered here, there is no frictional damping. As a result, the amplitude at resonance grows without limit, as the solution $-(A/2k)t \cos kt$ at resonance shows. The linear dependence on A of the solutions shows that the amplitude of oscillation is twice as large when the system is driven twice as hard. This arises because the system is linear; that is, it is a consequence of the superposition principle shown in Eq. (4.14).

As another example, let us use the Dirac δ function

$$R(x) = \delta(x - x')$$

in Eq. (4.21). Equation (4.36) now gives

$$\begin{aligned} v_1(x) &= \begin{cases} 0, & x < x' \\ -y_2(x')/W(x'), & x > x' \end{cases} \\ &= -[y_2(x')/W(x')]\Theta(x - x'), \end{aligned} \tag{4.40a}$$

where $\Theta(t)$ is a step function. Similarly

$$v_2(x) = [y_1(x')/W(x')]\Theta(x - x'). \tag{4.40b}$$

The solution of an inhomogeneous DE with a δ-function inhomogeneity is called a *Green function*. From Eqs. (4.30) and (4.33), we find that it has the general form

$$G(x,x') = G_p(x,x') + c_1 y_1(x) + c_2 y_2(x), \tag{4.41}$$

where

$$G_p(x,x') = \{[-y_1(x)y_2(x') + y_1(x')y_2(x)]/W(x')\}\Theta(x - x'). \quad (4.42)$$

As usual, there are two arbitrary constants that can be chosen to fit two boundary conditions, either at the same point or at two different points.

Problems

4.5.1. Verify Eqs. (4.34b) and (4.35).

4.5.2. Obtain the solution of the homogeneous LDE

$$\left(\frac{d^2}{dx^2} + k^2\right)y(x) = 0$$

satisfying the boundary conditions

$$y(a) = \tfrac{1}{2}, \qquad y'(a) = \tfrac{1}{4}k \qquad \text{at} \qquad x = a = \frac{\pi}{2k}.$$

4.5.3. A frictionless simple harmonic oscillator driven by a force $F(t)$ satisfies the inhomogeneous LDE

$$\left(\frac{d^2}{dt^2} + \omega_0^2\right)x(t) = F(t).$$

If the driving force is $F(t) = a \cos \omega t$, show that a particular solution is

$$x_p(t) = A \cos \omega t/(\omega_0^2 - \omega^2).$$

Describe the motion of this driven oscillation when the *initial conditions* at $t = 0$ are $x(0) = 1$, $\dot{x}(0) = \frac{1}{2}\omega_0$.

4.5.4. Consider the differential equation

$$x^2y'' + (1 - 2a)xy' + a^2y = x^b, \qquad a \neq b.$$

(a) Show that one solution of the homogeneous equation is

$$y_1(x) = x^a.$$

(b) Obtain the linearly independent second solution of the homogeneous equation.

(c) Obtain the solution of the inhomogeneous equation satisfying the boundary conditions

$$y(1) = c, \qquad y'(1) = d.$$

4.5.5. Obtain a particular solution of the equation for the driven harmonic oscillator with damping

$$\left(\frac{d^2}{dt^2} + 2\beta\frac{d}{dt} + \omega_0^2\right)y(t) = A\,\delta(t - t')$$

by the method of variation of constants.

4.6. Green functions

We have seen in Section 3.6 that given a solution of the DE

$$\mathscr{D}(x)G(x,x') = \delta(x-x') \tag{4.43}$$

satisfying specified boundary conditions, the solution of the general inhomogeneous DE

$$\mathscr{D}(x)y(x) = R(x) \tag{4.44}$$

satisfying the *same* boundary conditions can be written in the integral form

$$y(x) = \int_a^b G(x,x')R(x')\,dx', \tag{4.45}$$

where $[a,b]$ is the interval for the variable x. The *Green function* $G(x,x')$ is independent of the inhomogeneity function $R(x)$, so that every solution of Eq. (4.44) satisfying the *same* boundary conditions is given by Eq. (4.45) involving the *same* Green function even though $R(x')$ itself might change.

We would like to show that $G(x,x')$ is made up of suitable homogeneous solutions joined together at $x=x'$ in a well-defined way.

If $x \neq x'$, the δ function is identically zero. Equation (4.43) is then a homogeneous equation. Therefore

$$G(x,x') = \begin{cases} G_<(x,x'), & x<x' \\ G_>(x,x'), & x>x' \end{cases} \tag{4.46}$$

are just solutions of the homogeneous equations. Suppose a full set of n boundary conditions (for an nth-order DE) has been specified, say at the lower limit a of the interval. $G_<(x,x')$ is now uniquely defined up to a multiplicative constant until we reach $x=x'$. On passing this point, we must switch over to another homogeneous solution $G_>(x,x')$ before continuing to the upper limit $x=b$. This second homogeneous solution must satisfy a full set of n boundary conditions, say at $x=b$. This does not completely determine G, however, because we have not yet determined if the normalizations of $G_<$ and $G_>$ can be chosen arbitrarily.

The answer is no. If G is a solution of Eq. (4.43), $2G$ is not; it is a solution of an equation with the inhomogeneity $2\,\delta(x-x')$. Thus the overall normalization is not arbitrary. This is because Eq. (4.43) is an inhomogeneous equation, not a homogeneous one.

We now show that the relative normalization between $G_<$ and $G_>$ is also not arbitrary. The δ function that appears in the nth-order inhomogeneous DE

$$y^{(n)}(x) = \frac{d^n}{dx^n}y(x) = f(x,y(x),\ldots,y^{(n-1)}(x)) + \delta(x-x') \tag{4.47}$$

can be considered the derivative of the unit step function $\Theta(x-x')$, because

$$\int_{x'-\varepsilon}^{x'+\varepsilon} \frac{d}{dx}\Theta(x-x')\,dx = \Theta(\varepsilon)-\Theta(-\varepsilon)=1.$$

Hence Eq. (4.47) will be satisfied if $y^{(n-1)}(x)$ has a unit step discontinuity at $x=x'$, while $y^{(n-2)}(x)$ is continuous but has a kink there. The δ function cannot be associated with $y^{(n-1)}(x)$ or one of the lower derivatives, for otherwise on differentiations in Eq. (4.47) it will generate derivatives of the δ function that are not present on the right-hand side. This shows that the only solution is one for which $y^{(n-1)}(x)$ has a unit step or jump discontinuity at $x=x'$. That is,

$$G_{>}^{(n-1)}(x',x') = G_{<}^{(n-1)}(x',x')+1,$$

while

$$G_{>}^{(k)}(x',x') = G_{<}^{(k)}(x',x'), \qquad \text{for} \qquad k<n-1. \tag{4.48}$$

To summarize, all derivatives $G^{(k)}(x,x')$ are smooth for $k \leqslant n$ except at $x=x'$, where $G^{(n-2)}$ has a kink, $G^{(n-1)}$ has a jump discontinuity, and $G^{(n)}$ has a δ-function discontinuity.

The n boundary conditions needed to specify a Green function for an nth-order DE do not have to be specified all at one point. They may be specified at two or more points in the interval $[a,b]$. For example, the Green function for a second-order DE may be specified by one boundary condition at a and another at b.

Example 4.6.1. Obtain the Green function for the DE

$$\left(\frac{d^2}{dx^2}+k^2\right)y(x) = \delta(x-x') \tag{4.49}$$

describing the motion of a driven string of length L with fixed ends at $x=0$ and L.

Solutions of the homogeneous DE are sine and cosine functions. Since the string ends are fixed [i.e., $y(0)=y(L)=0$], we must have

$$G_{<}(x,x') = a\sin kx, \qquad G_{>}(x,x') = b\sin k(x-L). \tag{4.50}$$

The unknown coefficients are now determined from Eq. (4.48):

$$a\sin kx' = b\sin k(x'-L),$$
$$ak\cos kx' + 1 = bk\cos k(x'-L).$$

These simultaneous equations have the unique solutions

$$\begin{aligned}\begin{pmatrix} a \\ b \end{pmatrix} &= \begin{pmatrix} \sin kx' & -\sin k(x'-L) \\ k\cos kx' & -k\cos k(x'-L) \end{pmatrix}^{-1} \begin{pmatrix} 0 \\ -1 \end{pmatrix} \\ &= \begin{pmatrix} \sin k(x'-L) \\ \sin kx' \end{pmatrix} (k\sin kL)^{-1}. \end{aligned} \tag{4.51}$$

From these results we can see that the Green function in Eq. (4.50) can be written more compactly in the form

$$G(x,x') = \sin kx_< \sin k(x_> - L)/(k \sin kL). \tag{4.52}$$

This possibility is a consequence of the type of boundary conditions it satisfies.

Example 4.6.2. Obtain a particular integral of the driven oscillator equation

$$\left(\frac{d^2}{dt^2} + 2\beta\frac{d}{dt} + \omega_0^2\right)y(t) = e^{\gamma t}, \qquad \beta > 0, \tag{4.53}$$

giving its motion as a function of the time t.

We shall look for a solution of the form

$$y_p(t) = \int_{-\infty}^{\infty} G(t,t')e^{\gamma t'}\, dt', \tag{4.54}$$

where the Green function satisfies the DE

$$\left(\frac{d^2}{dt^2} + 2\beta\frac{d}{dt} + \omega_0^2\right)G(t,t') = \delta(t-t'). \tag{4.55}$$

Since the boundary conditions (here called *initial conditions*) are not of interest, we may take the trivial solution

$$G_<(t,t') = 0, \qquad \text{for} \qquad t < t'. \tag{4.56}$$

This is actually a physically useful choice because we expect that nothing special happens for $t < t'$ before the impulsive driving force $\delta(t-t')$ acts on the system.

To obtain $g_>(t,t')$, we first calculate the two homogeneous solutions. These turn out to be $\exp(-\alpha t)$ with two different values of α

$$\alpha_{1,2} = \beta \pm (\beta^2 - \omega_0^2)^{1/2}. \tag{4.57}$$

Hence

$$G_>(t,t') = a_1 e^{-\alpha_1 t} + a_2 e^{-\alpha_2 t}. \tag{4.58}$$

It should satisfy the initial conditions at $t = t'$ obtainable from Eq. (4.48):

$$G_>(t',t') = 0, \qquad G_>^{(1)} = \frac{d}{dt'}G_>(t',t') = 1.$$

The linear coefficients a_i can now be calculated. They are readily found to be

$$a_1 = \frac{e^{\alpha_1 t'}}{\alpha_2 - \alpha_1}, \qquad a_2 = \frac{e^{\alpha_2 t'}}{\alpha_1 - \alpha_2}.$$

Thus

$$G_>(t,t') = (e^{-\alpha_1(t-t')} - e^{-\alpha_2(t-t')})/(\alpha_2 - \alpha_1), \qquad t > t'. \tag{4.59}$$

Finally y_p is calculated from Eq. (4.54)

$$\begin{aligned} y_p(t) &= \int_{-\infty}^{t} G_>(t,t')e^{\gamma t'}\,dt' \\ &= \frac{1}{\alpha_2-\alpha_1}\left(\frac{1}{\alpha_1+\gamma}-\frac{1}{\alpha_2+\gamma}\right)e^{\gamma t} \\ &= \frac{e^{\gamma t}}{(\alpha_1+\gamma)(\alpha_2+\gamma)} = \frac{e^{\gamma t}}{\gamma^2+2\beta\gamma+\omega_0^2}. \end{aligned} \tag{4.60}$$

One can check by inspection that this is indeed a solution of Eq. (4.53).

For DEs with constant coefficients such as Eq. (4.53), y_p itself can be obtained more readily by other means. This is not the case with the more general Eq. (4.21), for which the present approach is very useful. The reader can check that when this is done one recovers the same result as Eq. (4.42), obtained in the previous section by the method of separation of constants. (See Problem 4.6.4).

Problems

4.6.1. Calculate the Green function for the following differential equations in the interval $(-\infty,\infty)$:

(a) $\dfrac{d^2}{dx^2}y(x) = R(x), \qquad$ with $y(0)=0,\ y^{(1)}(0)=0$;

(b) $\left(\dfrac{d^2}{dx^2}-\lambda^2\right)y(x) = R(x), \qquad$ with $y(-\infty)=0,\ y(\infty)=0$.

4.6.2. Calculate the Green function for the differential equation

$$\left(\frac{d}{dx^2}+\frac{2}{x}\frac{d}{dx}-\frac{l(l+1)}{x^2}\right)y(x) = R(x)$$

in the interval $(0,\infty)$ satisfying the boundary conditions $y(0)=y(\infty)=0$.

4.6.3. Solve the driven oscillator equation

$$\left(\frac{d^2}{dt^2}+2\beta\frac{d}{dt}+\omega_0^2\right)y(t) = F(t)$$

satisfying the initial conditions $y(0)=0,\ \dot{y}(0)=v$.

4.6.4. Obtain Eq. (4.42) by using the method of this section.

4.7. Series solution of the homogeneous second-order linear differential equation

We have seen in Section 4.5 that, given one solution of a homogeneous second-order LDE, a second solution linearly independent of the first can be obtained by integration. If we are given both solutions, a particular solution of the inhomogeneous LDE can be calculated by integration. It remains for us to obtain at least one solution of the homogeneous equation.

It is often possible and useful to obtain a solution $y(x)$ of the homogeneous LDE $\mathscr{L}y(x) = 0$ in the form of a power series in powers of x. According to Frobenius, the power series should have the general form

$$y(x) = x^s(a_0 + a_1 x + a_2 x^2 + \cdots) = \sum_{\lambda=0}^{\infty} a_\lambda x^{\lambda+s}, \qquad a_0 \neq 0, \quad (4.61)$$

where the power s is not necessarily zero. (If $s = 0$, the series is a Taylor series.) By direct differentiation, we get

$$\begin{aligned} dy(x)/dx &= x^{s-1}[sa_0 + (s+1)a_1 x + (s+2)a_2 x^2 + \cdots] \\ &= \sum_{\lambda'=0}^{\infty} a_{\lambda'}(\lambda' + s)x^{\lambda'+s-1}, \end{aligned} \quad (4.62)$$

$$\begin{aligned} \frac{d^2y(x)}{dx^2} &= x^{s-2}[s(s-1)a_0 + (s+1)sa_1 x + (s+2)(s+1)a_2 x^2 + \cdots] \\ &= \sum_{\lambda'=0}^{\infty} a_{\lambda'}(\lambda' + s)(\lambda' + s - 1)x^{\lambda'+s-2}. \end{aligned} \quad (4.63)$$

Substitution of these into the LDE shows that the DE can be satisfied if the coefficient of each power of x vanishes. This set of conditions is often sufficient to define a solution $y(x)$, as the following two examples will demonstrate.

Example 4.7.1. Obtain a Frobenius-series solution for Eq. (4.27) describing harmonic oscillations

$$\left(\frac{d^2}{dx^2} + k^2\right)y(x) = 0. \quad (4.27)$$

Direct substitution of Eqs. (4.62) and (4.63) gives

$$\begin{aligned} 0 &= x^{s-2}[s(s-1)a_0 + (s+1)sa_1 x + (s+2)(s+1)a_2 x^2 + \cdots \\ &\quad + k^2x^2(a_0 + a_1 x + a_2 x^2 + \cdots)] \\ &= a_0 s(s-1)x^{s-2} + a_1(s+1)sx^{s-1} + [a_0 k^2 + (s+2)(s+1)a_2]x^s \\ &\quad + \cdots \\ &\quad + [a_\lambda k^2 + (s+\lambda+2)(s+\lambda+1)a_{\lambda+2}]x^{s+\lambda} \\ &\quad + \cdots. \end{aligned} \quad (4.64)$$

The coefficient of the lowest power x^{s-2} must vanish; that is,

$$a_0 s(s-1) = 0. \quad (4.65)$$

Since $a_0 \neq 0$, this requires that $s = 0$ or 1. Equation (4.65), which determines the power s of the Frobenius series (4.61) is called an *indicial equation*.

The coefficient of the next power x^{s-1} must also vanish:

$$a_1(s+1)s = 0. \tag{4.66}$$

This expression shows that if $s = 0$, a_1 can be nonzero, while $a_1 = 0$ is required if $s = 1$.

In a similar way, the coefficient of a higher power $x^{s+\lambda}$ vanishes if

$$a_{\lambda+2} = -\frac{k^2}{(s+\lambda+2)(s+\lambda+1)} a_\lambda. \tag{4.67}$$

In this way, the coefficients a_λ of the Frobenius series can all be determined step by step. Equation (4.67) is called a *recurrence relation.* When Eq. (4.67) is used to generate unknown coefficients a_λ with increasing index λ, the recurrence relation is said to be used in the *forward* direction.

The recurrence relation (4.67) contains the special feature that it steps up λ by 2. As a result, the coefficients a_λ are separated into two disjoint groups—one with even λ and another with odd λ. For the even chain, we find

$$\text{a. for } s = 0: \qquad a_2 = -\frac{k^2}{2} a_0, \qquad a_4 = -\frac{k^2}{(4)(3)} a_2 = \frac{k^4}{4!} a_0, \ldots,$$

$$\text{b. for } s = 1: \qquad a_2 = -\frac{k^2}{(3)(2)} a_0, \qquad a_4 = -\frac{k^2}{(5)(4)} a_2 = \frac{k^4}{5!} a_0, \ldots \tag{4.68}$$

The resulting solutions are, respectively,

$$\text{a.} \qquad y(x) = y_{\text{even}}(x) = 1 - \frac{k^2}{2!} x^2 + \frac{k^4}{4!} x^4 - \cdots,$$

$$\text{b.} \qquad y(x) = y_{\text{odd}}(x) = x\left(1 - \frac{k^2}{3!} x^2 + \frac{k^4}{5!} x^4 - \cdots\right), \tag{4.69}$$

where we have used $a_0 = 1$.

What about the odd chain of coefficients? There is no solution for $s = 1$, since $a_1 = 0$ and therefore all higher odd coefficients vanish by virtue of Eq. (4.67). For $s = 0$, $a_1 \neq 0$; hence from Eq. (4.67)

$$a_3 = -\frac{k^2}{(3)(2)} a_1, \qquad a_5 = -\frac{k^2}{(5)(4)} a_3 = \frac{k^4}{5!} a_1, \ldots.$$

The resulting solution for $a_1 = 1$ is just $y_{\text{odd}}(x)$. Thus this solution is not new. This is not unexpected, because Eq. (4.27) has only two linearly independent solutions, and they are already given in Eq. (4.69).

The formal solutions in Eq. (4.69) are meaningful only if these infinite series converge. To determine their convergence, we examine the ratio of successive terms, and find

$$\frac{a_{\lambda+2}}{a_\lambda}x^2 = -k^2\frac{x^2}{(s+\lambda+2)(s+\lambda+1)} \xrightarrow[\lambda\to\infty]{} 0$$

for any x. Hence both solutions converge to well-behaved functions of x for all finite values of x. Indeed, Eqs. (4.69) are just the Taylor expansions about $x=0$ of the the functions

$$y_{\text{even}}(x) = \cos kx, \qquad y_{\text{odd}}(x) = \frac{1}{k}\sin kx,$$

which may be seen by inspection to be the two linearly independent solutions of Eq. (4.27).

4.7.1. *Parity property*

According to Eq. (4.69)

$$y_{\text{even}}(-x) = y_{\text{even}}(x), \qquad y_{\text{odd}}(-x) = -y_{\text{odd}}(x) \tag{4.70}$$

are, respectively, an even and an odd function of x. The function $y_{\text{even}}(y_{\text{odd}})$ is then said to have even (odd) *parity*.

For a solution of a LDE to have a definite parity, the linear operator $\mathscr{L}$ must be even (or invariant) under the *parity operation* $x \to -x$; that is,

$$\mathscr{L}(-x) = \mathscr{L}(x). \tag{4.71}$$

To see this, we first note that if $y_1(x)$ is a solution of the DE, then $y_1(-x)$ must be a solution of the DE $\mathscr{L}(-x)y(-x)=0$. However, if Eq. (4.71) is also satisfied, we have

$$0 = \mathscr{L}(-x)y_1(-x) = \mathscr{L}(x)y_1(-x),$$

so that $y_2(x) = y_1(-x)$ is also a solution of the DE $\mathscr{L}(x)y(x)=0$. furthermore, if $y_1(-x)$ is linearly independent of $y_1(x)$, then the following two solutions of definite parity can be constructed:

$$\begin{aligned} y_{\text{even}}(x) &= \tfrac{1}{2}[y_1(x) + y_1(-x)] \\ y_{\text{odd}}(x) &= \tfrac{1}{2}[y_1(x) - y_1(-x)]. \end{aligned} \tag{4.72}$$

Thus it is the parity invariance of the differential operator in Eq. (4.27) that permits the extraction of solutions of definite parities shown in Eq. (4.69).

Physical fields (describing physical properties) in space are also said to have even (odd) parity if they are even (odd) functions of x. If they satisfy an LDE $\mathscr{L}(x)y(x)=0$, then the linear operator involved must be parity even. An LDE whose solution is a physical field is called a *field equation* or an *equation of state,* in the sense that its solution describes a state of existence of the physical property in space. Physical systems whose field equations contain only parity-even operators are said to be *parity conserving* because states (i.e., solutions) of definite parities can be constructed.

On the other hand, there are states of physical systems with unavoidably mixed parities. This feature is a consequence of the lack of parity invariance in the differential operators appearing in their field equations. Such systems are said to be *parity nonconserving* or *parity violating*. An example of a DE whose solutions cannot have a definite parity is the equation (for a classical oscillator with damping if $x = t$)

$$\left(\frac{d^2}{dx^2} + 2\beta\frac{d}{dx} + \omega_0^2\right)y(x) = 0.$$

The recurrence relation for a parity-nonconserving equation such as this does not break up into two disjoint chains of different parities.

That the use of recurrence relations may require some ingenuity is illustrated by the following example.

Example 4.7.2. Solve the *Bessel equation*

$$\left(x^2\frac{d^2}{dx^2} + x\frac{d}{dx} - (x^2 - \mu^2)\right)y(x) = 0. \tag{4.73}$$

This differential operator is parity invariant, so that there are separate even and odd chains of coefficients. The DE itself may be written with the help of Eqs. (4.61)–(4.63) in the form

$$\sum_{\lambda=\lambda_{\min}}^{\infty} b_\lambda x^{\lambda+s} = 0, \tag{4.74}$$

where $\lambda_{\min} = 0$. The conditions for a solution are therefore

$$b_\lambda = a_\lambda[(\lambda + s)^2 - \mu^2] - a_{\lambda-2} = 0, \qquad \lambda \geqslant 0. \tag{4.75}$$

(The derivation of the expression for b_λ is left as an exercise in Problem 4.7.1.)

The coefficient of the lowest power ($\lambda = 0$) gives the indicial equation

$$a_0(s^2 - \mu^2) = 0, \qquad s = \pm\mu. \tag{4.76}$$

The coefficient of the next power ($\lambda = 1$) determines the coefficient a_1 of the odd chain:

$$a_1[(s + 1)^2 - \mu^2] = a_1(2s + 1) = 0, \tag{4.77}$$

where use has been made of Eq. (4.76). Equation (4.77) is satisfied

1. For $s \neq -\frac{1}{2}$, if $a_1 = 0$;
2. For $a_1 \neq 0$, if $s = -\frac{1}{2}$. (4.78)

For $\lambda \geq 2$, Eqs. (4.75) and (4.76) can be used to simplify the recursion formula to

$$a_\lambda = -\frac{a_{\lambda-2}}{\lambda(\lambda + 2s)}. \tag{4.79}$$

However, trouble might develop when $\lambda + 2s$ vanishes in the denominator. This occurs when

$$s = -\tfrac{1}{2}\lambda(\lambda \geq 2) = -1, -\tfrac{3}{2}, -2, \ldots. \tag{4.80}$$

Two situations can be distinguished: (1) If $s = -\frac{3}{2}, -\frac{5}{2}, \ldots$ is a negative half-integer, trouble might occur at $\lambda = 3, 5, \ldots$, that is, in the odd chain. (2) If $s = -1, -2, \ldots$, is a negative integer, trouble might develop at $\lambda = 2, 4, \ldots$, that is, in the even chain.

To see how we can get out of these problems, we first consider the trouble in the odd chain. Suppose $2s = -m$, $m = 3, 5, \ldots$. Since $s \neq -\frac{1}{2}$, $a_1 = 0$ must hold. The next higher coefficients also vanish one by one until we reach $\lambda = m$, when

$$a_m = -\frac{a_{m-2}}{m(m-m)} = \frac{0}{0} = \text{anything}.$$

If we now take $a_m \neq 0$ and go on to higher coefficients with Eq. (4.79), we find the Frobenius series

$$y_{s=-m/2}(x) = x^{-m/2}(a_1 + \cdots + a_m x^m + a_{m+2}x^{m+2} + \cdots) \tag{4.81}$$

actually contains zeros before the term with a_m. Removing these "ghosts," we find a series with an effective index $s_{\text{eff}} = \frac{1}{2}m$:

$$y_{s=-m/2}(x) = x^{m/2}(a_m + a_{m+2}x^2 + \cdots).$$

Furthermore, the recursion formula for the coefficients beyond a_m may be written as

$$a_{m+\lambda} = -\frac{a_{m+\lambda-2}}{(m+\lambda)\lambda}. \tag{4.82}$$

This is actually identical to the recursion formula for the Frobenius solution with $s = \frac{1}{2}m$. The only difference is in the subscripts of the coefficients due to the counting of the ghosts in Eq. (4.81). We therefore conclude that the Frobenius solution for $s = -\frac{1}{2}m$ "collapses" into the solution for $s = \frac{1}{2}m$, and is not linearly independent of the latter.

The same result can also be obtained for the even series when s is a negative integer. (The demonstration is left as an exercise.) In both cases we conclude that the Frobenius method does not necessarily give two linearly independent solutions even when s has two distinct solutions. This is all right, because the linearly independent second solution can always be obtained from the first by the method of Section 4.5.

Problems

4.7.1. Verify Eq. (4.75).

4.7.2. Show that the Frobenius-series solution of the Bessel equation (4.73) for $s = -m$, m any positive integer, is proportional to the solution for $s = m$. If we denote the solution of Eq. (4.73) as $J_s(x)$, then we may write

$$J_{-m}(x) = (-1)^m J_m(x),$$

where the choice of the proportionality constant $(-1)^m$ is a matter of convention.

4.7.3. Obtain the power-series solutions of each of the following homogeneous differential equations:
(a) $y' + 2y = 0$;
(b) $y'' + 4y' - 5y = 0$;
(c) $[x(1-x)y'' - xy' - y] = 0$;
(d) $xy' + (\sin x)y = 0$;
(e) $(\cos x)y'' + xy' + ay = 0$.

4.8. Differential eigenvalue equations and orthogonal functions

The *Legendre equation*

$$\left((1-x^2)\frac{d^2}{dx^2} - 2x\frac{d}{dx} + c\right)y(x) = 0 \tag{4.83}$$

has the unusual feature that at $x = \pm 1$ the first term disappears, leaving a first-order DE. These points $x = \pm 1$ are said to be *singular points* of the DE. Something interesting happens at these points, as we shall show in this section.

Let us first discuss its Frobenius-series solutions in a systematic way. The differential operator in Eq. (4.83) is parity invariant, as happens so often in DEs of interest to physicists. Thus there are as a rule two linearly independent solutions $y_{\text{even}}(x)$ and $y_{\text{odd}}(x)$, one of each parity. The coefficients of the Frobenius series for the entire DE is of the form of Eq. (4.74) with

$$b_\lambda = a_{\lambda+2}(\lambda + s + 2)(\lambda + s + 1) - a_\lambda[(\lambda + s)(\lambda + s + 1) - c] = 0. \tag{4.84}$$

For $\lambda = -2$ and -1, we find the same result as for the DE (4.27), namely

$$a_0 s(s-1) = 0, \qquad a_1(s+1)s = 0.$$

Hence $a_0 \neq 0$ is permitted if $s = 0$ or 1, while $a_1 \neq 0$ occurs only for $s = 0$.

The recursion formula from Eq. (4.84) is

$$a_{\lambda+2} = a_\lambda \frac{(\lambda + s)(\lambda + s + 1) - c}{(\lambda + s + 2)(\lambda + s + 1)}. \tag{4.85}$$

The subscript is stepped up by 2, so that all terms of the Frobenius series have the same parity. For $s = 0$, we have both an even and an odd chain of coefficients:

1. Even chain:

$$a_0 \to a_2 = a_0\left(-\frac{c}{2}\right) \to a_2 = a_2\left(\frac{6-c}{12}\right) \to a_6 \cdots,$$

2. Odd chain:

$$a_1 \to a_3 = a_1\left(\frac{2-c}{6}\right) \to a_5 = a_3\left(\frac{12-c}{20}\right) \to a_7 \cdots. \tag{4.86}$$

These yield the solutions

$$y_{\text{even}}(x) = 1 - \frac{c}{2}x^2 + \frac{c}{2}\left(\frac{c-6}{12}\right)x^4 + \cdots, \tag{4.87a}$$

$$y_{\text{odd}}(x) = x + \frac{2-c}{6}x^3 + \frac{2-c}{6}\left(\frac{12-c}{20}\right)x^5 + \cdots, \tag{4.87b}$$

where we have set $a_0 = 1$ and $a_1 = 1$, respectively. For $s = 1$, only an even chain is allowed. The solution turns out to be just $y_{\text{odd}}(x)$, as expected.

Next, we must determine if the Frobenius solutions (4.87) converge. The ratio of successive terms is

$$\frac{a_{\lambda+2}}{a_\lambda}x^2 = \frac{(\lambda+s)(\lambda+s+1)-c}{(\lambda+s+2)(\lambda+s+1)}x^2 \xrightarrow[\lambda\to\infty]{} x^2.$$

Hence the series converges if $|x| < 1$ and diverges if $|x| > 1$. (It turns out to be divergent also for $|x| = 1$.)

An important special case occurs when

$$c = c_l \equiv l(l+1), \qquad l = 0, 1, 2, \ldots$$
$$= \text{any non-negative integer}. \tag{4.88}$$

For each of these special values, one of the two chains of coefficients for $s = 0$ shown in Eq. (4.86) terminates at a_l, because

$$a_{l+2} = a_l \frac{l(l+1) - c_l}{(l+2)(l+1)} = 0 = a_{l+4} = \cdots. \tag{4.89}$$

In particular, the even chain terminates when l is even, while the odd chain terminates when l is odd. The corresponding Frobenius solution in Eq. (4.87) then simplifies to a polynomial. It is called the *Legendre polynomial* of degree l, or $P_l(x)$. (The degree of a polynomial is the value of its highest power.) Legendre polynomials occur so frequently in mathematics that the choice of the arbitrary multiplicative constant

a_l giving $P_l(x=1)=1$ has become standard:

$$\begin{aligned} &P_0(x)=1, && P_1(x)=x, \\ &P_2(x)=\tfrac{1}{2}(3x^2-1), && P_3(x)=\tfrac{1}{2}(5x^3-3x), \\ &P_4(x)=\tfrac{1}{8}(35x^4-30x^2+3), && \text{etc.} \end{aligned} \tag{4.90}$$

Legendre polynomials, being finite power series, are finite for all finite values of x. In particular, they are also defined at the singular points $x=\pm 1$ of the DE. This nice behavior at $x=\pm 1$ is the main reason for considering the special case $c=c_l$, because when $c \neq c_l$ the function is infinite at $x=\pm 1$ and therefore cannot be used to describe bounded physical properties at these points.

The other chain of coefficients in Eq. (4.86) does not terminate because its parity is opposite to that of the chain that does, and no coefficient in its chain ever vanishes. The corresponding Frobenius solution will remain an infinite series. As discussed earlier, this series converges to a well-defined function for $|x|<1$, and is called a *Legendre function of the second kind*, $Q_l(x)$. This function is also defined for $|x|>1$ if by $Q_l(x)$ we mean the linearly independent second solution of Eq. (4.83). However, it is not given by the Frobenius solution (4.61), which does not converge for $|x|>1$. It may be calculated either by the method of Section 4.5 or by a Frobenius series in inverse powers of x. At the regular points $x=\pm 1$, the function $Q_l(x)$ turns out to be infinite.

The special nature of the Legendre polynomials $P_l(x)$ at the singular points $x=\pm 1$ of Eq. (4.83) has a very useful description in the language of eigenvalue problems. An eigenvalue equation in mathematics involves a general equation for an unknown quantity $y(c)$ of the form

$$\mathscr{L}y(c)=cy(c), \tag{4.91}$$

where $\mathscr{L}$ is a linear operator that satisfies the linearity property of Eq. (4.10) and c is a scalar constant. Sometimes Eq. (4.91) does not have a solution unless $c=c_i$, $i=1, 2, \ldots, n$, is the one of a set of eigenvalues. That is the situation for the matrix eigenvalue equation (2.49).

If $\mathscr{L}$ is a differential operator in the variable x, the solutions are of course functions of x, so that Eq. (4.91) should be written more specifically as

$$\mathscr{L}(x)y(x;c)=cy(x;c). \tag{4.92}$$

It is possible that $y(x;c)$ for any c is finite everywhere except at a number of singular points of the equation where it is finite only when $c=c_l$ $(l=0, 1, 2, \ldots)$ is one of a set of eigenvalues. This is the situation for the Legendre DE. It is also possible that $y(x;c)$ is defined everywhere for all c, but only for a set of eigenvalues $c=c_i$ will the function $y(x;c=c_i)$ have a certain desirable property. This is the situation for the trigonometric functions appearing in the Fourier series for the interval $-\pi \leqslant x \leqslant \pi$. In this case, we are only interested in eigenfunctions that are periodic in x with the same period of 2π. A DE such as Eq. (4.92), in which the solution has been further selected by imposing an additional eigenvalue condition leading to the choice $c = c_i$, $i = 1, 2, \ldots$, is called a *differential eigenvalue equation*. Each of the allowed values c_i is called an *eigenvalue* and the corresponding solution $y(x;c_i)$ is its *eigenfunction*.

Since the eigenvalues of both the Legendre polynomials and the Fourier-series eigenfunctions are real, the corresponding differential operators must be Hermitian. The concept of Hermiticity for a differential operator turns out to be much more complicated than that for matrices. For example, it depends on the boundary conditions satisfied by the eigenfunctions, as well as on the operator itself. Nevertheless, it is also true that the eigenfunctions of a Hermitian differential operator are orthogonal or can be orthogonalized. (The orthogonality between two functions is defined in terms of an inner product that is an integral of a product of the functions. We have come across such inner products in Chapter 3.) For this reason we call fourier-series functions *orthogonal functions* and the Legendre polynomials *orthogonal polynomials*. An earlier discussion of orthogonal systems has been given in Section 3.10.

Orthogonal functions are useful in the expansion of arbitrary functions, especially in connection with the solution of partial differential equations. This application will be described in the next section.

Problem

4.8.1. Obtain a solution of the Legendre differential equation, Eq. (4.83), with $c = 0$, which is linearly independent of $P_0(x)$. [Answer: $Q_0(x) = \frac{1}{2} \ln[(1 + x)/(1 - x)]$.)

4.9. Partial differential equations of physics

In physics, there are differential equations of motion that describe the response of systems to external disturbances. There are also differential equations of states, or field equations, whose solutions give the space–time dependence of physical properties. These are, of course, *partial differential equations* (PDE) in the four variables x, y, z, and t. Some common PDEs of physics are shown in Table 4.1.

All these equations are linear. They are of second order in the space variables x,y,z (or x_1,x_2,x_3) and of first or second order in time. The use of differential operators guarantees invariance with respect to space translations and time displacements. The scalar differential operator ∇^2 is actually the simplest operator that will also respect invariance under rotation and the parity operation $x_i \to -x_i$.

It is also amusing to note that certain equations, such as the wave equation, is second order in time, so that they are invariant under time reversal, that is, the transformation $t \to -t$. A movie of a wave

Table 4.1. Some common partial differential equations in physics

Equation	PDE	Physical applications
Wave	$\nabla^2 u(\mathbf{r},t) = \frac{1}{c^2}\frac{\partial^2}{\partial t^2}(u(\mathbf{r},t)$	Wave motion
Helmholtz	$\nabla^2 u(\mathbf{r}) = -\frac{\omega^2}{c^2}u(\mathbf{r})$	Wave motion of frequency ω
Laplace	$\nabla^2 \phi(\mathbf{r}) = 0$	Electrostatic potential in free space
Poisson	$\nabla^2 \phi(\mathbf{r}) = \rho(\mathbf{r})$	Electrostatic potential with charge density
Diffusion	$\nabla^2 u(\mathbf{r},t) = \frac{1}{D}\frac{\partial}{\partial t}u(\mathbf{r},t)$	Diffusion and heat conduction
Schrödinger wave	$\left(-\frac{\hbar^2}{2m}\nabla^2 + V(\mathbf{r})\right)\psi(\mathbf{r}) = E\psi(\mathbf{r})$	Equation of state for a quantum-mechanical system

propagating to the left run backwards looks just like a wave propagating to the right. In diffusion or heat conduction, the field equation (for the concentration or temperature field) is only first order in time. The equation does not, and should not, satisfy time-reversal invariance, since heat is known to flow from a high-temperature region to a low-temperature region, never the other way around. A movie of a pool of water solidifying into a block of ice on a hot day has obviously been run backwards.

We have not included in Table 4.1 the famous Maxwell's equations [see Eq. (1.115)] of electromagnetism. These are a set of four coupled first-order PDEs for two vector fields—the electric field $\mathbf{E}$ and the magnetic induction $\mathbf{B}$. These equations are actually equivalent to two wave equations, one for each of these vector fields. Indeed, solutions to Maxwell's equations can be obtained from knowledge of the solutions of the wave equation.

4.10. Separation of variables and eigenfunction expansions

Under certain circumstances the solution of a PDE may be written as a sum of terms, each of which is a product of functions of one of the variables. This is called a solution by a *separation of variables*. The method is conveniently described by an example.

Let us consider first the one-dimensional wave equation describing for example the transverse vibrations of a string:

$$\frac{\partial^2}{\partial x^2}u(x,t) = \frac{1}{c^2}\frac{\partial^2}{\partial t^2}u(x,t). \tag{4.93}$$

It can be verified by direct substitution that the general form of the solution is

$$u(x,t) = f(x - ct) + g(x + ct), \tag{4.94}$$

where the two linearly independent terms on the right-hand side (RHS) represent waves propagating along the $+x$ and $-x$ direction, respectively.

The method of separation of variables does not require such insights into the nature of the solution right from the beginning. One simply looks for the possibility of solutions with the separable form

$$u(x,t) = X(x)T(t). \tag{4.95}$$

A direct substitution of Eq. (4.95) into the wave equation gives

$$\frac{\partial^2}{\partial x^2} u(x,t) = T(t) \frac{d^2X(x)}{dx^2} = \frac{1}{c^2} X(t) \frac{d^2T(t)}{dt^2}.$$

Thus

$$\frac{1}{X(x)} \frac{d^2X(x)}{dt^2} = \frac{1}{c^2T(t)} \frac{d^2T(t)}{dt^2} = \lambda, \tag{4.96}$$

where the left-hand side (LHS) is a function of x only, while the RHS is a function of t only. If the original PDE is to be satisfied, these two sides must be equal. This is possible only if each side is equal to a function *neither* of x *nor* of t. That is, each side must be equal to a constant, say λ. We have as a result two *separated* ordinary DEs:

$$\frac{d^2X(x)}{dx^2} = \lambda X(x), \tag{4.97a}$$

$$\frac{d^2T(t)}{dt^2} = c^2\lambda T(t). \tag{4.97b}$$

The separated ordinary DEs are not completely independent of each other, however, because the same *separation constant* λ must appear in both. The solution $X(x)$ of Eq. (4.97a) has the general solution:

$$X(x) = A \cos(x\sqrt{-\lambda}) + B \sin(x\sqrt{-\lambda})/\sqrt{-\lambda}, \tag{4.98}$$

while the solution for $T(t)$ is rather similar and contains the constant $c^2\lambda$ instead of λ.

As a rule, all possible values of the separation constant λ are allowed unless explicitly forbidden. Certain values of λ could be forbidden when the corresponding solution $X(x)$, which depends on λ as shown in Eq. (4.98), does not have the right properties. The properties in question are the boundary conditions that select one or more solutions from the infinitely many possibilities contained in the general solution. It can happen that one or more of these boundary conditions can be satisfied only when the separation constant takes on one of a set of special values. This set then contains the only permissible values, or *eigenvalues,* for the problem. The corresponding solutions are called their *eigenfunctions*.

The following example illustrates this point. Suppose, as in the case of the Fourier series, we are interested in solutions with a period of 2π, that is, the solutions 1, $\cos n\pi$, and $\sin n\pi$, where n is a positive integer. Then the only permissible separation constants are

$$\lambda_n = -n^2, \qquad n = 0, 1, 2, \ldots . \tag{4.99}$$

for $\lambda = \lambda_n$, we have the wave function

$$X_n(x)T_n(t) = a_n \cos nx \cos cn. + b_n \sin nx \cos cnt$$
$$+ c_n \cos nx \sin cnt + d_n \sin nx \sin cnt. \tag{4.100}$$

Since the one-dimensional wave equation (4.93) is linear, the general solution periodic in x with period 2π is the linear superposition

$$u(x,t) = \tfrac{1}{2}a_0 + \sum_{n=1}^{\infty} X_n(x)T_n(t) \tag{4.101}$$

of all the possible solutions. The double series defined by Eqs. (4.100) and (4.101) is called a *double Fourier series.*

In a similar way, the general solution of the three-dimensional wave equation that is periodic in $\mathbf{r}$ with a period of 2π is the quadruple Fourier series

$$u(\mathbf{r},t) = \tfrac{1}{2}a_0 + \sum_{l,m,n=1}^{\infty} X_l(x)Y_m(y)Z_n(z)T_{lmn}(t). \tag{4.102}$$

The derivation of this expression is left as an exercise.

Now that we have obtained a general solution (4.101) of the one-dimensional wave equation (4.93), it would be nice to extract from it the physical insight contained in Eq. (4.94), namely that there is a wave traveling along $\mathbf{e}_x$ and a wave traveling along $-\mathbf{e}_x$. This is readily done by rewriting Eq. (4.100) in the form

$$X_n(x)T_n(t) = A_n \cos n(x - ct) + B_n \sin n(x - ct)$$
$$+ C_n \cos n(x + ct) + D_n \sin n(x + ct), \tag{4.103}$$

where the coefficients A_n, B_n, C_n, and D_n can be related to those in Eq. (4.100). Thus we have gained the physical insight contained in Eq. (4.94) by the method of separation of variables when the situation might not have been obvious at the beginning.

To recapitulate, we note that in the method of separation of variables we look for a solution that is made up of a sum of products of functions of single variables. One or more of the boundary or other conditions can be absorbed into the general solution, which now involves eigenfunctions of one or more separated, ordinary differential equations. For this reason, the method is also referred to as an *eigenfunction expansion.*

Problems

4.10.1. Derive the general solution shown in Eq. (4.102) of the three-dimensional wave equation whose solutions are periodic in $\mathbf{r}$ with a period of 2π.

4.10.2. Obtain the separated equations of the Schrödinger wave equation (Table 4.1) in rectangular coordinates when the potential function $V(\mathbf{r})$ is

$$V(\mathbf{r}) = \tfrac{1}{2}m\omega^2 r^2 = \tfrac{1}{2}m\omega^2(x^2 + y^2 + z^2).$$

4.11. Boundary and initial conditions

The complete determination of a solution of a PDE requires the specification of a suitable set of boundary and initial conditions. The boundaries may not be just points, but, depending on the dimension, they can be lines or surfaces. Just what constitutes a suitable set of boundary conditions is a rather complicated question for PDEs. The answer depends on the nature of the PDE, the nature of the boundaries, and the nature of the boundary conditions. It is not our intention to describe this complicated situation. Rather, we would like to illustrate how boundary conditions can be imposed on functions of more than one variable. We shall restrict ourselves to the wave equation solved in rectangular coordinates.

4.11.1. Vibrations of a string

Let us consider first the one-dimensional vibrations of a string rigidly attached to a support at the points $x = 0$ and L. They are described by the one-dimensional wave equation (4.93) or the separated ordinary differential equations (4.97). The boundary conditions at $x = 0$ and L are satisfied if

$$X_n(x) = \sin\left(\frac{n\pi}{L}x\right), \tag{4.104}$$

that is, if

$$\lambda = \lambda_n = -\left(\frac{n\pi}{L}\right)^2, \qquad n = 1, 2, \ldots, \tag{4.105}$$

are used for the separation constant. The eigenfunction (4.104) of Eq. (4.97a) belonging to the eigenvalue λ_n is said to describe the nth *normal mode* (or eigenmode) of the vibration of the string. Figure 4.1 shows the first three normal modes. We note that there are points

$$r_m = \frac{mL}{n}, \qquad m = 1, \ldots, n-1,$$

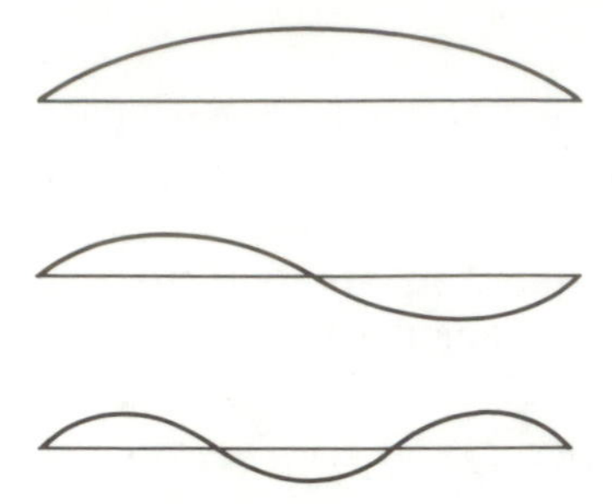

Fig. 4.1. The first three normal modes of a vibrating string with fixed ends.

at which the displacement is always zero; that is,

$$X_n(r_m) = \sin(m\pi) = 0.$$

These are called *nodal points*.

The time factor $T_n(t)$ associated with $X_n(x)$ is the solution of Eq. (4.97b) for the same separation constant λ_n of Eq. (4.105). Since $\lambda_n < 0$, we can see with the help of Eq. (4.98) that

$$T_n(t) = C_n \cos \omega_n t + D_n \sin \omega_n t, \tag{4.106}$$

where $\omega_n = n\pi c/L$ is the frequency of vibration of the nth normal mode. Hence the wave function of a vibrating string fixed at $x = 0$ and L has the general eigenfunction expansion

$$\begin{aligned} u(x,t) &= \sum_{n=1}^{\infty} X_n(x)T_n(t) \\ &= \sum_{n=1}^{\infty} \sin\left(\frac{n\pi}{L}x\right)(C_n \cos \omega_n t + D_n \sin \omega_n t). \end{aligned} \tag{4.107}$$

If we pluck a string at time $t = 0$, which normal modes will be excited? How strongly? The answer depends of course on how we pluck the string, that is, on the *initial conditions* at $t = 0$. Since the differential equation (4.97b) is second order in time, we need two initial conditions—the displacement $u_0(x)$ and the velocity $v_0(x)$ of the string at $t = 0$. Thus

$$u(x,t=0) = u_0(x) = \sum_{n=1}^{\infty} \sin\left(\frac{n\pi}{L}x\right)C_n.$$

This is a Fourier sine series for which the Fourier coefficients C_n can be shown to be

$$C_n = \frac{2}{L}\int_0^L \sin\left(\frac{n\pi}{L}x\right)u_0(x)\,dx. \tag{4.108}$$

Similarly

$$\frac{\partial}{\partial t}u(x,t)\bigg|_{t=0} = v_0(x) = \sum_{n=1}^{\infty} D_n\omega_n \sin\left(\frac{n\pi}{L}x\right),$$

so that

$$D_n = \frac{2}{\omega_n L} \int_0^L \sin\left(\frac{n\pi}{L} x\right) v_0(x)\, dx. \tag{4.109}$$

These linear coefficients describe the excitation strength of various normal modes.

4.11.2. Vibrations of a rectangular drum

The vibrations of a two-dimensional membrane fixed at the boundaries $x=0$, $x=a$, $y=0$, and $y=b$ can be described in a similar manner. The resulting eigenfunction expansion is

$$u(x,y,t) = \sum_{m,n=1}^{\infty} \sin\left(\frac{m\pi}{a} x\right) \sin\left(\frac{n\pi}{b} y\right)(C_{mn} \cos \omega_{mn} t + D_{mn} \sin \omega_{mn} y), \tag{4.110}$$

where

$$\omega_{mn} = \left(\frac{m^2}{a^2} + \frac{n^2}{b^2}\right)^{1/2} \pi c \tag{4.111}$$

is the frequency of the (m,n) normal mode. Figure 4.2 shows a few examples of the normal modes of vibration of a rectangular drum. The broken lines mark *nodal lines* where the displacement is zero at all times.

The strength with which various normal modes are excited depends on the initial conditions

$$u(x,y,t=0) = u_0(x,y), \qquad \frac{\partial}{\partial t} u(x,y,t)\bigg|_{t=0} = v_0(x,y). \tag{4.112}$$

From these functions the coefficients of the eigenfunction expansion (4.110) can readily be calculated:

$$\begin{Bmatrix} C_{mn} \\ D_{mn} \end{Bmatrix} = \frac{4}{ab} \int_0^a dx \int_0^b dy \sin\left(\frac{m\pi}{a} x\right) \sin\left(\frac{n\pi}{b} y\right) \begin{Bmatrix} u_0(x,y) \\ v_0(x,y) \end{Bmatrix}. \tag{4.113}$$

It should be pointed out, however, that, while the method of eigenfunction expansion is completely general, it tends to be rather cumbersome. In simple situations, it might be possible to solve a problem without using it at all. An example of this is provided by the following problem.

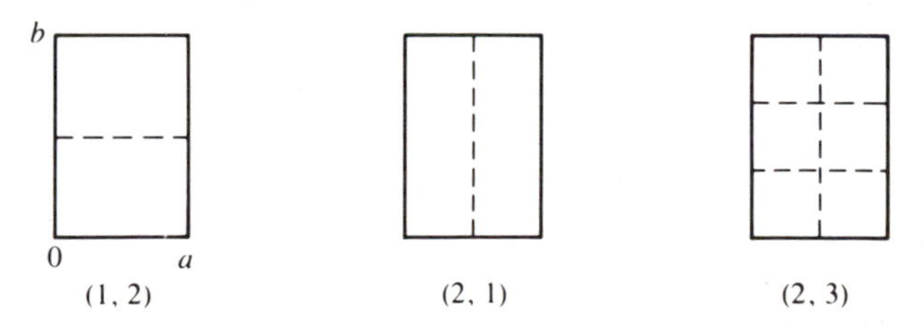

Fig. 4.2. A few normal modes of vibration of a rectangular drum.

Example 4.11.1. Obtain the solution of the one-dimensional wave equation satisfying the following boundary conditions:

$$u(x,0) = u_0(x)$$

$$\frac{\partial}{\partial t} u(x,0) = 0.$$

According to D'Alembert, the general form of the one-dimensional wave function is Eq. (4.94)

$$u(x,t) = f(x - ct) + g(x + ct).$$

At $t = 0$

$$\begin{aligned} u(x,0) &= u_0(x) = f(x) + g(x) \\ \frac{\partial}{\partial t} u(x,0) &= 0 = -cf'(x) + cg'(x). \end{aligned} \tag{4.114a}$$

That is,

$$g'(x) = \frac{d}{dx} g(x) = f'(x).$$

This means that

$$g(x) = f(x) + C, \tag{4.114b}$$

where C is an integration constant. Equations (4.114) have the solution

$$\begin{aligned} f(x) &= \tfrac{1}{2}u_0(x) - \tfrac{1}{2}C \\ g(x) &= \tfrac{1}{2}u_0(x) + \tfrac{1}{2}C \\ \therefore u(x,t) &= \tfrac{1}{2}u_0(x - ct) - \tfrac{1}{2}C + \tfrac{1}{2}u_0(x + ct) + \tfrac{1}{2}C \\ &= \tfrac{1}{2}u_0(x - ct) + \tfrac{1}{2}u_0(x + ct). \end{aligned} \tag{4.115}$$

Problems

4.11.1. An infinite string has an initial displacement

$$u_0(x) = \exp(-x^2)$$

and zero initial velocity. Obtain the wave function $u(x, t)$ for all times.

4.11.2. An infinite string has an initial displacement $u_0(x)$ and an initial velocity

$$v(x) = u'(x,0) = v_0 \exp(-ax).$$

Use D'Alembert's method to show that the wave function at all times is

$$u(x,t) = \tfrac{1}{2}[u_0(x - ct) + u_0(x + ct)] - \frac{v_0}{2ac}(e^{-a(x-ct)} - e^{-a(x+ct)}).$$

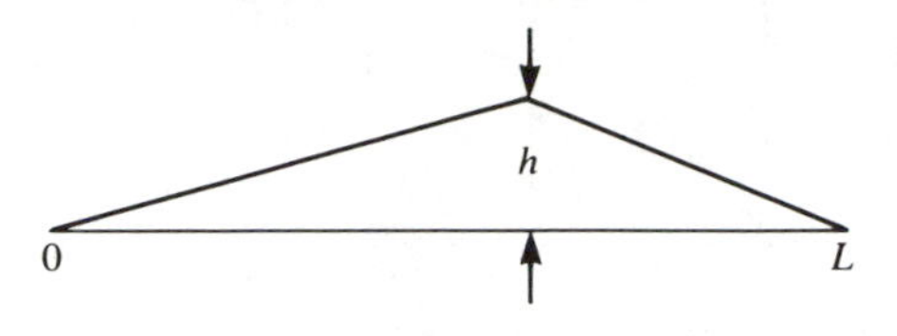

Fig. 4.3.

4.11.3. A stretched string of length L is fixed at each end. Suppose it has the initial displacement shown in Fig. 4.3

$$u_0(x) = \begin{cases} h\left(\frac{x}{a}\right), & x < a, \\ h\left(\frac{L-x}{L-a}\right), & x \geqslant a, \end{cases}$$

and zero initial velocity. Show that its motion is described by the function

$$u(x,t) = \frac{2hL^2}{\pi^2 a(L-a)} \sum_{n=1}^{\infty} \left(\frac{1}{n^2} \sin \frac{n\pi a}{L}\right) \sin \frac{n\pi x}{L} \cos \frac{n\pi c}{L} t$$

at subsequent times.

4.11.4. (a) A one-dimensional rod of length a originally at temperature T_0 has its ends placed in contact with a heat reservoir at temperature $T = 0$. Show that its subsequent temperature is

$$T(x,t) = T_0 \sum_{l=\text{odd integer}} \left(\frac{4}{l\pi}\right) \sin\left(\frac{l\pi}{a} x\right) e^{-(l\pi/a)^2 \kappa t},$$

where κ is the thermal conductivity.

(b) A cube of side a originally at temparature T_0 is suddenly placed in a heat reservoir at temperature $T = 0$. Show that its subsequent temperature is

$$T(x,t) = T_0 \sum_{\text{odd } l,m,n} \frac{64}{lmn\pi^3} \sin \frac{l\pi}{a} x \sin \frac{m\pi}{a} y \sin \frac{n\pi}{a} z$$
$$\times e^{-(l^2+m^2+n^2)(\pi/a)^2 \kappa t}.$$

(c) What would be the answers if the temperature of the heat reservoir had been T_1?

4.11.5. Consider the vibrations of a square drum of side a.

(a) Show that the (m,n) and (n,m) modes are degenerate in frequency; that is, $\omega_{mn} = \omega_{nm}$.

(b) Obtain the eigenfunctions $u_{mn}(x,y)$ of the (m,n) mode.

(c) Show that the nodal line of the hybrid mode described by each of the following spatial functions is as sketched in Fig. 4.4.

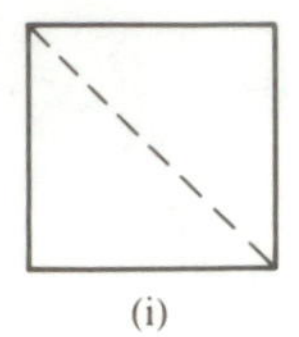
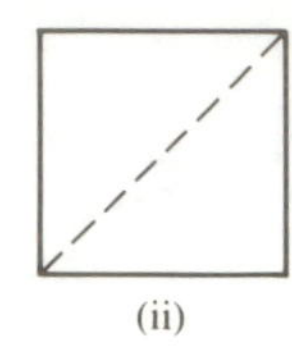
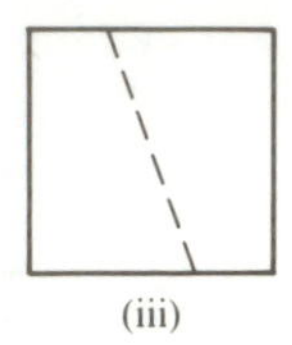
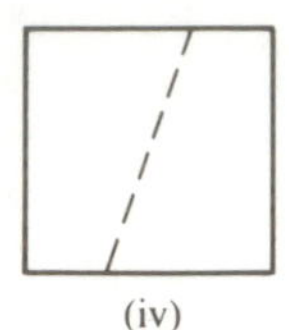

(i) (ii) (iii) (iv)

Fig. 4.4.

$$\begin{aligned}
\text{(i)}\ u(x,y) &= [u_{12}(x,y) + u_{21}(x,y)]/\sqrt{2},\\
\text{(ii)}\ u(x,y) &= [u_{12}(x,y) - u_{21}(x,y)]\sqrt{2},\\
\text{(iii)}\ u(x,y) &= [u_{12}(x,y) + 2u_{21}(x,y)]/\sqrt{5},\\
\text{(iv)}\ u(x,y) &= [u_{12}(x,y) - 2u_{21}(x,y)]/\sqrt{5}.
\end{aligned}$$

4.12. Separation of variables for the Laplacian

After separating out the time dependence, most of the PDE of Table 4.1 can be reduced to the form

$$(\nabla^2 + k^2)u(\mathbf{r}) = 0, \tag{4.116}$$

where the constant k^2 is zero for the Laplace equation, positive for the Helmholtz equation, and negative for the diffusion equation. The only exception is the Schrödinger equation (Table 4.1), where an additional term $-v(\mathbf{r})\phi(\mathbf{r})$ appears with

$$v(\mathbf{r}) = -\frac{2m}{\hbar^2} V(\mathbf{r}), \tag{4.117}$$

proportional to a potential function $V(\mathbf{r})$.

The Laplacian ∇^2, which appears in Eq. (4.116), is the simplest scalar differential operator of $\mathbf{r}$ that is invariant under rotation and parity transformations. It has a simple form when expressed in terms of the following common coordinate systems:

$$\begin{aligned}
\nabla^2 &= \frac{\partial^2}{\partial x^2} + \frac{\partial^2}{\partial y^2} + \frac{\partial^2}{\partial z^2}, \qquad \text{rectangular coordinates,}\\
&= \frac{1}{\rho}\frac{\partial}{\partial \rho}\left(\rho \frac{\partial}{\partial \rho}\right) + \frac{1}{\rho^2}\frac{\partial^2}{\partial \phi^2} + \frac{\partial^2}{\partial z^2}, \qquad \text{cylindrical coordinates,}\\
&= \frac{1}{r^2 \sin\theta}\left[\sin\theta \frac{\partial}{\partial r}\left(r^2 \frac{\partial}{\partial r}\right) + \frac{\partial}{\partial \theta}\left(\sin\theta \frac{\partial}{\partial \theta}\right)\right.\\
&\quad \left. + \frac{1}{\sin\theta}\frac{\partial^2}{\partial \phi^2}\right], \qquad \text{spherical coordinates.}
\end{aligned} \tag{4.118}$$

The Laplacian ∇^2 in rectangular coordinates is just a sum of the separate partial differential operators. Hence the development of Section 4.10 can be used to show that the solution of Eq. (4.116) is separable in rectangular coordinates. Even the Schrödinger equation is separable in rectangular coordinates if

$$v(\mathbf{r}) = v_1(x) + v_2(y) + v_3(z) \tag{4.119}$$

can be written as a sum of functions of x, y, or z individually. This is the case for a harmonic-oscillator potential for which $v(\mathbf{r})$ is proportional to r^2.

To separate Eq. (4.116) in cylindrical coordinates, we assume the separable form

$$u(\mathbf{r}) = u(\rho, \phi, z) = R(\rho)\Phi(\phi)Z(z) \tag{4.120}$$

for the solution u. The z part of the PDE then separates out cleanly as

$$\frac{d^2}{dz^2} Z(z) = \alpha^2 Z(z) \tag{4.121a}$$

with the help of the separation constant α^2, leaving the PDE

$$\left[\frac{1}{\rho}\frac{\partial}{\partial\rho}\left(\rho\frac{d}{d\rho}\right) + \frac{1}{\rho^2}\frac{\partial^2}{\partial\phi^2} + k^2 + \alpha^2\right]R(\rho)\Phi(\phi) = 0. \tag{4.122}$$

This may be written as

$$\frac{1}{R(\rho)}\left[\rho\frac{d}{d\rho}\left(\rho\frac{d}{d\rho}\right) + \rho^2(k^2 + \alpha^2)\right]R(\rho) = -\frac{1}{\Phi(\phi)}\frac{d^2}{d\phi^2}\Phi(\phi) = m^2. \tag{4.123}$$

The separation of $R(\rho)$ from $\Phi(\phi)$ requires that both sides of Eq. (4.123) be equal to the same constant. In this way, another separation constant m^2 is introduced, leading to the separated ordinary DE:

$$\frac{d^2}{d\phi^2}\Phi(\phi) = -m^2\Phi(\phi), \tag{4.121b}$$

$$\left(\rho^2\frac{d^2}{d\rho^2} + \rho\frac{d}{d\rho} + (\beta^2\rho^2 - m^2)\right)R(\rho) = 0, \qquad \beta^2 = k^2 + \alpha^2. \tag{4.121c}$$

The separation in spherical coordinates proceeds in a very similar way when the separable form

$$u(\mathbf{r}) = u(r, \theta, \phi) = R(r)\Theta(\theta)\Phi(\phi) \tag{4.124}$$

is substituted into Eq. (4.116). This calculation is left as an exercise. The results are the three separated ordinary DEs:

$$\frac{d^2}{d\phi^2}\Phi(\phi) = -m^2\Phi(\phi), \tag{4.125a}$$

$$\left[\frac{1}{\sin\theta}\frac{d}{d\theta}\left(\sin\theta\frac{d}{d\theta}\right) - \frac{m^2}{\sin^2\theta} + c\right]\Theta(\theta) = 0, \tag{4.125b}$$

$$\left[\frac{1}{r^2}\frac{d}{dr}\left(r^2\frac{d}{dr}\right) + k^2 - \frac{c}{r^2}\right]R(r) = 0, \tag{4.125c}$$

where m^2 and c are the separation constants.

The separated ordinary DE for $\Phi(\phi)$ of the azimuthal angle ϕ turns out to be the same in both cylindrical and spherical coordinates:

$$\left(\frac{d^2}{d\phi^2}+m^2\right)\Phi(\phi)=0.$$

This is the DE for harmonic oscillations. We note that, although the angle ϕ can increase monotonically, the point $\mathbf{r}$ in reality always returns to the same position after one or more complete turns of 2π. Since $\phi(\mathbf{r})$ is supposed to be a single-valued function of $\mathbf{r}$, the angular function $\Phi(\phi)$ must have a period of 2π. That is,

$$\Phi(\phi+2\pi)=\Phi(\phi). \tag{4.126}$$

This *periodicity condition* is satisfied if

$$\Phi(\phi)=\Phi_m(\phi)=A_m\cos m\phi+b_m\sin m\phi, \tag{4.127}$$

where m is an integer. In other words, Eq. (4.127) gives the two eigenfunctions of the eigenvalue equation (4.125a) belonging to the same eigenvalue m^2.

Equation (4.125b) is also an interesting differential eigenvalue equation. If we eliminate θ in favor of $x=\cos\theta$ in this equation, we find

$$\left(\frac{d}{dx}(1-x^2)\frac{d}{dx}-\frac{m^2}{1-x^2}+c\right)X(x)=0.$$

That is,

$$\left[(1-x^2)\frac{d^2}{dx^2}-2x\frac{d}{dx}-\left(c-\frac{m^2}{1-x^2}\right)\right]X(x)=0, \qquad |x|\leqslant 1. \tag{4.128}$$

This equation is called an *associated Legendre equation,* which contains the Legendre equation (4.83) as the special case of $m=0$. Equation (4.128) can be solved by the Frobenius method. The infinite Frobenius series turns out to be convergent only for $|x|<1$. The solution at $|x|=1$ turns out to be infinite as a rule, except when $c=c_l=l(l+1)$, $l=0, 1, 2, \ldots$. At these eigenvalues the Frobenius series terminates as a polynomial, the *associated Legendre polynomial* $P_l^m(x)$.

The points $|x|=|\cos\theta|=1$ are, of course, the singular points of the DE (4.128). The reason its solutions tend to be ill behaved at these points is that they are the north and south poles of the sphere. Since every meridian passes through both poles, the longitude (or azimuthal angle ϕ) at the poles can have any value ($-\infty\leqslant\phi\leqslant\infty$). This is, of course, a purely artificial result since the poles can be chosen to be elsewhere. Physical fields on the sphere must be bounded everywhere. This is possible only if we use the eigenfunction $P_l^m(x)$ of Eq. (4.128) belonging to the eigenvalue $c=c_l$, l being a non-negative integer.

Equations (4.128), (4.121c), and (4.125c) are among the most common nonelementary differential equations we need in mathematical physics.

They are so important that it is worthwhile to rewrite them in a standard form:

$$\left[(1-x^2)\frac{d^2}{dx^2}-2x\frac{d}{dx}+\left(l(l+1)-\frac{m^2}{1-x^2}\right)\right]P_l^m(x)=0, \qquad x=\cos\theta; \tag{4.129}$$

$$\left(z^2\frac{d^2}{dz^2}+z\frac{d}{dz}+(z^2-m^2)\right)J_m(z)=0, \qquad z=\beta\rho; \tag{4.130}$$

$$\left(z^2\frac{d^2}{dz^2}+2z\frac{d}{dz}+[z^2-l(l+1)]\right)j_l(z)=0, \qquad z=kr. \tag{4.131}$$

Equation (4.130) is called a (cylindrical) *Bessel equation,* while Eq. (4.131) is a *spherical Bessel equation.* The corresponding solutions $J_m(z)$ and $j_l(z)$ are the (cylindrical) Bessel function of order m and the spherical Bessel function of order l. The function $J_m(z)$ is that solution of Eq. (4.130) that behaves like $(z/2)^m$ for small values of z, while

$$j_l(z)=\left(\frac{\pi}{2z}\right)^{1/2}J_{l+1/2}(z).$$

(See Problem 4.12.1.) These Bessel functions are among the most useful of the so-called *higher transcendental functions.* (A transcendental function is one that is not algebraic. An algebraic function of x is one that can be generated from x by a *finite* number of algebraic operations $+$, $-$, $\times$, $\div$. For example, a polynomial is algebraic, but $\log x$ is not. The elementary transcendental functions are the exponential, logarithmic, circular, and hyperbolic functions.)

The linearly independent second solution of Eq. (4.129) is denoted by the symbol $Q_l^m(x)$, and is called an *associated Legendre function of the second kind.* Similarly, there is the Bessel function of the second kind $N_m(z)$ and the spherical Bessel function of the second kind $n_l(z)$.

The most general form of the solution of a linear PDE is, according to Eq. (4.13), a superposition of all permissible solutions. In the case of cylindrical coordinates this superposition is

$$u(\mathbf{r})=\sum_i u_i(\rho,\phi,z), \tag{4.132}$$

where i denotes collectively the two separation constants α and m in Eqs. (4.121). There are three ordinary DEs in Eqs. (4.121), each with two linearly independent solutions. Hence each $u_i(\rho,\phi,z)$ is a sum of eight terms:

$$\begin{aligned} u_i(\rho,\phi,z) &= u_{\alpha m}(\rho,\phi,z) \\ &= [a_1J_m(\beta\rho)+a_2N_m(\beta\rho)](b_1\cos m\phi+b_2\sin m\phi) \\ &\quad \times[c_1Z_1(z,\alpha)+c_2Z_2(z,\alpha)], \end{aligned} \tag{4.133}$$

where we have made use of the fact that Eq. (4.121c) is a Bessel equation. Similarly we have in spherical coordinates

$$u(\mathbf{r}) = \sum_i u_i(r, \theta, \phi), \tag{4.134}$$

where

$$\begin{aligned} u_i(r, \theta, \phi) &= u_{lm}(r, \theta, \phi) \\ &= [a_1 j_l(kr) + a_2 n_l(kr)][b_1 P_l^m(\cos\theta) + b_2 Q_l^m(\cos\theta)] \\ &\quad \times (c_1 \cos m\phi + c_2 \sin m\phi). \end{aligned} \tag{4.135}$$

Here we have used the result that Eq. (4.125c) is a spherical Bessel equation.

Since we are not yet familiar with some of the higher transcendental functions appearing in Eqs. (4.133) and (4.135), we shall not use these results until Chapter 5 except for one very special case. This involves the Laplace equation

$$\nabla^2 u(\mathbf{r}) = 0,$$

for which the separated radial equation (4.125c) simplifies because $k = 0$. With $c = l(l+1)$ we can readily verify that the two linearly independent solutions are r^l and r^{-l-1}. The function Q_l^m is known to be singular at $\cos\theta = \pm 1$, that is, at the north and south poles of the sphere. Hence physically interesting solutions defined everywhere on the sphere do not contain Q_l^m. Thus $u_i(r, \theta, \phi)$ of Eq. (4.135) simplifies to

$$u_i(r, \theta, \phi) = (a_1 r^l + a_2 r^{-l-1})(c_1 \cos m\phi + c_2 \sin m\phi) P_l^m(\cos\theta).$$

The simplest of such solutions are those with axial symmetry, that is solutions independent of the aximuthal angle ϕ. This occurs if $m = 0$ so that $\cos m\phi = 1$, $\sin m\phi = 0$. The functions $P_l^{m=0}$ are just the Legendre polynomials of Eq. (4.90). Thus we have shown that axially symmetric solutions of the Laplace equation can be written as

$$u(\mathbf{r}) = \sum_{l=0} (a_1 r^l + a_2 r^{-l-1}) P_l(\cos\theta). \tag{4.136}$$

As an example, we note that spherically symmetric solutions are independent of θ and therefore must involve the $l = 0$ terms only. Furthermore, if the solution is also finite as $r \to \infty$, then $a_1 = 0$ must be satisfied. Thus by successive eliminations we are left with only one function r^{-1}. This is just the Coulomb potential due to a point charge at the origin.

We must postpone applications of these eigenfunction expansions to Chapter 5 because such applications require the calculation of the linear coefficients a_i. This cannot be done without some knowledge of the orthogonality properties of these special functions.

We also note that PDE (4.116) is known to be separable in the eight other coordinate systems shown in Table 1.2 in addition to the three common ones shown in Eq. (4.118). Discussions of these less common coordinate systems can be found in Morse and Feshbach (Chapter 5) and in Arfken (Second Edition, Chapter 2).

Problems

4.12.1. Starting from Eqs. (4.130) and (4.131) show that

$$j_l(z) = \text{const} \times \frac{1}{\sqrt{z}} J_{l+1/2}(z).$$

4.12.2. Verify that the ordinary differential equations shown in Eq. (4.125) are obtained on separating the Helmholtz equation in spherical coordinates.

4.13. Green functions for partial differential equations

Green functions for partial DEs are more complicated than those for ordinary DEs. The main reason is that the boundary conditions necessary and sufficient to define a unique solution are much more complicated. For example, they depend also on the nature of the DE itself. However, in many cases of physical interest, simple expressions can be obtained readily without a detailed knowledge of the theory of PDEs, as the following examples show.

We shall first consider the Green functions for the Poisson equation in two and three dimensions:

$$\nabla_1^2 G(\mathbf{r}_1,\mathbf{r}_2) = \delta(\mathbf{r}_1 - \mathbf{r}_2). \tag{4.137}$$

The three-dimensional case is perhaps more familiar. The three-dimensional δ function can be expressed simply in rectangular coordinates as

$$\delta(\mathbf{r}_1 - \mathbf{r}_2) = \delta(x_1 - x_2)\, \delta(y_1 - y_2)\, \delta(z_1 - z_2),$$

or more generally in terms of the integral

$$\int \delta(\mathbf{r}_1 - \mathbf{r}_2)\, d^3r_1 = 1 \qquad \text{(or 0)} \tag{4.138}$$

if the integration region includes (excludes) $\mathbf{r}_2$. Thus $\mathbf{r}_2$ is a very special point of the equation. If the DE is defined over the entire three-dimensional space, nothing is changed by putting $\mathbf{r}_2$ at the origin. Then the solution must be spherically symmetric in

$$\mathbf{r} = \mathbf{r}_1 - \mathbf{r}_2;$$

that is, $G(\mathbf{r}_1,\mathbf{r}_2)$ must be a function of r only.

Suppose we integrate the DE over the volume of a sphere of radius r. We get

$$\int \nabla \cdot \nabla G(r)\, d^3r = 1.$$

This may be written, with the help of Gauss's theorem, as

$$1 = \int \nabla G(r) \cdot d\boldsymbol{\sigma} = \left(\frac{d}{dr} G(r)\right) 4\pi r^2, \tag{4.139}$$

where $4\pi r^2$ is the area of the sphere. This gives $(d/dr)G(r)$, which can next be integrated directly to give

$$\begin{aligned} G(r) &= \int^r \left(\frac{d}{dr} G(r)\right) dr = \int^r \frac{1}{4\pi r^2}\, dr \\ &= -\frac{1}{4\pi r} + C, \end{aligned} \tag{4.140}$$

where C is an integration constant. This shows that it is possible to make G vanish as r approaches infinity by choosing $C = 0$.

Equation (4.140) appears in physics as the familiar Coulomb field surrounding a unit point charge at $\mathbf{r}_2$. We should note that, while the Coulomb field is spherically symmetric about the source at $\mathbf{r}_2$, it is not spherically symmetric in a coordinate system in which the source is not located at the origin:

$$G(|\mathbf{r}_1 - \mathbf{r}_2|) = -\frac{1}{4\pi\, |\mathbf{r}_1 - \mathbf{r}_2|}. \tag{4.141}$$

The problem in two dimensions differs from the above only because the two-dimensional δ-function relation

$$\int \delta(\mathbf{r}_1 - \mathbf{r}_2)\, d^2 r_1 = 1 \tag{4.142}$$

involves a two-dimensional integration. Equation (4.139) now reads

$$1 = \left(\frac{d}{dr} G(r)\right) 2\pi r,$$

where $2\pi r$ is the circumference of the circle around $\mathbf{r}_2$ (which is also located at the origin). As a result

$$G(r) = \int^r \frac{1}{2\pi r}\, dr = \frac{1}{2\pi} \ln r + C. \tag{4.143}$$

This shows that a two-dimensional $G(r)$ *cannot* be finite as r approaches infinity.

We now know what a function should look like near a point source where its Laplacian is a δ function. This information can be used to obtain the Green functions of many second-order PDEs. Consider, for example, the Helmholtz equation

$$(\nabla^2 + k^2) G_k(r) = \delta(\mathbf{r}). \tag{4.144}$$

If $r \neq 0$, $G_k(r)$ satisfies the homogeneous equation

$$(\nabla^2 + k^2)G_k(r) = 0.$$

Hence $G_k(r)$ must be that solution that behaves at the origin like $(-4\pi r)^{-1}$ in three dimensions, or $(2\pi)^{-1} \ln r$ in two dimensions, as $k \to 0$.

Let us work out the three-dimensional case. Since $G_k(r)$ depends only on r, we may "separate" the Laplacian in spherical coordinates to yield (for $r \neq 0$) the spherical Bessel equation (4.131) with $l = 0$:

$$\left(r^2 \frac{d^2}{dr^2} + 2r \frac{d}{dr} + (kr)^2\right) j_0(kr) = 0. \tag{4.145}$$

Although this equation does not look too familiar, the equation for $u(r) = rG(R)$ can easily be shown by direct substitution into Eq. (4.145) to be

$$\left(\frac{d^2}{dr^2} + k^2\right) u(r) = 0.$$

Thus the two linearly independent solutions for $G(r)$ are $(-4\pi r)^{-1} \cos kr$ or $(-4\pi r)^{-1} \sin kr$. The second solution, called a *regular solution,* is well behaved at the origin and does not give rise to a δ function when operated on by ∇^2. It is the *irregular solution*

$$G(R) = (-4\pi r)^{-1} \cos kr, \tag{4.146a}$$

which is the Green function for the Helmholtz equation.

This Green function can be written in two other alternative forms

$$G_{\pm k}(r) = -\frac{1}{4\pi r} e^{\pm ikr}, \tag{4.146b}$$

where the factors $\exp(\pm ikr)$ describe spherical waves since the wave fronts of constant phase ($kr = \text{const}$) are spheres. The direction of the wave motion cannot be determined in the absence of a time dependence. It is customary to add the time factor of $\exp(-i\omega t)$, so that $\exp(ikr)$ describes an *outgoing* spherical wave, while $\exp(-ikr)$ describes an *ingoing* spherical wave.

An equivalent convention leading to the same result is obtained by examining the eigenvalue of the *propagation operator*

$$\mathbf{K} = -i\nabla,$$

which is also the infinitesimal generator of space translation, as we have discussed in Section 2.9. Thus

$$\mathbf{K} e^{\pm ikr} = \pm k \mathbf{e}_r e^{\pm ikr}. \tag{4.147}$$

The wave $\exp(ikr)$ thus propagates outward. It is therefore called an outgoing spherical wave.

The solution $(-4\pi r)^{-1} \cos kr$ thus contains an equal linear combination of an outgoing and an ingoing spherical wave. These two waves in opposite directions will interfere to form a standing-wave pattern in the same way two waves running in opposite directions in a violin string do. Hence $(-4\pi r)^{-1}\cos kr$ is called a *standing-wave Green function*.

To change the boundary conditions, we can add the homogeneous solution $\pm i \sin kr/(-4\pi r)$ to get back the *spherical-wave Green functions*. These Green functions are said to satisfy the outgoing or ingoing spherical-wave boundary conditions.

Problems

4.13.1. Show that the three-dimensional δ function can be written explicitly in spherical coordinates as

$$\delta(\mathbf{r}_1 - \mathbf{r}_2) = \frac{1}{r_1^2}\,\delta(r_1 - r_2)\,\delta(\cos\theta_1 - \cos\theta_2)\,\delta(\phi_1 - \phi_2).$$

4.13.2. Express the three-dimensional δ function explicitly in an orthogonal curvilinear coordinate system using the notation of Section 1.10.

4.13.3. Construct a one-dimensional function $f(x)$ whose Laplacian $(d^2/dx^2)f(x)$ is the δ function $\delta(x)$.

4.13.4. Verify that the outgoing-wave Green function for the two-dimensional Helmholtz equation is $-(i/4)H_0^{(1)}(kr)$, where $H_0^{(1)}$ is the Hankel function of the first kind and of order 0. It satisfies the Bessel equation (4.130) for $m = 0$, behaves like $(2i/\pi)\ln kr$ near $r = 0$, and is proportional to $(kr)^{-1/2}e^{ikr}$ as $r \to \infty$.

4.14. Introduction to nonlinear systems

Systems far away from states of equilibrium are intrinsically nonlinear. Recent studies of nonlinear properties of physical systems have emphasized the concepts of catastrophe, bifurcation, chaos, and strange attractors. This section gives a brief introduction to these interesting ideas, and is based in part on a review by Martens [*Physics Reports* **115,** 315 (1985).]

4.14.1. Catastrophe and hysteresis

Catastrophe refers to a sudden change in the response of a system to a small change in the value of a parameter. This behavior is exhibited by the following first-order nonlinear differential equation:

$$\dot{x} = \frac{d}{dt}x(t) = bx - x^3 + p. \tag{4.148}$$

For a given positive b, this system is in equilibrium (with zero velocity $\dot{x} = 0$) along the cubic curve shown in Fig. 4.5:

$$p = x_e^3 - bx_e. \tag{4.149}$$

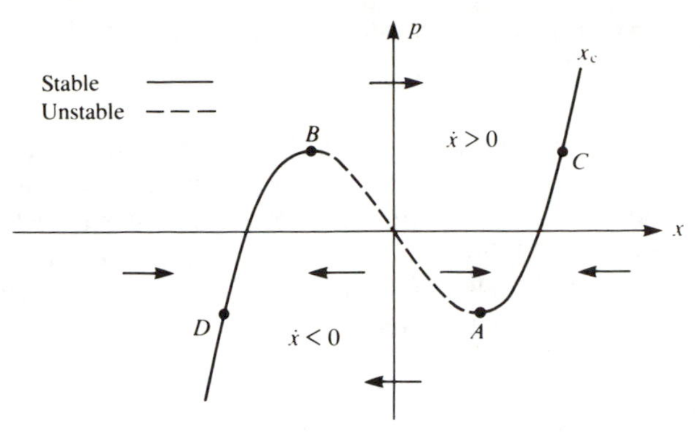

Fig. 4.5. The equilibrium or static solution of Eq. (4.149).

The equilibrium curve by its nature separates the px diagram into two regions, each with a definite sign for the velocity $\dot{x}$. When $p \gg bx > 0$, $\dot{x} \approx p - x^3$; the negative velocity region therefore occurs to the right of x_e, as shown by the direction of arrows in Fig. 4.5.

Starting from a large and positive value of x_e, we see that the equilibrium or static solution is stable down to the local minimum A at $(x_A, p_A) = (\sqrt{b/3}, -(2b/3)\sqrt{b/3})$. It is unstable from A to the local maximum B at $(-x_A, -p_A)$, as indicated by the broken curve in the figure. Below B, the equilibrium is stable again.

Suppose b is fixed and p is decreased slowly from a large and positive value. The system will follow the equilibrium curve until the point A is reached. When p is further decreased, the system will jump to the point D on the left half of the stability curve, and then continues on to more negative values of p.

If p is now increased from this large and negative value, the system does not trace its original path exactly. It goes all the way up to the local maximum B instead, then jumps over to the point C on the other stability curve before continuing upwards. The jumps $A \to D$ and $B \to C$, where the system changes abruptly in response to quite small changes in p, are called *catastrophes*, while the history-dependent behavior described above is referred to as *hysteresis*.

Given values for b and p and an arbitrary initial position $x(0)$, the system will move to the stable equilibrium point on the nearer of the two stability curves. Such a stable point is called a *point attractor*, because it "attracts" the system for a variety of initial conditions. Another rather familiar example of a point attractor is an oscillating spring that comes to rest as a result of dissipation.

4.14.2. Bifurcation

At $p = 0$, Fig. 4.5 shows two stable static solutions, at $x_e = \pm\sqrt{b}$. These solutions approach each other as $b \to 0$, collapsing into a common solution at $b = 0$. For negative values of b, the differential equation

$$\dot{x} = bx - x^3 = b(x - x_e) - (x^3 - x_e^3) \tag{4.150}$$

has only one static solution. Since $\dot{x} < 0$ if $x > x_e$, this unique solution is stable. This situation is described graphically by the xb diagram of Fig. 4.6, where the equilibrium curves define regions in which the velocities have a common sign. The branching of static solutions caused by a change of parameter is called a *bifurcation*. Successive bifurcations play a crucial role in turbulence and chaos.

4.14.3. *Chaos*

Many adult insect populations do not overlap from season to season. The population x_{n+1} in one season depends only on the population x_n of the last season. An interesting example of a system of this type is described by the first-order nonlinear difference equation

$$x_{n+1} = bx_n(1 - x_n). \tag{4.151}$$

Since $x(1 - x)$ has a maximum of $\frac{1}{4}$ at $x = \frac{1}{2}$, it is convenient to restrict the parameter b to $1 < b < 4$, so that a population value of $0 \leq x_n \leq 1$ will be mapped into the same interval. Equation (4.151) itself is called a one-dimensional *nonlinear map*.

Equilibrium solutions $x_{n+1} = x_n = x_e$ of this equation satisfy the equation

$$bx_e^2 + (1 - b)x_e = 0. \tag{4.152}$$

This has a trivial solution $x_e = 0$, and a nontrivial solution $x_e = 1 - b^{-1}$, which goes from 0 to $\frac{3}{4}$ as b increases from 1 to 4. (The nontrivial solutions with $b < 1$ are not allowed because they involve negative populations. Solutions for $b > 4$ will diverge to $-\infty$. A negative population means extinction.)

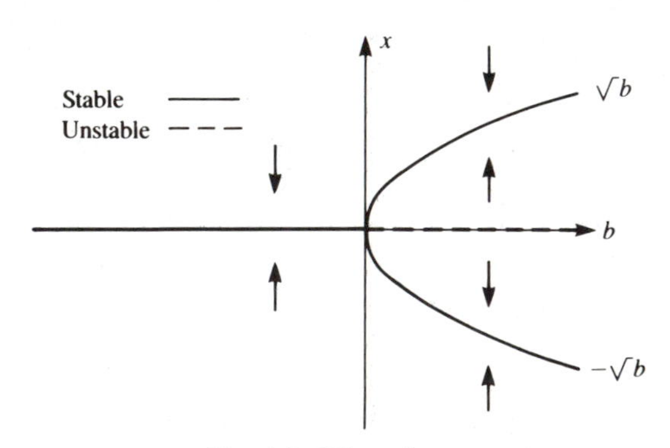

Fig. 4.6. Bifurcation.

Not all equilibrium solutions are stable. Near x_e

$$\begin{aligned} x_{n+1} - x_e &= b[x_n(1-x_n) - x_e(1-x_e)] \\ &= (x_n - x_e)[2 - b - b(x_n - x_e)] \\ &\approx (2-b)(x_n - x_e). \end{aligned} \tag{4.153}$$

The solutions are therefore stable for $1 < b < 3$, but unstable if $b \geqslant 3$.

For $3 < b < 3.45$, one can show by explicit calculation that the solution becomes periodic with a period of two; that is, the system oscillates between two distinct populations in succeeding seasons. This "period-doubling" bifurcation is shown in Fig. 4.7, which summarizes the behavior of the system as b increases: The system undergoes repeated period-doubling bifurcations at appropriate values of b. These occur more and more frequently with an accumulation point at $b = 3.57$, where the number of periods becomes infinite. Beyond this point, the behavior is so complicated that periodicity is lost and the system becomes aperiodic. The behavior is now sensitive to the initial conditions in such a way that two slightly different initial conditions no matter how close will eventually give rise to very different population evolutions after a sufficiently large number of seasons. The system is then said to be unpredictable and *chaotic,* even though it is still completely deterministic. A detailed description of Eq. (4.151) and its solutions has been given by May [*Nature* **261,** 459 (1976)].

4.14.4. *Strange attractor*

An attractor is so called because the system is attracted to it even when the initial conditions are very different. Sensitivity to initial conditions is just the opposite of this: Two points on opposite sides of a point of unstable equilibrium will end up in very different places. Such a point of unstable equilibrium is called a *repellor.*

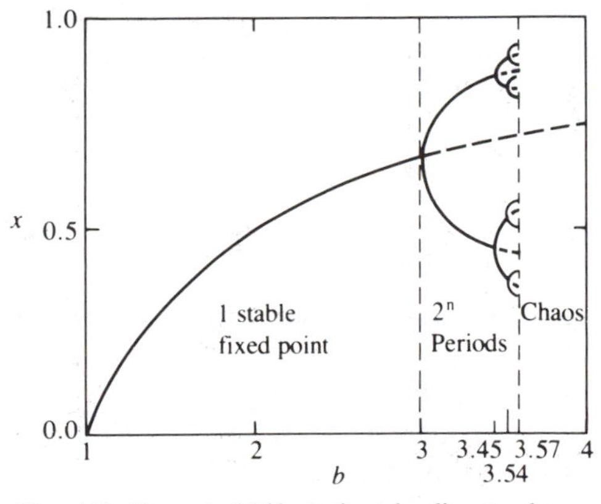

Fig. 4.7. Repeated bifurcations leading to chaos.

A *strange attractor* is an attractor that is also sensitive to initial conditions. That is to say, it is also a repellor. How such contrary attributes could coexist can be illustrated by a simple example taken from Martens: A good labyrinth or trap is an attractor in that people who go in do not come out. It is also a repellor since the people inside may not even see one another, although they all got in originally through the same gate. In other words, there is a trapping region with complicated instabilities inside.

Strange attractors were discovered quite recently (in 1963). One simple example appears in the two-dimensional map

$$x_{n+1} = y_n + 1 - ax_n, \qquad y_{n+1} = bx_n, \tag{4.154}$$

studied by Hénon. The trapping property is just the self-mapping of a region already discussed in one dimension for Eq. (4.151). Indeed, the x part of the Hénon map is described by the second-order nonlinear difference equation

$$x_{n+1} = 1 - ax_n^2 + bx_{n-1}. \tag{4.155}$$

The situation in two dimensions is more interesting: A point attractor in one dimension becomes a curve that in computer experiments appears to be a normal (i.e., dense) one-dimensional curve. A strange attractor is made up of an infinity of such curves that appear to be parallel to each other when examined at sufficiently small scales. These curves do not cover the entire xy plane. They have instead the unusual property that they look the same at every magnification. Mathematical objects having such scale-invariant geometrical properties are called *fractals*. Fractals are characterized by well-defined fractional dimensions. The strange attractor of the Hénon map has a dimension somewhere between 1 and 2.

The first strange attractor was discovered in meteorological studies of turbulence in the atmosphere. Turbulence is characterized by a highly unpredictable behavior with nonetheless well-defined average properties. Strange attractors appear to satisfy these requirements better than chaos.

Problem

4.14.1. (a) Plot the two functions

$$y_{\mathrm{L}}(x) = x, \qquad y_{\mathrm{R}}(x) = bx(1-x)$$

appearing in the nonlinear map of Eq. (4.151) for $b = 2$. Starting with $x_1 = 0.5$, show on your graph how the mapping proceeds step by step until the behavior of the system becomes clear to you. Describe this behavior.

(b) Repeat this graphical solution for $b = 3.2$.

(c) Repeat this graphical solution for $b = 5$.

Appendix 4. Tables of mathematical formulas

1. Linearity property

If $\mathscr{L}$ is a linear operator in x and

$$\mathscr{L}y_1(x)=0, \qquad \mathscr{L}y_2(x)=0, \qquad \mathscr{L}y_p(x)=R(x),$$

then

$$\mathscr{L}[y_p(x)+c_1y_1(x)+c_2y_2(x)]=R(x).$$

2. Linear differential equations

The first-order linear differential equations

$$\left(\frac{d}{dx}+p(x)\right)y_h(x)=0,$$

$$\left(\frac{d}{dx}+p(x)\right)y_p(x)=R(x)$$

have solutions

$$y_h(x)=y_h(a)\exp\left(-\int_a^x p(x')\,dx'\right),$$

and

$$y_p(x)=\left\{\int_a^x \left(\frac{R(x')}{y_h(x')}\right)dx'\right\}y_h(x).$$

The second-order linear differential equation

$$\mathscr{L}y_i(x)\equiv\left(\frac{d^2}{dx^2}+P(x)\frac{d}{dx}+Q(x)\right)y_i(x)=0, \qquad i=1,2$$

has two linearly independent solutions $y_1(x)$ and $y_2(x)$ with a nonzero Wronskian

$$W(x)=\begin{vmatrix} y_1(x) & y_2(x) \\ y_1'(x) & y_2'(x) \end{vmatrix}$$

$$=W(a)\exp\left(-\int_a^x P(x')\,dx'\right).$$

If $y_1(x)$ is known, the differential equations

$$\mathscr{L}y_2(x)=0,$$

$$\mathscr{L}y_p(x)=R(x)$$

have the solutions

$$y_2(x)=\left\{\int_b^x \left[\exp\left(-\int_b^t P(t')\,dt'\right)\Big/ y_1^2(t)\,dt\right\}y_1(x),$$

and

$$y_p(x)=v_1(x)y_1(x)+v_2(x)y_2(x),$$

where

$$v_1(x) = -\int_a^x \frac{y_2(t)R(t)}{W(t)}\,dt, \qquad v_2(x) = \int_a^x \frac{y_1(t)R(t)}{W(t)}\,dt.$$

The homogeneous differential equation

$$\mathscr{L}y(x) = 0$$

often has one or more solutions of the form

$$y(x) = \sum_{\lambda=0}^{\infty} a_\lambda x^{\lambda+s}.$$

The coefficients a_λ can be obtained by requiring that the coefficients b_μ of the power series

$$\mathscr{L}y(x) = \sum_\mu b_\mu x^{\mu+s}$$

are identially zero, that is, $b_\mu = 0$.

3. Green functions

If $G(x,x')$ is a solution of the differential equation

$$\mathscr{D}G(x,x') = \delta(x - x'),$$

then the solution of the differential equation

$$\mathscr{D}y(x) = R(x)$$

is

$$y(x) = \int_a^b G(x,x')R(x')\,dx'.$$

This solution $y(x)$ satisfies the same boundary conditions as $G(x,x')$.

If $\mathscr{D}$ is an nth-order differential operator, the kth derivative

$$G^{(k)}(x,x') = \frac{d^k}{dx^k} G(x,x')$$

for $k \leqslant n$ is smooth everywhere except at $x = x'$, where $G^{(n-2)}$ has a kink, $G^{(n-1)}$ has a jump discontinuity, and $G^{(n)}$ has a δ-function discontinuity.

4. Separation of variables and eigenfunction expansion

In certain coordinate systems, a linear partial differential equation might be separable into a number of ordinary differential equations of the general form

$$\mathscr{L}f(x) = \lambda f(x).$$

The separation constant λ appears in more than one equation, thus permitting the relationship between different coordinates to be maintained. Each ordinary differential equation must satisfy certain boundary or regularity conditions. As a result, only a selection or *spectrum* of separation constants is allowed. These are called *eigenvalues*, while the corresponding solutions $f(x)$ are their *eigenfunctions*.

Under these circumstances, the linear partial differential equation has solutions in the form of a product of eigenfunctions, one for each separated coordinate. The most general solution is therefore a linear combination of such products, the linear coefficients being determined by fitting to suitable boundary and initial conditions in the manner of a Fourier series.

4(a). One-dimensional wave equation

$$\frac{\partial^2}{\partial x^2} u(x,t) = \frac{1}{c^2} \frac{\partial^2}{\partial t^2} u(x,t)$$

$$\left(\frac{d^2}{dx^2} - \lambda_n\right) X_n(x) = 0$$

$$\left(\frac{d^2}{dt^2} - c^2\lambda_n\right) T_n(t) = 0$$

$$X_n(x) = A_n \cos(x\sqrt{-\lambda_n}) + B_n \sin(x\sqrt{-\lambda_n})/\sqrt{-\lambda_n}$$

$$T_n(t) = C_n \cos(ct\sqrt{-\lambda_n}) + D_n \sin(ct\sqrt{-\lambda_n})/\sqrt{-\lambda_n}$$

$$u(x,t) = \sum_{n=1} X_n(x) T_n(t).$$

This eigenfunction expansion can be written symbolically as

$$u(x,t) = \sum_n A_n \begin{Bmatrix} \cos(x\sqrt{-\lambda_n}) \\ \sin(x\sqrt{-\lambda_n})/\sqrt{-\lambda_n} \end{Bmatrix} \begin{Bmatrix} \cos(ct\sqrt{-\lambda_n}) \\ \sin(ct\sqrt{-\lambda_n})/\sqrt{-\lambda_n} \end{Bmatrix}.$$

4(b). Helmholtz equation

$$(\nabla^2 + k^2) u(\mathbf{r}) = 0$$

$$u(x,y,z) = \sum_{m,n} A_{mn} \begin{Bmatrix} \cos(x\sqrt{-\lambda_m}) \\ \sin(x\sqrt{-\lambda_m})/\sqrt{-\lambda_m} \end{Bmatrix} \begin{Bmatrix} \cos(y\sqrt{-\lambda_n}) \\ \sin(y\sqrt{-\lambda_n})/\sqrt{-\lambda_n} \end{Bmatrix}$$
$$\times \begin{Bmatrix} \cos(z\sqrt{\lambda_m + \lambda_n + k^2}) \\ \sin(z\sqrt{\lambda_m + \lambda_n + k^2})/\sqrt{\lambda_m + \lambda_n + k^2} \end{Bmatrix}$$

$$u(\rho,\theta,z) = \sum_{m,n} A_{mn} \begin{Bmatrix} J_m(\sqrt{k^2 + \lambda_n}\,\rho) \\ N_m(\sqrt{k^2 + \lambda_n}\,\rho) \end{Bmatrix} \begin{Bmatrix} \cos m\phi \\ \sin m\phi \end{Bmatrix} \begin{Bmatrix} \cos(z\sqrt{-\lambda_n}) \\ \sin(z\sqrt{-\lambda_n})/\sqrt{-\lambda_n} \end{Bmatrix}$$

$$u(r,\theta,\phi) = \sum_{l,m} A_{lm} \begin{Bmatrix} j_l(kr) \\ n_l(kr) \end{Bmatrix} \begin{Bmatrix} P_l^m(\cos\theta) \\ Q_l^m(\cos\theta) \end{Bmatrix} \begin{Bmatrix} \cos m\phi \\ \sin m\phi \end{Bmatrix}.$$

The special case $k = 0$ gives the solutions to the Laplace equation. We note the simplification

$$\begin{Bmatrix} j_l(kr) \\ n_l(kr) \end{Bmatrix} \to \begin{Bmatrix} r^l \\ r^{-l-1} \end{Bmatrix}.$$

5. Green functions for partial differential equations

5(a). Laplace equation

$$G(\mathbf{r}_1 - \mathbf{r}_2) = \frac{1}{2\pi} \ln |\mathbf{r}_1 - \mathbf{r}_2| \qquad \text{in two dimensions,}$$

$$= -\frac{1}{4\pi |\mathbf{r}_1 - \mathbf{r}_2|} \qquad \text{in three dimensions.}$$

5(b). Helmholtz equation

The outgoing-wave Green functions are

$$G(\mathbf{r}_1 - \mathbf{r}_2) = -\frac{i}{4} H_0^{(1)}(k |\mathbf{r}_1 - \mathbf{r}_2|) \qquad \text{in two dimensions,}$$

$$= -\frac{e^{ik |\mathbf{r}_1 - \mathbf{r}_2|}}{4\pi |\mathbf{r}_1 - \mathbf{r}_2|} \qquad \text{in three dimensions.}$$

5

SPECIAL FUNCTIONS

5.1. Introduction

A function is a rule defining the relation between one variable and another. Among the oldest functions is the sine function, which relates the length of a chord to the half-angle of the arc on a circle. A table of chords could already be found in a book compiled by Ptolemy of Alexandria in the second century. Most functions are of relatively recent origin, however. The logarithmic function was invented in the early, and the exponential function in the late, seventeenth century. The idea of a power is older, but its algebraic representation began with Descartes in the early seventeenth century.

Indeed our modern understanding of functions can be said to begin with Descartes. It gained much cohesion with the publication of the first treatise on functions by Euler in 1748. Lagrange, Fourier, Cauchy, and Riemann, among others, gave the subject its modern form in the early and mid-nineteenth century.

Special functions are functions in frequent use in the mathematical and physical sciences. Functions of particular interest in physics are those that describe simple systems. They are almost always the solutions of differential equations—the equations of motion or of state of simple systems. Perhaps not surprisingly, the most useful functions are those associated with the description of wave motion.

5.2. Generating function for Legendre polynomials

We have seen in Eq. (4.136) that an axially symmetric solution of the Laplace equation

$$\nabla^2 u = 0$$

regular at the poles of the sphere and finite at infinity is just the "Coulomb" potential r^{-1}. In Section 4.13 we have seen that this is actually the solution of the inhomogeneous Poisson equation

$$\nabla^2 u(\mathbf{r}) = \rho(\mathbf{r})$$

for a δ-function, or point, source

$$\rho(\mathbf{r}) = -4\pi\,\delta(\mathbf{r}). \tag{5.1}$$

Suppose the source is placed not at the origin, but at the point $\mathbf{r}_s = (0,0,1)$ on the z axis. The resulting Coulombic potential

$$u(\mathbf{r}) = \frac{1}{|\mathbf{r}-\mathbf{r}_s|} = \frac{1}{|\mathbf{r}-\mathbf{e}_z|} = \sum_{l=0}^{\infty} A_l r^l P_l(\cos\theta) \qquad (5.2)$$

will still be axially symmetric about the z axis, but not spherically symmetric about the origin. It is finite at the origin where $u(0)=1$; hence near the origin it can be written in the form of Eq. (5.2), according to the method of separation of variables.

To determine the coefficients A_l, we examine the field along the z axis. Since $\cos\theta = 1$ and all $P_l(1) = 1$ along the z axis, we have

$$u(\mathbf{r} = r\mathbf{e}_z) = \sum_{l=0}^{\infty} A_l r^l.$$

Now the field along the z axis between the origin and the source is just $1/(1-r)$, according to one of the middle expressions in Eq. (5.2). Therefore

$$\begin{aligned} u(\mathbf{r}) &= \frac{1}{1-r} \\ &= 1 + r + r^2 + \cdots = \sum_{l=0}^{r} r^l. \end{aligned} \qquad (5.3)$$

This shows that $A_l = 1$ for all l.

The series shown in Eq. (5.3) is convergent for $r<1$. At $r=1$, that is, at the *source* position $\mathbf{r} = \mathbf{r}_s$, it becomes infinite or singular. We shall see in Chapter 6 that such a singular point, or *singularity*, of a function is a *source* of the function all over space, in the same sense that a point charge is a source of the electrostatic potential.

The series shown in Eq. (5.3) is divergent for $r>1$. This means that the power series in r shown on the right-hand side of Eq. (5.2) is valid only for $r<1$, although the Coulomb potential on the left-hand side is valid for all r.

Now that A_l is determined, we have for $r<1$ and any angle θ

$$(1 - 2r\cos\theta + r^2)^{-1/2} = \sum_{l=0}^{\infty} r^l P_l(\cos\theta).$$

A function like this, written in the form

$$G(x,t) = (1 - 2xt + t^2)^{-1/2} = \sum_{l=0}^{\infty} t^l P_l(x), \qquad t<1, \qquad (5.4)$$

in which a set of functions, here $P_l(x)$, appears as the coefficients of the power-series expansion in t is called their *generating function* (GF). The name arises from the fact that the function $P_l(x)$ can be generated from

$G(x,t)$ by differentiation

$$P_l(x) = \frac{1}{l!}\left(\frac{\partial^l}{\partial t^l} G(x,t)\right)_{t=0}. \tag{5.5}$$

A GF is very useful because it can be used to derive many properties of the functions it generates.

For example, the GF in Eq. (5.4) has the property

$$G(-x,t) = (1 + 2xt + t^2)^{-1/2} = G(x,-t).$$

As a result,

$$\sum_{l=0}^{\infty} t^l P_l(-x) = \sum_{l=0}^{\infty} (-t)^l P_l(x) = \sum_{l=0}^{\infty} t^l (-1)^l P_l(x).$$

Hence

$$P_l(-x) = (-1)^l P_l(x). \tag{5.6}$$

Thus Legendre polynomials of even (odd) degrees are said to have even (odd) parity.

Legendre polynomials of different degrees are related to each other. For example, since $P_0(x) = 1$, we have

$$P_1(x) = x = xP_0(x). \tag{5.7}$$

Relations between these functions of different degrees, or between functions and their derivatives, are called *recursion formulas,* or *recurrence relations*. These can be derived readily from the GF in Eq. (5.4). As illustrations, we shall derive the following two recursion formulas:

$$(2l + 1)xP_l(x) = lP_{l-1}(x) + (l + 1)P_{l+1}(x), \tag{5.8}$$

$$lP_l(x) = xP_l'(x) - P_{l-1}'(x), \tag{5.9}$$

where the primes denote differentiation with respect to x.

Equation (5.8) involves no derivatives of P_l, but there are changes in the degree of the polynomials, that is, in the power of t in Eq. (5.4). This suggests that we have to differentiate $G(x,t)$ with respect to t:

$$\frac{\partial G}{\partial t} = \frac{x - t}{(1 - 2xt + t^2)^{3/2}} = \sum_{l'=0}^{\infty} l' t^{l'-1} P_{l'}(x).$$

That is,

$$(x - t)G(x,t) = (\underline{\underline{1}} - 2\underline{x}t + \underline{t^2}) \sum_{l'=0}^{\infty} t^{l'-1} P_{l'}(x)$$

$$= (\underline{x} - \underline{\underline{t}}) \sum_{l'=0}^{\infty} t^{l'} P_{l'}(x).$$

We now rearrange the power series so that they will eventually carry the same power t^l. As a visual aid, terms that will be added together are underlined the same way:

$$\begin{aligned}\sum_{l=0} (2l+1)\underline{x}P_l t^l &= \sum_{l'=0}^{\infty} (l'+1)P_{l'}\underline{\underline{t}}^{l'+1} + \sum_{l'=0}^{\infty} l'P_{l'}\underline{\underline{\underline{t}}}^{l'-1} \\ &= \sum_{l=1}^{\infty} lP_{l-1}t^l + \sum_{l=0}^{\infty} (l+1)P_{l+1}t^l \\ &= P_1 + \sum_{l=1}^{\infty} [lP_{l-1} + (l+1)P_{l+1}]t^l.\end{aligned}$$

Equating coefficients of t^l, we obtain Eq. (5.7) for $l=0$, and Eq. (5.8) for $l>0$.

Equation (5.8) is very useful for calculating the numerical values of a Legendre polynomial. This is because, if we know $P_{l-1}(x)$ and $P_l(x)$ for a given value of x, the next Legendre polynomial must be

$$P_{l+1}(x) = [(2l+1)xP_l(x) - lP_{l-1}(x)]/(l+1). \tag{5.10}$$

Since we already know $P_0(x)=1$ and $P_1(x)=x$, all other Legendre polynomials can thus be generated by successive applications of Eq. (5.10). This is particularly easy to do on a digital computer. When a recursion formula is used like this in the direction of increasing degree, it is called a *forward* recursion formula. (Occasionally a recursion formula is used backwards in the direction of decreasing degrees in order to improve numerical precision by minimizing roundoff errors in a computer.)

For Eq. (5.9), it is obvious that a differentiation with respect to x is needed:

$$\frac{\partial}{\partial x}G(x,t) = \frac{t}{(1-2xt+t^2)^{3/2}} = \frac{t}{x-t}\frac{\partial}{\partial t}G(x,t),$$

or

$$t\frac{\partial}{\partial t}G(x,t) = (x-t)\frac{\partial}{\partial x}G(x,t).$$

In terms of the power series in t^l, this equation reads

$$\begin{aligned}\sum_{l=0}^{\infty} P_l l t^l &= (x-t)\sum_{l=0}^{\infty} P_l' t^l \\ &= \sum_{l=0}^{\infty} xP_l' t^l - \sum_{l=1}^{\infty} P_{l-1}' t^l.\end{aligned}$$

Hence Eq. (5.9) follows for $l>0$. For $l=0$, the relation obtained is

$$xP_0' = 0,$$

which can be considered to be a special case of Eq. (5.9) if $P_{-1}(x)=0$ is used.

The Legendre polynomials form a family of orthogonal polynomials satisfying the orthogonality relation

$$\int_{-1}^{1} P_l(x)P_{l'}(x)\,dx = \frac{2}{2l+1}\,\delta_{ll'}. \tag{5.11}$$

This relation can also be derived from the GF by squaring it and integrating with respect to x:

$$\begin{aligned}\int_{-1}^{1} G^2(x,t)\,dx &= \int_{-1}^{1} (1+t^2-2tx)^{-1}\,dx \\ &= \frac{1}{t}[\ln(1+t)-\ln(1-t)] \\ &= \sum_{l=0}^{\infty} \frac{2}{2l+1}\,t^{2l},\end{aligned} \tag{5.12}$$

where use has been made of the Taylor expansion

$$\ln(1+t) = \sum_{n=1}^{\infty} (-1)^{n+1}\frac{t^n}{n}.$$

The same integral can be expressed in terms of integrals of the Legendre polynomials, if one uses the right-hand side of Eq. (5.4):

$$\int_{-1}^{1} G^2(x,t)\,dx = \sum_{l=0}^{\infty}\sum_{l'=0}^{\infty} t^{l+l'}\int_{-1}^{1} P_l(x)P_{l'}(x)\,dx. \tag{5.13}$$

The double sum in Eq. (5.13) can be made equal to the single sum in Eq. (5.12) only if the integral of two Legendre polynomials is proportional to $\delta_{ll'}$. Equation (5.11) then follows on equating coefficients.

Finally, we should note that explicit formulas for some $P_l(x)$ can be found in Section 3.9. The polynomials P_2 to P_5 are plotted in Fig. 5.1.

Problems

5.2.1. Use the generating function of Eq. (5.4) to derive the Legendre differential equation.

5.2.2. Obtain the recursion formula

$$P'_{l+1}(x) - P'_{l-1}(x) = (2l+1)P_l(x)$$

by using
(a) Eq. (5.9); and
(b) Eq. (5.4).

5.2.3. Obtain the recursion formula

$$(1-x^2)P'_l(x) = l[P_{l-1}(x) - xP_l(x)].$$

5.2.4. Calculate $P_4(x)$ at $x = 0.1,\ 0.5,\ -0.1$ by using a forward recursion formula.

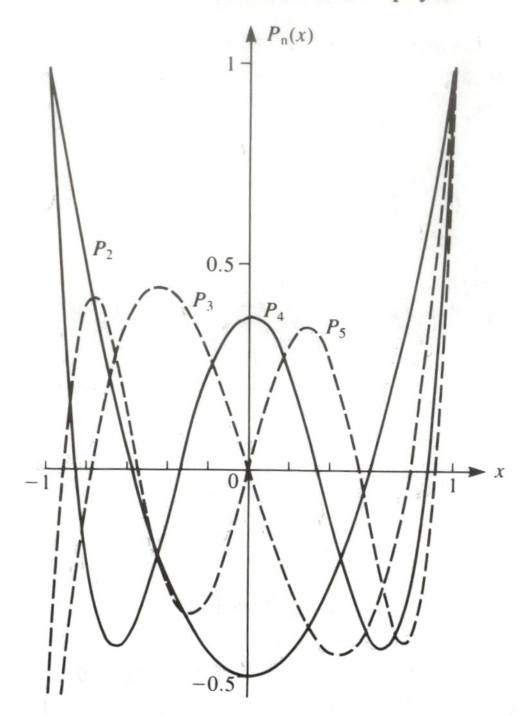

Fig. 5.1. Legendre polynomials $P_n(x)$, $n = 2$–5 (from Abramowitz and Stegun).

5.3. Hermite polynomials and the quantum oscillator

A family of functions can be defined by the differential equations they satisfy (Section 4.12). They can be defined by means of a generating function (Section 5.1). Functions can also be defined in terms of a differential formula called a *Rodrigues formula.* For example, the Rodrigues formula for Hermite polynomials is

$$H_n(x) = (-1)^n e^{x^2} \frac{d^n}{dx^n} e^{-x^2}, \quad n = 0, 1, 2, \ldots. \tag{5.14}$$

Direct evaluation of this formula gives

$$H_0(x) = 1, \qquad H_1(x) = 2x, \qquad H_2(x) = 4x^2 - 2, \ldots.$$

The generating function for $H_n(x)$ can itself be deduced from Eq. (5.14). This can be done by first introducing a "dummy" variable t into Eq. (5.14) to generate the necessary powers of t:

$$H_n(x) = e^{x^2}\left(\frac{\partial^n}{\partial t^n} e^{-(x-t)^2}\right)_{t=0} = \left(\frac{\partial^n}{\partial t^n}(e^{2x-t^2})\right)_{t=0}.$$

There is thus a Taylor expansion about $t = 0$ of the function

$$e^{2xt-t^2} = \sum_{n=0}^{\infty} H_n(x)\frac{t^n}{n!} = G(x,t). \tag{5.15}$$

This simple GF can now be used to derive recurrence and orthogonality relations such as the following:

$$H_{n+1}(x) = 2xH_n(x) - 2nH_{n-1}(x), \tag{5.16}$$

$$H'_n(x) = 2nH_{n-1}(x), \tag{5.17}$$

$$H_{n+1}(x) - 2xH_n(x) + H'_n(x) = 0, \tag{5.18}$$

$$\int_{-\infty}^{\infty} e^{-x^2} H_m(x) H_n(x)\, dx = 2^n n!\, \sqrt{\pi}\, \delta_{mn} = h_n\, \delta_{mn}. \tag{5.19}$$

The actual derivation of these relations using the GF in Eq. (5.15) will be left as exercises. Equation (5.18) itself can also be obtained directly from the Rodrigues formula (5.14) as follows:

$$\begin{aligned} H'_n(x) &= (-1)^n (2xe^{x^2}) \frac{d^n}{dx^n} e^{-x^2} + (-1)^n e^{x^2} \frac{d^{n+1}}{dx^{n+1}} e^{-x^2} \\ &= 2xH_n(x) - H_{n+1}. \end{aligned}$$

On differentiating Eq. (5.18) once more, we find

$$H''_n = 2H_n + 2xH'_n - H'_{n+1}.$$

A differential equation for H_n can be obtained by eliminating H'_{n+1} in favor of H_n with the help of Eq. (5.17):

$$H''_n(x) - 2xH'_n(x) + 2nH_n(x) = 0. \tag{5.20}$$

This is just the Hermite differential equation, which can also be used to define these Hermite polynomials.

Hermite polynomials appear in the quantum-mechanical description of conservative systems near equilibrium. The statement of energy conservation

$$\frac{p^2}{2m} + V(\mathbf{r}) = E, \tag{5.21}$$

for a particle of mass m and momentum p that is experiencing a potential energy $V(\mathbf{r})$ at the position $\mathbf{r}$ holds in quantum mechanics as in classical mechanics. However, atomic systems do not behave in the same way as familiar systems in classical mechanics. It was Niels Bohr who first realized that an electron bound electrostatically to an atomic nucleus in a circular orbit could exist only in certain "stationary" states, while other rather similar orbits were not allowed. Bohr discovered that the states permitted by nature satisfied the "quantum condition" that the orbital angular momentum of the electron was an integral multiple of $\hbar = h/2\pi$, the reduced Planck's constant. When quantum mechanics was eventually formulated by Heisenberg, Born, and Jordon, and by Schrödinger, it was realized that this selectivity was a consequence of a change of meaning of the momentum p in Eq. (5.21), namely that it is not a number, but the differential operator

$$\mathbf{p} = -i\hbar\nabla. \tag{5.22}$$

In quantum mechanics, even the potential energy $V(\mathbf{r})$ should be treated as an operator. We have in fact discussed in Section 3.8 why this should be the case. It is useful to refer to the operator nature of the left-hand side of Eq. (5.21) in quantum mechanics by giving it the special name of *Hamiltonian* and denoting it by the symbol H:

$$H = \frac{p^2}{2m} + V(\mathbf{r}). \tag{5.23}$$

The statement (5.21) for energy conservation now becomes $H = E$ in quantum mechanics. But surely this cannot be literally true, because an operator H cannot be equal to a number such as E. It is conceivable, however, that the operator H operating on a certain function $\psi(\mathbf{r})$ could give a result equal to the number E times the same function $\psi(\mathbf{r})$. That is, energy conservation in quantum mechanics might take the form of a differential equation

$$\left(\frac{p^2}{2m} + V(\mathbf{r}) - E\right)\psi(\mathbf{r}) = 0. \tag{5.24}$$

This is the famous Schrödinger wave equation. The wave function $\psi(\mathbf{r})$ that solves the equation provides a physical description of the system.

The mathematical structure of Eq. (5.24) reminds us of the rotation of a rigid body. The angular momentum of a rigid body, given by the matrix product of its 3×3 inertial matrix and its angular velocity vector, can actually be parallel to its angular velocity, but this occurs only when the angular velocity is along one of its three principal axes of inertia. The situation is described by Eq. (2.48), which can be seen to be similar to Eq. (5.24) here. For the same reason, only for very special functions will the Schrödinger equation be satisfied. The energies E at which this occurs are called its *energy eigenvalues,* while the wave functions $\psi(\mathbf{r})$ so obtained are called its *energy eigenfunctions.*

Consider now a one-dimensional quantum system near a potential minimum at $x = 0$. Near this equilibrium point, the force is linear restoring and the potential is quadratic

$$V(x) \simeq \tfrac{1}{2}m\omega^2x^2. \tag{5.25}$$

The Schrödinger equation can then be simplified to

$$\left[-\frac{1}{2}\left(\frac{d^2}{dx^2} - \frac{x^2}{x_0^4}\right) - \frac{e}{x_0^2}\right]\psi(x) = 0, \tag{5.26}$$

where

$$x_0 = \left(\frac{\hbar}{m\omega}\right)^{1/2}$$

has the dimension of a distance. It gives a characteristic size to the system, while

$$e = \frac{E}{\hbar\omega} \tag{5.27}$$

is a "dimensionless" energy, that is, energy in units of the characteristic energy $\hbar\omega$. Equation (5.26) can be further simplified by writing it as an equation for the dimensionless distance $\xi = x/x_0$:

$$\left[-\frac{1}{2}\left(\frac{d^2}{d\xi^2} - \xi^2\right) - e\right]\psi(\xi) = 0. \tag{5.28}$$

It is now appropriate to ask what exactly is the physical information contained in the wave function $\psi(\xi)$. The answer first proposed by Born is that $|\psi(\xi)|^2$ gives the probability density for finding the system at the position ξ. That is, the integral

$$P = x_0 \int_{-\infty}^{\infty} |\psi(\xi)|^2 \, d\xi \tag{5.29}$$

gives the total probability of finding the system somewhere. If we know for a certainty that the system is oscillating about the point of equilibrium, this total probability must be 1. This means that the wave function $\psi(\xi)$ must be square integrable. To be square integrable, $\psi(\xi)$ must vanish sufficiently rapidly for large values of $|\xi|$.

At large $|\xi|$, the Schrödinger equation (5.26) simplifies to

$$\left(\frac{d^2}{d\xi^2} - \xi^2\right)\psi(\xi) = 0. \tag{5.30}$$

Its solutions can be calculated easily:

$$\psi(\xi) \underset{\xi\to\infty}{\sim} e^{\pm(1/2)\xi^2}. \tag{5.31}$$

The solution $\exp(\frac{1}{2}\xi^2)$ is not square integrable; so we are left with only $\exp(-\frac{1}{2}\xi^2)$. This suggests that there might be a solution of the complete Eq. (5.26) having the form

$$\psi(\xi) = H(\xi)e^{-(1/2)\xi^2}. \tag{5.32}$$

The DE satisfied by $H(\xi)$ can be obtained by direct substitution. We find

$$\begin{aligned}
\psi'(\xi) &= -\xi\psi + e^{-(1/2)\xi^2}H', \\
\psi''(\xi) &= -\psi - \xi(-\xi\psi + e^{-(1/2)\xi^2}H') - \xi e^{-(1/2)\xi^2}H' + e^{-(1/2)\xi^2}H'' \\
&= e^{-(1/2)\xi^2}(H'' - 2\xi H' - H + \xi^2 H).
\end{aligned}$$

The result is

$$\left(\frac{d^2}{d\xi^2} - 2\xi\frac{d}{d\xi} + (2e - 1)\right)H(\xi) = 0. \tag{5.33}$$

Comparison with Eq. (5.20) shows that the solutions are just the Hermite polynomials $H_n(\xi)$ if $2e-1=2n$. That is, the dimensionless energy eigenvalues are

$$e = n + \tfrac{1}{2}. \tag{5.34}$$

What happens when $e \neq n + \frac{1}{2}$? One can show by doing Problem 5.3.4 that Eq. (5.33) then has solutions that are infinite series rather than polynomials. These solutions all behave like $\exp(\xi^2)$ at large ξ. The resulting wave functions are then $\exp(\xi^2/2)$ at large distances. They are not square integrable, and are therefore not realizable as physical states.

The requirement of square integrability thus selects certain functions $H_n(\xi)$ out of the original set $H(\xi)$. Certain quantum characteristics of atomic systems have their origin in the discreteness of their eigenvalues caused by this selection.

The wave functions

$$\psi_n(\xi) = c_n H_n(\xi) e^{-(1/2)\xi^2} \tag{5.35}$$

of the quantum oscillator are plotted in Fig. 5.2. The normalization

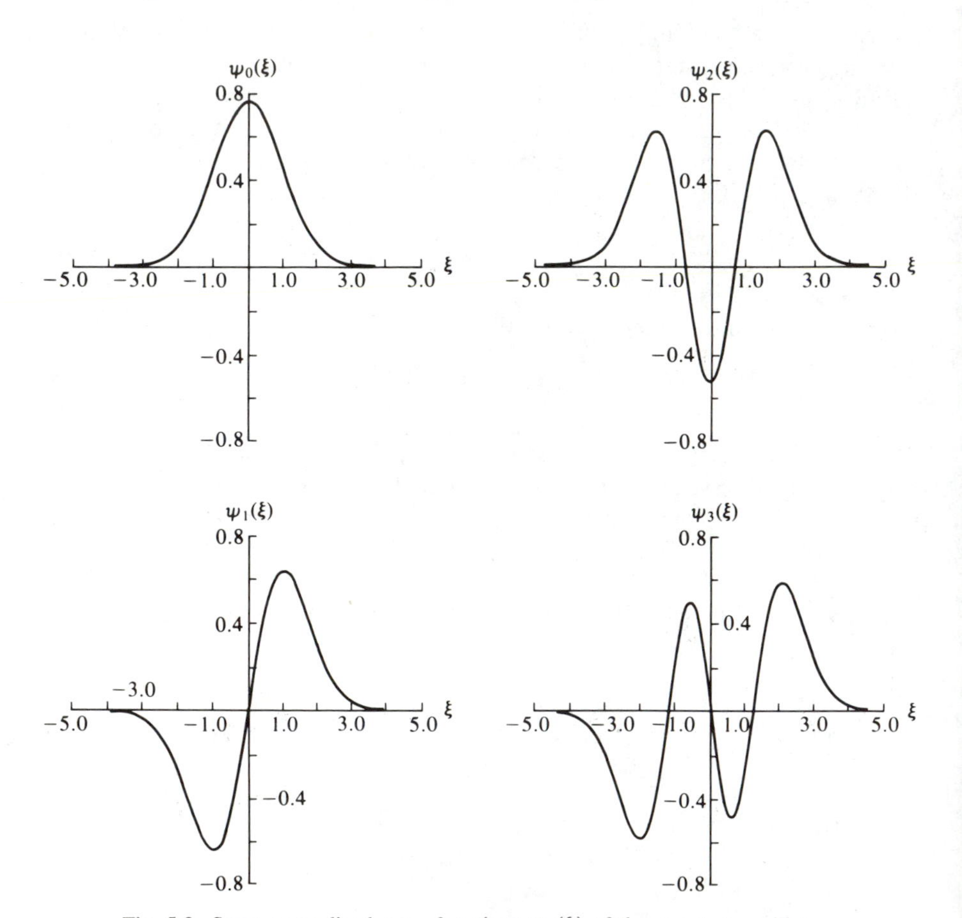

Fig. 5.2. Some normalized wave functions $\psi_n(\xi)$ of the quantum oscillator.

constants c_n are chosen so that

$$x_0 \int_{-\infty}^{\infty} \psi_n^2(\xi)\, d\xi = 1. \tag{5.36}$$

Problems

5.3.1. Show that $H_n(-x) = (-1)^n H_n(x)$.

5.3.2. Use the generating function shown in Eq. (5.15) for Hermite polynomials to derive the recursion formulas Eq. (5.16)–(5.18).

5.3.3. Use the generating function shown in Eq. (5.15) for Hermite polynomials to derive their orthogonality relation Eq. (5.19).

5.3.4. Solve the Hermite differential equation (5.20) by looking for a Frobenius-series solution of the form

$$H(\xi) = \xi^s(a_0 + a_1\xi + a_2\xi^2 + \cdots),$$

with $a_0 \neq 0$ and $s \geqslant 0$. Show explicitly that

(a) The Frobenius series terminates as a polynomial when $e = n + \frac{1}{2}$; and

(b) The infinite series can be summed to $\exp(\xi^2)$ for large ξ when $e \neq n + \frac{1}{2}$.

5.3.5. If $\psi_n(\xi) = (2^n n!\sqrt{\pi})^{-1/2} H_n(\xi) e^{-(1/2)\xi^2}$ are the square-integrable solutions of the oscillator Eq. (5.28), show that

(a) $\left(\frac{d}{d\xi} + \xi\right)\psi_0(\xi) = 0;$

(b) $\frac{1}{\sqrt{2}}\left(\frac{d}{d\xi} + \xi\right)\psi_n(\xi) \equiv a\psi_n(\xi) = \sqrt{n}\,\psi_{n-1}(\xi);$

(c) $\frac{1}{\sqrt{2}}\left(\frac{d}{d\xi} - \xi\right)\psi_n(\xi) = -a^\dagger\psi_{n+1} = -\sqrt{n+1}\,\psi_{n+1}(\xi);$

(d) $[a, a^\dagger] = aa^\dagger - a^\dagger a = 1;$

(e) $H = -\frac{1}{2}\left(\frac{d^2}{d\xi^2} - \xi^2\right) = a^\dagger a + \frac{1}{2};$

(f) $(H - e_n)\psi_n(\xi) = 0$, with $e_n = n + \frac{1}{2}$, is the oscillator equation (5.28) for physically realizable states;

(g) $[H, a^\dagger] = a^\dagger$, $(H, a] = a$.

(h) If the solution $\psi_n(\xi)$ is said to have n oscillator "quanta," show that $a^\dagger\psi_n$ has $n+1$ quanta and $a\psi_n$ has $n-1$ quanta.

The first-order differential operator $a(a^\dagger)$ is called a *destruction* (*creation*) operator for oscillator quanta. It is also called a *lowering* (*raising*) operator, or a *step-down* (*step-up*) operator, or a *ladder* operator. H is the Hamiltonian of the quantum oscillator, and e_n its energy eigenvalue in the physical state with n oscillator quanta.

5.4. Orthogonal polynomials

We have seen in Section 3.10 that a family of polynomials $\{f_n(x), n = 0, 1, \ldots, \infty\}$, where

$$f_n(x) = k_n x^n + k'_n x^{n-1} + \cdots \tag{5.37}$$

is a polynomial of degree n, is said to be *orthogonal* on the closed interval $[a,b]$, with respect to the *weight function* $w(x)$ if

$$\int_a^b w(x) f_m(x) f_n(x)\, dx = h_n\, \delta_{mn}. \tag{5.38}$$

Although the f_n in different families are all polynomials of degree n, they have different coefficients $k_n, k'_n, \ldots$. These differences arise from differences in the weight functions and in the intervals of integration. Tables of a, b, $w(x)$, and h_n for four common orthogonal polynomials (Legendre, Laguerre, Hermite polynomials, and Chebyshev polynomials of the first kind) have been given in Table 3.1.

For example, the Legendre and Chebyshev polynomials are defined in the same interval $x = -1$ to 1, but differ in their weight functions. As a result, orthogonality means different things to different families of orthogonal polynomials, even for those defined in the same interval. This point was emphasized in Section 3.10. A table of the first four polynomials of each family has been given in Table 3.2.

These orthogonal polynomials satisfy a generalized Rodrigues formula of the same general form

$$f_n(x) = \frac{1}{a_n w(x)} \frac{d^n}{dx^n} \{w(x)[s(x)]^n\}, \tag{5.39}$$

where $w(x)$ is the weight function of Eq. (5.38), $s(x)$ is a polynomial of degree ≤ 2, and a_n is a normalization constant. These quantities are shown in Table 5.1.

Table 5.1. Rodrigues' formula for several orthogonal polynomials

$$f_n(x) = \frac{1}{a_n w(x)} \frac{d^n}{dx^n} \{w(x)[s(x)]^n\}$$

$f_n(x)$	Name	$w(x)$	$s(x)$	a_n
$P_n(x)$	Legendre	1	$1 - x^2$	$(-1)^n 2^n n!$
$L_n(x)$	Laguerre	e^{-x}	x	$n!$
$H_n(x)$	Hermite	e^{-x^2}	1	$(-1)^n$
$T_n(x)$	First Chebyshev	$(1 - x^2)^{-1/2}$	$1 - x^2$	$(-1)^n 2^{n+1} \dfrac{\Gamma(n + \frac{1}{2})}{\sqrt{\pi}}$

N.B.: $\Gamma(n + \frac{1}{2}) = (n - \frac{1}{2})\Gamma(n - \frac{1}{2})$, $\Gamma(\frac{1}{2}) = \sqrt{\pi}$.

Table 5.2. Generating functions $G(x,t) = \sum_{n=0}^{\infty} b_n f_n(x)t^n$, for $|t| < 1$.

$f_n(x)$	$G(x,t)$	b_n	Note
$P_n(x)$	R^{-1}	1	$\|x\| < 1$
$L_n(x)$	$\frac{1}{1-t}\exp\left(\frac{xt}{t-1}\right)$	1	
$H_n(x)$	$\exp(2xt - t^2)$	$\frac{1}{n!}$	
$T_n(x)$	$(1 - xt)/R^2$	1	$\|x\| < 1$

N.B.: $R = \sqrt{1 - 2xt + t^2}$.

Table 5.2 gives the generating functions

$$G(x,t) = \sum_{n=0}^{\infty} b_n f_n(x)t^n, \qquad \text{for} \qquad |t| < 1. \tag{5.40}$$

From these one can derive recursion formulas of the form

$$f_{n+1}(x) = (b_n x + c_n)f_n - d_n f_{n-1}, \tag{5.41}$$

where the coefficients b_n, c_n, and d_n are given in Table 5.3. One can also obtain the differential relations

$$p_2(x)f'_n(x) = p_1(x)f_n(x) + p_0(n)f_{n-1}(x), \tag{5.42}$$

where $f'_n(x) = df(x)/dx$. The functions $p_i(x)$ are shown in Table 5.4.

Finally, the orthogonal polynomials satisfy second-order differential equations of the form

$$g_2(x)f''_n(x) + g_1(x)f'_n(x) + g_0(n)f_n(x) = 0, \tag{5.43}$$

where the functions $g_i(x)$ are given in Table 5.5.

Plots of some Legendre polynomials have already been given in Fig. 5.2. Some of the other polynomials are shown in Fig. 5.3.

Orthogonal polynomials can be used to represent arbitrary functions, including solutions of differential equations, in their interval. This is done by writing the function to be represented as

$$R(x) = \sum_{n=0}^{\infty} c_n f_n(x). \tag{5.44}$$

Table 5.3. Recursion formulas for orthogonal polynomials $f_{n+1}(x) = (b_n x + c_n)f_n - d_n f_{n-1}$

$f_n(x)$	b_n	c_n	d_n
$P_n(x)$	$(2n+1)/(n+1)$	0	$n/(n+1)$
$L_n(x)$	$-1/(n+1)$	$(2n+1)/(n+1)$	$n/(n+1)$
$H_n(x)$	2	0	$2n$
$T_n(x)$	4	-2	1

Table 5.4. Differential relation for orthogonal polynomials $p_2(x)f'_n(x) = p_1(x)f_n(x) + p_0 f_{n-1}(x)$

$f_n(x)$	$p_2(x)$	$p_1(x)$	p_0
$P_n(x)$	$1-x^2$	$-nx$	n
$L_n(x)$	x	n	$-n$
$H_n(x)$	1	0	$2n$
$T_n(x)$	$1-x^2$	$-nx$	n

The expansion coefficients c_n can be extracted by multiplying both sides by $f_m(x)w(x)$ and integrating from a to b. It is convenient to write the resulting integral in the compact notation of a functional inner product:

$$(f_m,R) \equiv \int_a^b f_m(x)w(x)R(x)\,dx$$

$$= \sum_{n=0}^{\infty} c_n(f_m,f_n), \tag{5.45}$$

where all quantities have been taken to be real. Since (f_m,f_n) is just Eq. (5.38), we have

$$(f_m,R) = \sum_{n=0}^{\infty} c_n h_n\,\delta_{mn} = c_m h_m. \tag{5.46}$$

Example 5.4.1. Expand $R(x) = e^{ax}$ in the Hermite series

$$R(x) = \sum_{n=0}^{\infty} c_n H_n(x). \tag{5.47}$$

According to Eq. (5.46), the coefficients are

$$c_n = h_n^{-1}(H_n,e^{ax}). \tag{5.48}$$

Table 5.5. The differential equation satisfied by orthogonal polynomials $g_2(x)f''_n + g_1(x)f'_n + g_0(n)f_n = 0$

$f_n(x)$	$g_2(x)$	$g_1(x)$	$g_0(n)$
$P_n(x)$	$1-x^2$	$-2x$	$n(n+1)$
$L_n(x)$	x	$1-x$	n
$H_n(x)$	1	$-2x$	$2n$
$T_n(x)$	$1-x^2$	$-nx$	n

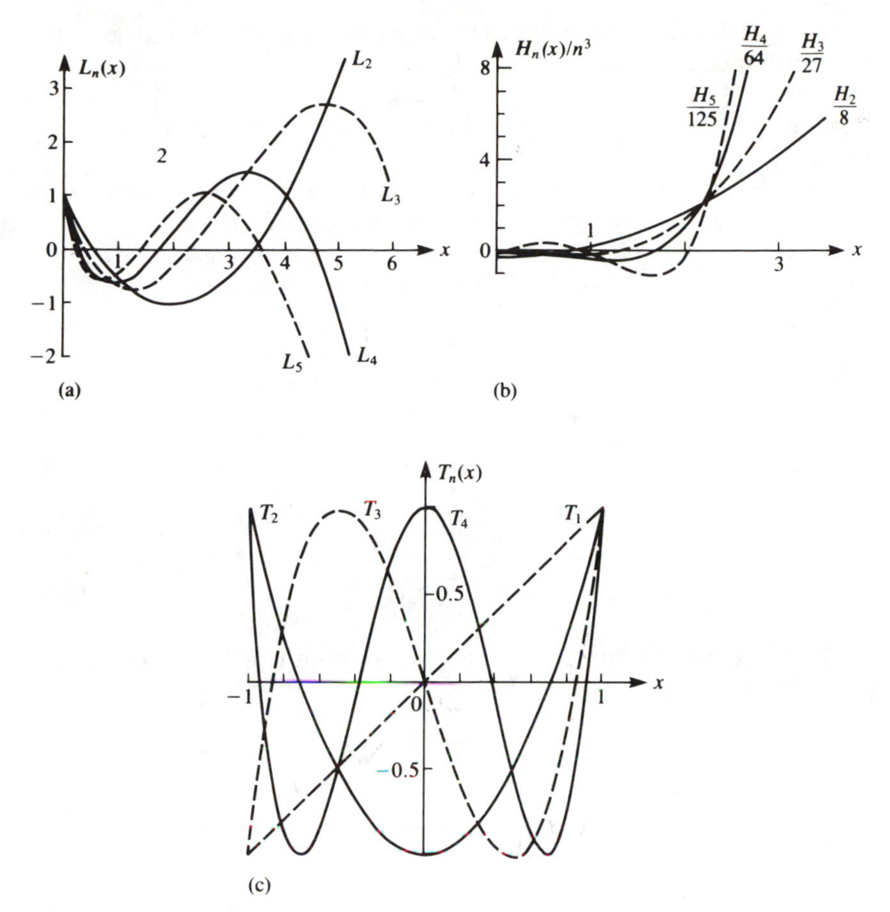

Fig. 5.3. (a) Laguerre polynomials $L_n(x)$, $n = 2$–5; (b) Hermite polynomials $H_n(x)/n^3$, $n = 2$–5; (c) Chebyshev polynomials $T_n(x)$, $n = 1$–4 (from Abramowitz and Stegun).

Using Tables 3.2 and 3.3, we find

$$c_0 = \frac{1}{\sqrt{\pi}} \int_{-\infty}^{\infty} e^{ax} e^{-x^2}\, dx$$

$$= \frac{1}{\sqrt{\pi}} e^{a^2/4} \int_{-\infty}^{\infty} e^{-[x-(1/2)a]^2}\, dx = e^{a^2/4}, \tag{5.49}$$

$$c_1 = \frac{1}{2\sqrt{\pi}} \int_{-\infty}^{\infty} 2x e^{ax} e^{-x^2}\, dx$$

$$= \frac{1}{\sqrt{\pi}} \frac{d}{da} \int_{-\infty}^{\infty} e^{ax} e^{-x^2}\, dx$$

$$= \frac{d}{da} e^{a^2/4} = \frac{a}{2} e^{a^2/4}, \tag{5.50}$$

and so on. The calculation increases in complexity when the degree n of the polynomial increases.

An equivalent method is to use the Rodrigues formula and integrate by parts.

In many cases, the calculation of c_n can be greatly simplified by using the generating function, as the following example illustrates.

Example 5.4.2. Calculate the Hermite coefficients c_n in Eq. (5.47) with the help of the generating function

$$G(x,t) = e^{2xt-t^2} = \sum_{n=0}^{\infty} \frac{t^n}{n!} H_n(t).$$

We have by direct substitution

$$\begin{aligned} (G,R) &= \sum_{n=0}^{\infty} \frac{t^n}{n!} (H_n,R) \\ &= \sum_{n=0}^{\infty} \frac{t^n}{n!} h_n c_n. \end{aligned} \tag{5.51}$$

If (R,G) can be calculated and expanded in power of t^n, the coefficients c_n can be extracted. Now

$$\begin{aligned} (G,e^{ax}) &= \int_{-\infty}^{\infty} e^{-x^2} e^{2xt-t^2} e^{ax}\, dx \\ &= e^{-t^2} \int_{-\infty}^{\infty} e^{-x^2+2(t+a/2)x}\, dx \\ &= e^{-t^2+(t+a/2)^2} \int_{-\infty}^{\infty} e^{-[x-(t+a/2)]^2}\, dx \\ &= \sqrt{\pi}\, e^{at+a^2/4}. \end{aligned}$$

The expansion in powers of t turns out to be easy in this example:

$$(G,e^{ax}) = \sqrt{\pi}\, e^{a^2/4} \sum_{n=0}^{\infty} \frac{(at)^n}{n!}. \tag{5.52}$$

Comparison with Eq. (5.51) shows that

$$c_n = \sqrt{\pi}\, a^n e^{a^2/4}/h_n = (a/2)^n e^{a^2/4}/n!, \tag{5.53}$$

where h_n is given by Table 3.1. The generating function allows us to obtain all the coefficients "wholesale," thus avoiding the laborious calculation of coefficients one after another.

Problems

5.4.1. Use the generating functions to derive the recursion formulas shown in Table 5.3.

5.4.2. Use the generating functions to derive the differential relations shown in Table 5.4.

5.4.3. Use the generating functions to derive the differential equations shown in Table 5.5.

5.4.4. Give the Hermite representation [Eq. (5.47)] of
(a) x^3;
(b) x^4;
(c) $\exp(-a^2x^2)$.

5.4.5. Give the Laguerre representation of $\exp(-ax)$.

5.4.6. Show that

$$e^{t^2}\cos 2xt = \sum_{n=0}^{\infty} \frac{(-1)^n H_{2n}(x)}{(2n)!} t^{2n},$$

$$e^{t^2}\sin 2xt = \sum_{n=0}^{\infty} \frac{(-1)^n H_{2n+1}(x)}{(2n+1)!} t^{2n+1}.$$

5.5. Classical orthogonal polynomials

It turns out to be possible to define families of orthogonal polynomials by the generalized Rodrigues formula (5.39)

$$f_n(x) = \frac{1}{a_n w(x)} \frac{d^n}{dx^n} \{w(x)[s(x)]^n\} \tag{5.39}$$

if three conditions are imposed:

1. $f_1(x)$ is a polynomial of first degree;
2. $s(x)$ is a polynomial of degree $\leqslant 2$ and has real roots;
3. The weight function $w(x)$ is real, positive, and integrable in the interval $[a,b]$. It satisfies the boundary conditions (BC)

$$w(a)s(a) = w(b)s(b) = 0 \tag{5.54}$$

at the boundary points a and b.

We shall first show that the function $f_n(x)$ is a polynomial $p_{\leqslant n}$ of degree $\leqslant$n if f_1 is a first-degree polynomial satisfying the equation

$$\frac{d}{dx}(ws) = a_1 w f_1 = w p_{\leqslant 1}. \tag{5.55}$$

This can be done by first considering

$$\begin{aligned}\frac{d}{dx}(ws^n) &= s^{n-1}\frac{d}{dx}(ws) + ws(n-1)s^{n-2}\frac{ds}{dx}\\ &= ws^{n-1}p_{\leqslant 1} + (n-1)ws^{n-1}p_{\leqslant 1}\\ &= ws^{n-1}p_{\leqslant 1},\end{aligned} \tag{5.56}$$

where we have used the information that $s = p_{\leq 2}$. The different $p_{\leq 1}$ that appear are not the same function, but they all are polynomials of degree not exceeding one. Repeated differentiation or an inductive argument can be used to show that

$$\frac{d^m}{dx^m}(ws^n) = ws^{n-m}p_{\leq m}, \qquad m \leq n. \tag{5.57}$$

Equation (5.57) used with $m = n$ states that f_n is a polynomial of degree $\leq n$.

We next show that f_n is orthogonal to any polynomial p_l of degree $l < n$, and is therefore *not* of degree $< n$. The overlap between these two polynomials is from Eqs. (5.38) and (5.39)

$$\begin{aligned}(p_l, f_n) &\equiv \int_a^b w(x)p_l(x)f_n(x)\,dx \\ &= \frac{1}{a_n}\int_a^b p_l(x)\frac{d^n}{dx^n}(ws^n)\,dx.\end{aligned}$$

We now integrate by parts n times. Each time there appear boundary terms that vanish because of the BC (5.54). Hence

$$(p_l, f_n) = \frac{(-1)^n}{a_n}\int_a^b ws^n \frac{d^n}{dx^n}(p_l)\,dx. \tag{5.58}$$

This also vanishes for $l < n$.

Since f_n is a polynomial of degree $\leq n$ [from Eq. (5.57)], but not of degree $< n$ [from Eq. (5.58)], it must be a polynomial of degree exactly n. We have thus verified that given $s(x)$ and $w(x)$, the polynomials $f_n(x)$ defined by Eq. (5.42) form a family of orthogonal polynomials.

The three conditions given just before Eq. (5.54) are actually very restrictive. Only three distinct possibilities exist. They give rise to three classes of polynomials called *classical orthogonal polynomials*. These are obtained by choosing $s(x)$ to have one of the following forms:

$$s(x) = \begin{cases} \alpha, & (5.59\text{a}) \\ \beta(x - \alpha), & (5.59\text{b}) \\ \gamma(x - \alpha)(\beta - x), \quad \beta > \alpha, & (5.59\text{c}) \end{cases}$$

where α, β, and γ are constants. Once $s(x)$ is chosen, $w(x)$ can be determined by solving the first-order differential equation obtained from Eq. (5.55),

$$s\frac{dw}{dx} = w\left(a_1 f_1 - \frac{ds}{dx}\right).$$

subject to the BC (5.54). We can actually take $a_1 f_1$ to be $-x$, because the normalization turns out to be unimportant, while a constant term in f_1 can be removed by redefining x. Hence

$$s\frac{dw}{dx} = -w\left(x + \frac{ds}{dx}\right). \tag{5.60}$$

Equation (5.60) is easily solved for Eq. (5.59a). Here $ds/dx = 0$. Hence

$$w(x) = Ae^{-x^2/2\alpha}.$$

The normalization A can be taken to be 1, because Eq. (5.39) already contains a normalization a_n, while 2α can be taken to be 1 by redefining x. We therefore get the standard form

$$w(x) = e^{-x^2}. \tag{5.61}$$

To satisfy the BC (5.54), the boundary points should be taken to be $[0, \infty)$, $(-\infty, 0]$ or $(-\infty, \infty)$, the last being the standard form. (The points $\pm\infty$ are always considered to be outside the interval.)

The solutions for the other two cases are left as exercises. The results are summarized in Table 5.6. The polynomials constructed for Eq. (5.59b) are the generalized Laguerre polynomials. The Jacobi polynomials come from Eq. (5.59c) and include the Legendre and first Chebyshev polynomials of Table 5.1:

$$\begin{aligned} P_n(x) &= P_n^{(0,0)}(x) \\ T_n(x) &= \frac{n!\sqrt{\pi}}{\Gamma(n+\frac{1}{2})} P_n^{(-1/2,-1/2)}(x), \end{aligned} \tag{5.62}$$

with the usual choice of normalization.

Since $f_n(x)$ is a polynomial of degree n, $xf_n(x)$ must be a polynomial of degree $n+1$. We can therefore write

$$xf_n(x) = \sum_{l=0}^{n+1} a_{nl} f_l(x).$$

The fact [from (5.58)] that $(p_l, f_n) = 0$ if $l < n$ can now be used to show that

$$(xf_n, f_l) = (f_n, xf_l) \neq 0 \tag{5.63}$$

only if $l \geq n-1$. This means that there is a recursion formula of the form

$$f_{n+1}(x) = (b_n x + c_n) f_n(x) - d_n f_{n-1}(x). \tag{5.64}$$

Table 5.6. The three classes of classical orthogonal polynomials.

Eq.	$s(x)$	$w(x)$	Interval	Name
(5.59a)	1	e^{-x^2}	$(-\infty,\infty)$	Hermite $H_n(x)$
(5.59b)	x	$x^\nu e^{-x}(\nu > -1)$	$[0,\infty)$	Laguerre $L_n^{(\nu)}(x)$
(5.59c)	$1-x^2$	$(1-x)^\nu(1+x)^\mu(\mu,\nu > 1)$	$[-1,1]$	Jacobi $P_n^{(\nu,\mu)}(x)$

The reader can verify by direct construction and with the help of Eqs. (5.37) and (5.38) that

$$\begin{aligned} b &= k_{n+1}/k_n, \\ c_n &= (k_{n+1}/k_n)[(k'_{n+1}/k_{n+1}) - (k'_n/k_n)] \\ d_n &= (h_n/h_{n-1})k_{n+1}k_{n-1}/k_n^2. \end{aligned} \tag{5.65}$$

(This is Problem 5.5.2.)

Since $f_n(x)$ is a polynomial of degree n, $df_n(x)/dx$ must be a polynomial of degree $n-1$. Now Eq. (5.56) can be written in the more general form (see Problem 5.5.3)

$$\frac{d}{dx}(ws^n p_m) = ws^{n-1}p_{\leq m+1}. \tag{5.66}$$

This shows that

$$\begin{aligned} \frac{1}{w}\frac{d}{dx}\left(ws\frac{d}{dx}f_n\right) &= p_{\leq n}(x) \\ &= -\sum_{l=0}^{n} \lambda_{nl} f_l(x). \end{aligned} \tag{5.67}$$

The expansion coefficients on the right-hand side are just the overlaps

$$\lambda_{nl}h_l = -\left(f_l, \frac{1}{w}\frac{d}{dx}\left(ws\frac{d}{dx}f_n\right)\right) = -\int_a^b f_l \frac{d}{dx}\left(ws\frac{d}{dx}f_n\right) dx.$$

We now integrate twice by parts, using the boundary condition (5.54) to eliminate all boundary terms:

$$\begin{aligned} \lambda_{nl}h_l &= +\int_a^b \left(\frac{d}{dx}f_l\right) ws \left(\frac{d}{dx}f_n\right) dx \\ &= -\int_a^b f_n \frac{d}{dx}\left[\left(\frac{d}{dx}f_l\right) ws\right] dx \\ &= \lambda_{ln}h_n. \end{aligned} \tag{5.68}$$

Equation (5.67) can now be applied to the last integrand to get

$$\lambda_{ln}h_n = -\int_a^b f_n w p_{\leq l}\, dx.$$

According to Eq. (5.58), this vanishes if $l < n$. We have thus shown that the right-hand side of Eq. (5.67) contains only one term involving the coefficient λ_{nn}. In other words, the polynomial $f_n(x)$ satisfies the second-order differential equation

$$\frac{d}{dx}\left(ws\frac{d}{dx}f_n(x)\right) + w\lambda_{nn}f_n(x) = 0. \tag{5.69}$$

We may note with some satisfaction that both Eqs. (5.64) and (5.69) are derived by applying simple arguments cleverly.

The undaunted reader may want to derive the actual expression for λ_{nn}, which is

$$\lambda_{nn} = -n\left(a_1 \frac{df_1}{dx} + \tfrac{1}{2}(n-1)\frac{d^2 s}{dx^2}\right), \tag{5.70}$$

where a_1 is the normalization constant used in Eq. (5.39). Problem 5.5.4 gives a hint on how this can be done.

Problems

5.5.1. Verify the weight function $w(x)$ and the choice of interval shown in Table 5.6 for the generalized Laguerre and Jacobi polynomials.
5.5.2. Verify Eq. (5.65).
5.5.3. Use the method of Eqs. (5.55) and (5.56) to verify Eq. (5.66).
5.5.4. Verify Eq. (5.70) by first showing that

$$\frac{d}{dx}\left(sw\frac{df_n}{dx}\right) = \left(a_1 n\frac{df_1}{dx} + \tfrac{1}{2}n(n-1)\frac{d^2 s}{dx^2}\right)k_n x^n + \cdots,$$

where k_n is the coefficient shown in Eq. (5.37).

5.6. Associated Legendre polynomials and spherical harmonics

Other orthogonal polynomials can be obtained from the classical ones by additional manipulations. For example, the *associated Legendre polynomials* $P_l^m(x)$ of degree l and order m are related to the Legendre polynomials (order $m = 0$) by the differential relation

$$P_l^m(x) = (1-x^2)^{m/2}\frac{d^m}{dx^m}P_l(x), \qquad 0 \leq m \leq l. \tag{5.71}$$

It is possible to show (for example by induction) that the function

$$u_m(x) = \frac{d^m}{dx^m}P_l(x) \tag{5.72a}$$

satisfies the DE

$$\left((1-x^2)\frac{d^2}{dx^2} - 2(m+1)x\frac{d}{dx} + [l(l+1) - m(m+1)]\right)u_m(x) = 0, \tag{5.72b}$$

while P_l^m itself satisfies the associated Legendre DE:

$$\left[(1-x^2)\frac{d^2}{dx^2} - 2x\frac{d}{dx} + \left(l(l+1) - \frac{m^2}{1-x^2}\right)\right]P_l^m(x) = 0. \tag{5.73}$$

(This is Problem 5.6.1.)

These polynomials satisfy the orthogonality relation

$$(P_{l'}^m, P_l^m) \equiv \int_{-1}^{1} P_{l'}^m(x) P_l^m(x)\, dx = \frac{(l+m)!}{(l-m)!} \int_{-1}^{1} P_{l'}(x) P_l(x)\, dx \tag{5.74a}$$

$$= \frac{(l+m)!}{(l-m)!} \delta_{ll'} \frac{2}{2l+1}. \tag{5.74b}$$

This can be proved by using Eq. (5.71) in Eq. (5.74a) and by integrating by parts:

$$(P_{l'}^m, P_l^m) = \int_{-1}^{1} \frac{d^m P_{l'}}{dx^m} (1-x^2)^m \frac{d^m P_l}{dx^m}\, dx$$

$$= -\int_{-1}^{1} \frac{d^{m-1} P_{l'}}{dx^{m-1}} \frac{d}{dx}\left((1-x^2)^m \frac{d^m P_l}{dx^m}\right) dx,$$

where zero boundary terms have been discarded. The required differentiation in the integrand can be performed explicitly with the help of Eq. (5.72b)

$$\frac{d}{dx}\left((1-x^2)^m \frac{d^m P_l}{dx^m}\right) = \frac{d}{dx}\left((1-x^2)^m \frac{d}{dx} u_{m-1}\right)$$

$$= (1-x^2)^m \frac{d^2 u_{m-1}}{dx^2} - 2mx(1-x^2)^{m-1} \frac{du_{m-1}}{dx}$$

$$= -(1-x^2)^{m-1}(l+m)(l-m+1) u_{m-1}.$$

Hence

$$(P_{l'}^m, P_l^m) = (l+m)(l-m+1) \int_{-1}^{1} \frac{d^{m-1} P_{l'}}{dx^{m-1}} (1-x^2)^{m-1} \frac{d^{m-1} P_l}{dx^{m-1}}\, dx$$

$$= (l+m)(l-m+1)(P_{l'}^{m-1}, P_l^{m-1}). \tag{5.75}$$

Successive applications of this reduction formula for the order m yield Eq. (5.74a). Equation (5.74b) then follows from the orthogonality relation for Legendre polynomials.

Explicit formulas for $P_l^m(x)$ for $m > 0$ can be obtained directly from Eq. (5.71). They are conveniently expressed in terms of $\cos\theta = x$ and $\sin\theta = (1-x^2)^{1/2}$:

$$P_1^1(x) = (1-x^2)^{1/2} = \sin\theta$$

$$P_2^1(x) = 3x(1-x^2)^{1/2} = 3\cos\theta \sin\theta$$

$$P_2^2(x) = 3(1-x^2) = 3\sin^2\theta, \qquad \text{etc.} \tag{5.76}$$

Note that P_l^m always vanishes at $x = \pm 1$ for $m \neq 0$.

The generating function for P_l^m can be obtained by applying Eq. (5.71) to both sides of the GF (5.4) for P_l:

$$(1-x^2)^{m/2} \frac{(2m)!}{2^m m!\,(1-2tx+t^2)^{m+1/2}} = \sum_{l=m}^{\infty} P_l^m(x) t^{l-m}. \tag{5.77}$$

The resulting expression is often too complicated to be of much use. In many cases, it might be simpler to use Eq. (5.71) directly.

If we go back all the way to the Rodrigues formula for

$$P_l(x) = \frac{1}{2^l\, l!}\frac{d^l}{dx^l}(x^2-1)^l,$$

it is clear that associated Legendre polynomials of negative orders $m = -|m|$ can also be defined down to $m = -l$:

$$P_l^{-|m|}(x) = (1-x^2)^{-|m|/2}\frac{d^{l-|m|}}{dx^{l-|m|}}(x^2-1)^l\frac{1}{2^l l!} \tag{5.78}$$

with

$$P_l^{-l}(x) = (-1)^l(1-x^2)^{l/2}/(2^l l!).$$

This turns out, perhaps unexpectedly, to be proportional to P_l^m itself:

$$P_l^{-m}(x) = (-1)^m\frac{(l-m)!}{(l+m)!}P_l^m(x). \tag{5.79}$$

This result can be proved by showing that P_l^{-m} is not orthogonal to P_l^m. The key is the derivation of a reduction formula for the overlap $(P_{l'}^m, P_l^{-m})$ in the order m by integration by parts in a way similar to the derivation of Eq. (5.75):

$$\begin{aligned}(P_{l'}^m, P_l^{-m}) &= \int_{-1}^{1}\frac{d^m P_{l'}}{dx^m}\frac{d^{l-m}}{dx^{l-m}}(x^2-1)^l\frac{1}{2^l l!}\,dx \\ &= -\int_{-1}^{1}\frac{d^{m-1}P_{l'}}{dx^{m-1}}\frac{d^{l-m+1}}{dx^{l-m+1}}(x^2-1)^l\frac{1}{2^l l!}\,dx \\ &= -(P_{l'}^{m-1}, P_l^{-m+1}).\end{aligned} \tag{5.80}$$

(The boundary terms in the integration by parts vanish because P_l^{-m} vanishes at $x = \pm 1$.) Equation (5.80) can be used repeatedly to give

$$\begin{aligned}(P_{l'}^m, P_l^{-m}) &= (-1)^m(P_{l'}, P_l) \\ &= (-1)^m\frac{2}{2l+1}\delta_{ll'}.\end{aligned} \tag{5.81}$$

Comparison with Eq. (5.74b) gives Eq. (5.79).

Associated Legendre polynomials appear frequently in physics as parts of functions called *spherical harmonics*. *Harmonic functions* are functions that satisfy the Laplace equation $\nabla^2\psi = 0$. The spherical harmonics $Y_l^m(\theta,\phi)$ are those that satisfy it on the surface of a sphere. That is, the Laplace equation in spherical coordinates has solutions of the form

$$\psi(r,\theta,\phi) = \sum_{l=0}^{\infty}\sum_{m=-l}^{l} Y_l^m(\theta,\phi)(A_l r^l + B_l r^{-l-1}). \tag{5.82}$$

According to the results of Section 4.12,

$$Y_l^m(\theta,\phi) = \Phi(\phi)\Theta(\theta),$$

where

$$\Phi(\phi) = \Phi_{\pm m}(\phi) = \exp(\pm im\phi)/(2\pi)^{1/2} \tag{5.83}$$

satisfies Eq. (4.125a), while $\Theta(\theta)$ satisfies the associated Legendre equation Eq. (4.129) or Eq. (5.73). Hence the normalized spherical harmonics are

$$Y_l^m(\theta,\phi) = \left(\frac{(l-m)!}{(l+m)!}\frac{2l+1}{4\pi}\right)^{1/2} e^{im\phi}P_l^m(\cos\theta). \tag{5.84}$$

Spherical harmonics behave simply under the complex conjugation and parity operations:

$$\begin{aligned} Y_l^{m*}(\theta,\phi) &= \left(\frac{(l-m)!}{(l+m)!}\frac{2l+1}{4\pi}\right)^{1/2} e^{-im\phi}P_l^m(\cos\theta) \\ &= (-1)^m Y_l^{-m}(\theta,\phi), \end{aligned} \tag{5.85}$$

where use has been made of Eq. (5.79). Under the parity transformation $\mathbf{r}\to-\mathbf{r}$ or $(r,\theta,\phi)\to(r,\pi-\theta,\pi+\phi)$. Since $\cos\theta$ then transforms into $-\cos\theta$, we find

$$\begin{aligned} P_l^m(x)\to P_l^m(-x) &= (1-x^2)^{m/2}\frac{d^m}{d(-x)^m}P_l(-x) \\ &= (1-x^2)^{m/2}(-1)^m\frac{d^m}{dx^m}(-1)^lP_l(x) \\ &= (-1)^{l+m}P_l^m(x) \end{aligned}$$

with the help of Eq. (5.6). As a result,

$$\begin{aligned} Y_l^m(\theta,\phi)\to Y_l^m(\pi-\theta,\pi+\phi) &= \left(\frac{(l-m)!}{(l+m)!}\frac{2l+1}{4\pi}\right)^{1/2} e^{im(\pi+\phi)}(-1)^{l+m}P_l^m(x) \\ &= (-1)^l Y_l^m(\theta,\phi). \end{aligned} \tag{5.86}$$

Thus it has the same parity property as $P_l(x)$.

Explicit formulas for $Y_l^m(\theta,\phi)$ are listed below for the reader's convenience:

$$\begin{aligned} &Y_0^0 = \frac{1}{\sqrt{4\pi}} \\ &Y_1^0 = \sqrt{3/4\pi}\cos\theta, \qquad Y_1^{\pm1} = \mp\sqrt{3/8\pi}\sin\theta\, e^{\pm i\phi}, \\ &Y_2^0 = \sqrt{5/16\pi}\,(3\cos^2\theta - 1), \qquad Y_2^{\pm1} = \mp\sqrt{15/8\pi}\cos\theta\sin\theta\, e^{\pm i\phi}, \\ &\qquad Y_2^{\pm2} = \sqrt{15/32\pi}\sin^2\theta\, e^{\pm 2i\phi}, \ldots . \end{aligned} \tag{5.87}$$

Occasionally, the *solid spherical harmonics,* or (spherical) harmonic polynomials

$$\mathcal{Y}_l^m(\mathbf{r}) = r^l Y_l^m(\theta,\phi) \tag{5.88}$$

are used because they are the "tensor" products of the x, y, z coordinates:

$$\mathcal{Y}_0^0 = \frac{1}{\sqrt{4\pi}}$$
$$\mathcal{Y}_1^0 = \sqrt{3/4\pi}\, z, \qquad \mathcal{Y}_1^{\pm 1} = \mp\sqrt{3/8\pi}\,(x \pm iy),$$
$$\mathcal{Y}_2^0 = \sqrt{5/16\pi}\,(2z^2 - x^2 - y^2), \qquad \mathcal{Y}_2^{\pm 1} = \mp\sqrt{15/8\pi}\, z(x \pm iy),$$
$$\mathcal{Y}_2^{\pm 2} = \sqrt{15/32\pi}\,(x \pm iy)^2, \ldots. \tag{5.89}$$

The functions (5.83) satisfy the orthonormal relation

$$\int_0^{2\pi} \Phi_{m'}^*(\phi)\Phi_m(\phi)\, d\phi = \delta_{m'm}. \tag{5.90}$$

It therefore follows that the orthonormal relation for the spherical harmonics is

$$\int Y_{l'}^{m'*}(\theta,\phi) Y_l^m(\theta,\phi)\, d\cos\theta\, d\phi = \delta_{l'l}\,\delta_{m'm}. \tag{5.91}$$

We recall that the discussion of Section 4.12 makes clear that $Y_l^m(\theta,\phi)$ satisfies the periodicity condition

$$Y_l^m(\theta,\phi+2\pi) = Y_l^m(\theta,\phi),$$

everywhere and the regularity condition (Y_l^m finite) at the poles $\theta = 0$ and π. As a result, any function on the sphere satisfying these conditions can be expanded in spherical harmonics

$$F(\Omega) = \sum_{l=0}^{\infty} \sum_{m=-l}^{l} c_{lm} Y_l^m(\Omega), \tag{5.92}$$

where the notation $\Omega = (\theta, \phi)$ is used. The expansion coefficients can be calculated readily from $F(\Omega)$ with the help of Eq. (5.91):

$$c_{lm} = \int d\Omega\, Y_l^{m*}(\Omega) F(\Omega), \tag{5.93}$$

where

$$d\Omega = d\cos\theta\, d\phi,$$

as in Eq. (1.89). Substitution of Eq. (5.93) into Eq. (5.92)

$$F(\Omega) = \int d\Omega'\, F(\Omega') \sum_{lm} Y_l^{m*}(\Omega') Y_l^m(\Omega)$$

yields the *completeness relation*:

$$\sum_{lm} Y_l^{m*}(\Omega') Y_l^m(\Omega) = \delta(\Omega - \Omega') = \delta(\cos\theta - \cos\theta')\,\delta(\phi - \phi'), \tag{5.94}$$

where the Dirac δ function satisfies the usual properties discussed in Section 3.4.

$$\delta(\Omega - \Omega') = 0, \qquad \text{if} \qquad \Omega \neq \Omega'; \tag{5.95a}$$

$$\int \delta(\Omega)\, d\Omega = 1. \tag{5.95b}$$

The completeness relation (5.94) can be further simplified into a very useful identity known as the *addition theorem* for spherical harmonics. The possibility of a simplification arises because the angular difference $\Omega - \Omega'$ between two directions on a sphere can be characterized by a single angle α between them. If we choose one of these directions as the new z axis, any function of $\Omega - \Omega'$ is a function only of the new colatitude angle α and can therefore be expanded in terms of $P_l(\cos\alpha)$. In particular,

$$\delta(\Omega - \Omega') = \sum_l d_l P_l(\cos\alpha), \tag{5.96}$$

where

$$\begin{aligned} d_l &= \frac{2l+1}{2} \int d\cos\alpha\, \delta(\Omega - \Omega')\, P_l(\cos\alpha) \\ &= \frac{2l+1}{4\pi} \int d\Omega'\, \delta(\Omega - \Omega')\, P_l(\cos\alpha) \\ &= \frac{2l+1}{4\pi} P_l(1) = \frac{2l+1}{4\pi}. \end{aligned}$$

Comparison between this and Eq. (5.94) yields the desired result

$$P_l(\cos\alpha) = \frac{4\pi}{2l+1} \sum_{m=-l}^{l} Y_l^{m*}(\Omega') Y_l^m(\Omega). \tag{5.97}$$

Problems

5.6.1. Verify Eqs. (5.72b) and (5.73) with the help of the Legendre differential equation (4.83).

5.6.2. Obtain explicit expressions for $P_3^m(x)$, $m = -3, \ldots, 3$.

5.6.3. Derive Eq. (5.80).

5.6.4. Use the generating function given in Eq. (5.77) to show that

(a) $P_l^m(0) = 0, \qquad l - m = \text{odd},$

$$= (-1)^{(l-m)/2} \frac{(l+m-1)!!}{(l-m)!!}, \qquad l - m = \text{even};$$

(b) $P_l^m(1) = \delta_{m0}, \qquad Y_l^m(0, \phi) = \left(\dfrac{2l+1}{4\pi}\right)^{1/2} \delta_{m0}.$

5.6.5. Obtain the following recurrence relations with the help of Eqs. (5.8, 5.9, 5.71):

(a) $P_{l+1}^m(x) = (1 - x^2)^{1/2}(2l+1)P_l^{m-1}(x) + P_{l-1}^m(x)$;

(b) $(l + 1 - m)P_{l+1}^m(x) = (2l+1)xP_l^m(x) - (l+m)P_{l-1}^m(x)$.

Hint: $\dfrac{d^m}{dx^m}(xP_l) = x\dfrac{d^m}{dx^m} P_l + m\dfrac{d^{m-1}}{dx^{m-1}} P_l.$

5.6.6. Expand $f(x,y,z) = x + y + z$ in spherical harmonics.

5.6.7. Express the Coulomb field $u(\mathbf{r})$ in spherical coordinates [See Eq. (5.2)] at the field point $\mathbf{r}$ due to a source $-4\pi\delta(\mathbf{r}-\mathbf{r}_s)$ at an arbitrary source point $\mathbf{r}_s$ in terms of the spherical harmonics $Y_{lm}(\Omega_\mathbf{r})$ and $Y_{lm}(\Omega_{\mathbf{r}'})$.

5.7. Bessel functions

Orthogonal polynomials are well suited to the expansion of functions, or the description of physical systems, that are localized near the origin of coordinates. Such an expansion is quite similar to the Taylor expansion. The further we go away from the origin, the more terms we need for a reasonable approximation.

Orthogonal polynomials are not so useful if the system is not localized, but is spread out all over space. For example, solutions of the wave equation describing waves over all space are more conveniently expanded in terms of trigonometric functions (as in Fourier series) and other functions containing infinite series of powers rather than polynomials of finite degrees. Indeed, solutions of the wave equation (including the Helmholtz equation of Section 4.12) in various coordinate systems give a class of special functions particularly suited to the description of physical phenomena that can propagate to infinity.

The *Bessel function* $J_n(x)$ of an integral order n is the solution of the Bessel differential equation (4.130)

$$x^2J_n'' + xJ_n' + (x^2 - n^2)J_n(x) = 0 \tag{5.98}$$

obtained by separating the two-dimensional Helmholtz equation in circular coordinates, or the three-dimensional equation in cylindrical coordinates. These Bessel functions can also be defined in terms of the generating function

$$G(X,t) = \exp\left[\tfrac{1}{2}x\left(t - \frac{1}{t}\right)\right] = \sum_{n=-\infty}^{\infty} t^n J_n(x). \tag{5.99}$$

Both sides of this equation diverge at $t = 0$. The infinite series on the right-hand side is not a Taylor series, which has only positive powers. Infinite series containing negative powers like this are called *Laurent series*. They will be discussed in Section 6.8. For the present, it is sufficient to note that Bessel functions of negative integral orders are associated with negative powers of t in Eq. (5.99).

To generate explicit expansions for $J_n(x)$, we expand $G(x,t)$ in powers of t:

$$\exp\left[\left(\frac{x}{2}\right)t\right]\exp\left[-\left(\frac{x}{2}\right)\frac{1}{t}\right]=\left[\sum_{r=0}^{\infty}\left(\frac{x}{2}\right)^{r}\frac{t^{r}}{r!}\right]\left[\sum_{s=0}^{\infty}\left(-\frac{x}{2}\right)^{s}\frac{t^{-s}}{s!}\right]$$

$$=\sum_{r,s}(-1)^{s}\left(\frac{x}{2}\right)^{r+s}\frac{t^{r-s}}{r!\,s!}.$$

A change of the summation index to $n=r-s$ (or $r=n+s$, $r+s=n+2s$) yields

$$G(x,t)=\sum_{n=-\infty}^{\infty}\sum_{s=0}^{\infty}(-1)^{s}\left(\frac{x}{2}\right)^{n+2s}\frac{t^{n}}{(n+s)!\,s!}.$$

Comparison with Eq. (5.99) shows that

$$\begin{aligned}J_n(x)&=\sum_{s=0}^{\infty}\frac{(-1)^{s}}{(n+s)!\,s!}\left(\frac{x}{2}\right)^{n+2s}\\&=\left(\frac{x}{2}\right)^{n}\left(\frac{1}{n!}-\frac{x^2/4}{1!\,(n+1)!}+\frac{(x^2/4)^2}{2!\,(n+2)!}-\cdots\right),\end{aligned}\tag{5.100}$$

if $n\geqslant 0$.

The expression for $J_n(x)$ with $n<0$ is slightly more complicated. This is because the condition

$$r=n+s=s-|n|\geqslant 0$$

requires that $s\geqslant|n|$. Hence

$$J_{-|n|}(x)=\sum_{s=|n|}^{\infty}\frac{(-1)^{s}}{(s-|n|)!\,s!}\left(\frac{x}{2}\right)^{2s-|n|}.\tag{5.101}$$

If we now change variables to $s'=s-|n|$ or $s=s'+|n|$, we see that

$$J_{-|n|}(x)=\sum_{s'=0}^{\infty}\frac{(-1)^{s'+|n|}}{s'!\,(s'+|n|)!}\left(\frac{x}{2}\right)^{2s'+|n|}=(-1)^{n}J_{|n|}(x)\tag{5.102}$$

is proportional to $J_{|n|}(x)$. The mathematically facile reader will recognize this relation by simply inspecting Eq. (5.99) (Problem 5.7.1).

A number of recursion formulas can be obtained easily from $G(x,t)$. Two are of special interest (see Problem 5.7.2):

$$\tfrac{1}{2}(J_{n-1}+J_{n+1})=\frac{n}{x}J_n,\tag{5.103}$$

$$\tfrac{1}{2}(J_{n-1}-J_{n+1})=J_n'.\tag{5.104}$$

The difference and the sum of these equations give the *ladder operators* for J_n (*raising* operator R_n and *lowering* operator L_n):

$$\begin{aligned}R_nJ_n&=\left(\frac{n}{x}-\frac{d}{dx}\right)J_n=J_{n+1},\\L_nJ_n&=\left(\frac{n}{x}+\frac{d}{dx}\right)J_n=J_{n-1}.\end{aligned}\tag{5.105}$$

The expression $L_{n+1}R_nJ_n$ raising the order once and lowering it once must be equal to J_n itself. Hence

$$(L_{n+1}R_n - 1)J_n = 0. \tag{5.106}$$

This is just the Bessel equation (5.98).

One can easily verify by direct differentiation (in Problem 5.7.3) that Eq. (5.100) with n replaced by any real number ν defines a $J_\nu(x)$ that also satisfies the recursion formulas (5.103) and (5.104). Hence it is also a solution of the Bessel equation (5.98) with n replaced by ν.

We should note that the factorial $(\nu + s)!$ appearing in Eq. (5.100) has the usual property that

$$z! = z(z-1)!. \tag{5.107}$$

It follows from this that

$$\begin{aligned} 0! &= \frac{1!}{1} = 1 \\ (-1)! &= \frac{0!}{0} = \infty = (-2)!, \qquad \text{etc.} \end{aligned} \tag{5.108}$$

Thus $\mu!$ is finite except at the negative nonzero integers. Because of Eq. (5.107), we need to know $z!$ only over the range -1 to 0. A useful definition turns out to be the integral:

$$\begin{aligned} z! &\equiv \int_0^\infty e^{-t}t^z\,dt, \qquad \text{Re}\, z > -1 \\ &= \Gamma(z+1). \end{aligned} \tag{5.109}$$

It is actually valid for complex values of z provided that $\text{Re}\, z > -1$. The factorial function is also called a *gamma function,* but for argument $z + 1$, as shown in Eq. (5.109). One gamma function of fractional argument is particularly worthy of note

$$\Gamma(\tfrac{1}{2}) = (-\tfrac{1}{2})! = 2\int_0^\infty e^{-u^2}\,du = \sqrt{\pi}. \tag{5.110}$$

Equation (5.100) shows that $J_\nu(x)$ and $J_{-\nu}(x)$ are linearly independent if ν is nonintegral, because they differ in at least some terms. If ν becomes integral, their difference vanishes, and they become linearly dependent. This difference in behavior is a minor annoyance that can be avoided by using the *Neumann function*

$$N_\nu(x) = \frac{\cos \nu\pi\, J_\nu(x) - J_{-\nu}(x)}{\sin \nu\pi} \tag{5.111}$$

instead. It is obvious that N_ν is linearly independent of J_ν. When $\nu \to n$ becomes integral, both numerator and denominator vanish, but the quotient remains well defined, as one can verify by applying l'Hospital's rule. One finds (Problem 5.7.7) that $N_n(x)$ then takes the form

$$N_n(x) = \frac{1}{\pi}\left(\frac{\partial J_\nu(x)}{\partial \nu} - (-1)^n \frac{\partial J_{-\nu}(x)}{\partial \nu}\right)_{\nu=n}. \tag{5.112}$$

One can show (Problem 5.7.4) that this $N_n(x)$ also satisfies the recursion formulas (5.103) and (5.104). Hence it is a solution of the Bessel differential equation (5.106). That it it linearly independent of J_n can be demonstrated by first calculating the Wronskian for N_ν

$$W = J_\nu N'_\nu - J'_\nu N_\nu = (-J_\nu J'_{-\nu} + J'_\nu J_{-\nu})/\sin \nu\pi.$$

The result turns out to be

$$W = \frac{2}{\pi x}, \tag{5.113}$$

which is independent of ν. It should also hold for $\nu = n$, thus showing that N_n is linearly independent of J_n. The calculation for Eq. (5.113) is too long to reproduce here. The interested reader can find a derivation in Arfken (1985), p. 599.

Some Bessel and Neumann functions are shown in Fig. 5.4.

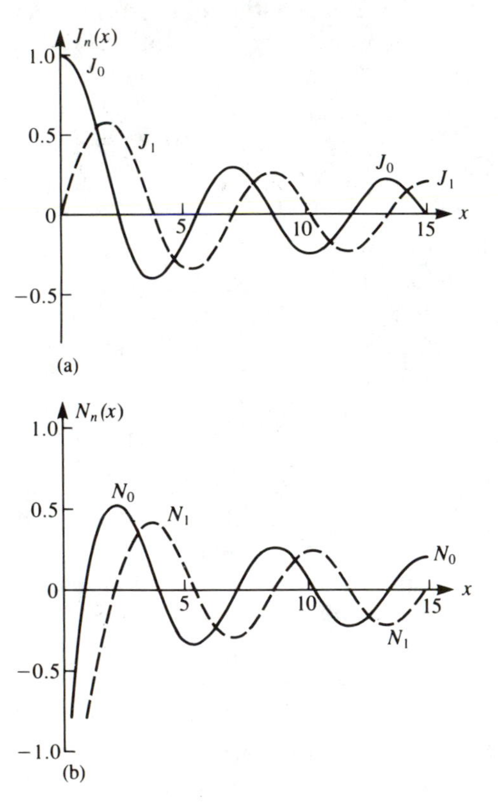

Fig. 5.4. (a) Bessel functions $J_0(x)$ and $J_1(x)$; (b) Neumann functions $N_0(x)$ and $N_1(x)$.

Separation of the three-dimensional Helmholtz equation in spherical coordinates yields the spherical Bessel equation (4.131)

$$\left(z^2 \frac{d^2}{dz^2} + 2z \frac{d}{dz}\right) + [z^2 - l(l+1)]\Big) f_l(z) = 0, \qquad (5.114)$$

where $z = kr$ is a dimensionless variable. Physical applications often involve integrals of the form

$$\int f_l(z) f_{l'}(az) z^2\, dz = \int F_l(z) F_{l'}(az) z\, dz,$$

where

$$F_l(z) = z^{1/2} f_l(z) \qquad (5.115)$$

is the "equivalent" function in cylindrical or circular coordinates. Indeed, a direct substitution of $f_l = z^{-1/2} F_l$ into Eq. (5.114) shows that F_l satisfies the Bessel equation

$$\left(z^2 \frac{d^2}{dz^2} + z \frac{d}{dz} + [z^2 - (l + \tfrac{1}{2})^2]\right) F_l(z) = 0. \qquad (5.116)$$

Hence F_l is proportional to $J_{l+1/2}$ or $N_{l+1/2}$. A standard choice of the proportional constant then gives the *spherical Bessel* and *Neumann functions*:

$$\begin{aligned} j_l(z) &= \left(\frac{\pi}{2z}\right)^{1/2} J_{l+1/2}(z), \\ n_l(z) &= \left(\frac{\pi}{2z}\right)^{1/2} N_{l+1/2}(z). \end{aligned} \qquad (5.117)$$

Equations (5.100) and (5.111) can now be used to obtain the power series

$$j_l(z) = z^l \left(\frac{1}{(2l+1)!!} - \frac{z^2/2}{1!(2l+3)!!} + \frac{(z^2/2)^2}{2!\,(2l+5)!!} - \cdots \right)$$

and

$$n_l(z) = -\frac{(2l-1)!!}{z^{l+1}} \left(1 - \frac{z^2/2}{1!\,(1-2l)} + \frac{(z^2/2)^2}{2!\,(1-2l)(3-2l)} - \cdots \right), \qquad (5.118)$$

where the double factorial of a positive integer n is

$$n!! = n(n-2) \times \cdots \times \begin{cases} 3.1, & \text{odd} \\ 4.2, & \text{even.} \end{cases} \qquad (5.119)$$

Recursion formulas for f_l follow from those for J_ν. For example, Eqs. (5.103) and (5.104) give

$$f_{l-1} + f_{l+1} = \frac{2l+1}{z} f_l \qquad (5.120)$$

and

$$l f_{l-1} - (l+1) f_{l+1} = (2l+1) f_l'. \qquad (5.121)$$

The resulting ladder operators are

$$R_l f_l = \left(\frac{l}{z} - \frac{d}{dz}\right) f_l = f_{l+1} \tag{5.122}$$

$$L_l f_l = \left(\frac{l+1}{z} + \frac{d}{dz}\right) f_l = f_{l-1}. \tag{5.123}$$

The spherical Bessel equation (5.114) for $l=0$ can be cast into a familiar equation for

$$Z(z) = z f_0(z).$$

The result (Problem 5.7.11) is

$$\left(\frac{d^2}{dz^2} + 1\right) Z(z) = 0. \tag{5.124}$$

This means that $Z(z)$ is proportional to $\sin z$ or $\cos z$. Equation (5.118) then shows that

$$j_0(z) = \frac{\sin z}{z}, \qquad n_0(z) = -\frac{\cos z}{z}. \tag{5.125}$$

Finally, functions of higher orders can be generated by using the raising operator of Eq. (5.122):

$$\begin{aligned} j_1 &= -j_0' = \frac{\sin z}{z^2} - \frac{\cos z}{z}, \\ j_2 &= \frac{j_1}{z} - j_1' = \left(\frac{3}{z^3} - \frac{1}{z}\right) \sin z - \frac{3}{z^2} \cos z; \end{aligned} \tag{5.126}$$

$$\begin{aligned} n_1 &= -\frac{\cos z}{z^2} - \frac{\sin z}{z}, \\ n_2 &= -\left(\frac{3}{z^3} - \frac{1}{z}\right) \cos z - \frac{3}{z^2} \sin z. \end{aligned} \tag{5.127}$$

Some spherical Bessel and Neumann functions are shown in Fig. 5.5.

Problems

5.7.1. Show that Eq. (5.102) follows from the symmetry property $G\left(x, -\frac{1}{t}\right) = G(x,t)$.

5.7.2. Derive the recurrence relations (5.103) and (5.104).

5.7.3. Show that the function J_ν defined by Eq. (5.100) with n replaced by any real number ν also satisfies Eqs. (5.103) and (5.104).

5.7.4. Show that the Neumann functions $N_n(x)$ defined by Eq. (5.112) satisfies Eqs. (5.103) and (5.104).

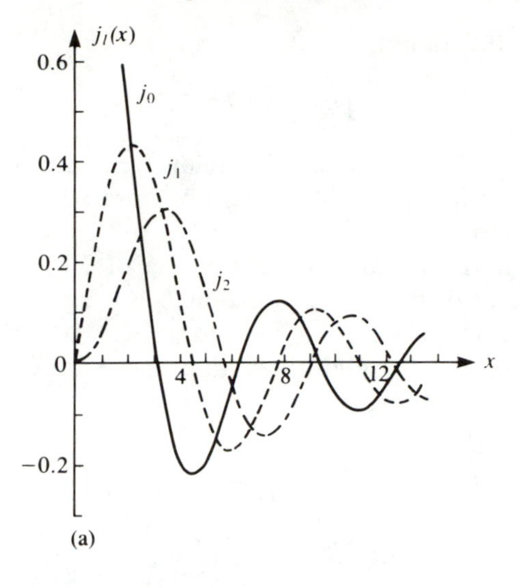

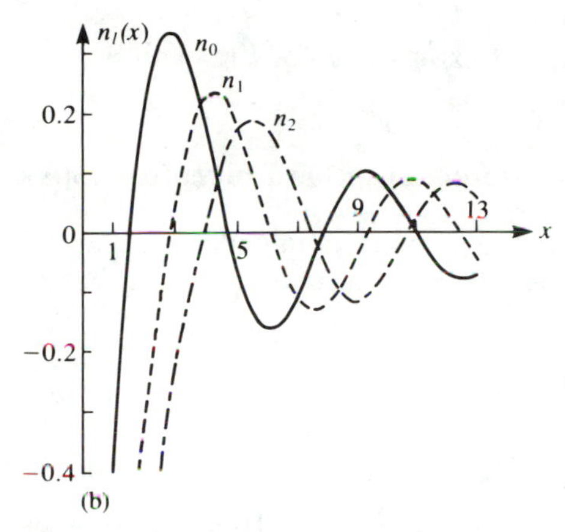

Fig. 5.5. (a) Spherical Bessel functions $j_l(x)$, $l = 0$–2; (b) Spherical Neumann functions $n_l(x)$, $l = 0$–2 (from Abramowitz and Stegun).

5.7.5. Use the product of generating functions

$$G(x + y, t) = G(x, t)G(y, t)$$

to derive the *addition theorem*

$$J_n(x + y) = \sum_{k=-\infty}^{\infty} J_k(x)J_{n-k}(y).$$

5.7.6. Use the product of generating functions

$$G(x, t)G(-x, t) = 1$$

to derive the identity

$$1 = J_0^2(x) + 2 \sum_{n=1}^{\infty} J_n^2(x).$$

5.7.7. Verify Eq. (5.112) for the Neumann function of integral order n by differentiating both numerator and demoninator of Eq. (5.111) before setting $\nu = n$.

5.7.8. Show that the Neumann function N_n of Eq. (5.112) is linearly independent of the Bessel function J_n.

5.7.9. Verify Eq. (5.116).

5.7.10. Verify Eq. (5.118) and show that these functions can also be expressed in the forms

$$j_l(z) = (2z)^l \sum_{n=0}^{\infty} \frac{(-1)^n (n+l)!}{n!\,(2n+2l+1)!} z^{2n}$$

$$n_l = 2(-2z)^{-l-1} \sum_{n=0}^{\infty} \frac{(-1)^n (n-l)!}{n!\,(2n-2l)!} z^{2n}.$$

5.7.11. Verify Eq. (5.124).

5.7.12. Obtain explicit expressions for $j_3(z)$ and $n_3(z)$.

5.8. Sturm–Liouville equation and eigenfunction expansions

Ordinary differential equations obtained by separating the Helmholtz equation $(\nabla^2 + k^2)\psi = 0$ in various coordinate systems can be written in the general form

$$\begin{aligned} \mathscr{L}\psi &\equiv \left[\frac{d}{dz}\left(p(z)\frac{d}{dz}\right) + q(z)\right]\psi \\ &= -\lambda w(z)\psi, \qquad a \leqslant z \leqslant b, \end{aligned} \tag{5.128}$$

which is called a *Liouville equation*. Here λ is the separation constant (Section 4.12). The function $p(z)$ could have zeros where the term involving ψ'' disappears. These points are called *singular points* of the equation. The behavior of the equation changes across a singular point. Hence it is convenient to separate a region into intervals in each of which $p(z)$ has the same sign. Thus, by design, singular points when they appear coincide with one or both of the boundary points a and b of Eq. (5.128).

The functions $p(z)$, $q(z)$, and $w(z)$ are determined by the chosen coordinate system, and are real and well behaved inside each interval. Furthermore, the *weight function* $w(z)$ by choice does not change sign in the interval, and can be taken to be positive definite. A detailed discussion of these properties can be found in Morse and Feshbach (Sections 5.1 and 6.3).

In the Sturm–Liouville theory, we study how λ affects the values of ψ and ψ' at certain points in the interval. Consider two DEs of the form

(5.128):

$$\begin{aligned} \mathscr{L}\psi_j &= -\lambda_j w \psi_j, \\ \mathscr{L}\psi_i^* &= -\lambda_i^* w \psi_i^*, \end{aligned} \tag{5.129}$$

where we have allowed for complex solutions of DEs containing only real functions $p(z)$, $q(z)$, and $w(z)$. Complex solutions are always possible because, if $\psi_{1,2}$ are two linearly independent real solutions of a DE, then $\psi_1 \pm i\psi_2$ are two linearly independent complex solutions.

These equations can be combined into the expression

$$\psi_i^* \mathscr{L}\psi_j - \psi_j \mathscr{L}\psi_i^* = (\lambda_i^* - \lambda_j) w \psi_i^* \psi_j.$$

Let us integrate both sides

$$(\lambda_i^* - \lambda_j) \int_a^b w \psi_i^* \psi_j \, dz = \int_a^b (\psi_i^* \mathscr{L}\psi_j - \psi_j \mathscr{L}\psi_i^*) \, dz. \tag{5.130}$$

The first integral on the right-hand side can be integrated twice by parts

$$\begin{aligned} &\int_a^b \psi_i^* \frac{d}{dx}\left(p \frac{d}{dz} \psi_j\right) dz \\ &\qquad = \psi_i^* p \psi_j' \big|_a^b - \int_a^b \left(\frac{d}{dz} \psi_i^*\right) p \frac{d}{dz} \psi_j \, dz \\ &\qquad = p(\psi_i^* \psi_j' - \psi_i^{*\prime} \psi_j) \big|_a^b + \int_a^b \left[\frac{d}{dz}\left(p \frac{d}{dz} \psi_i^*\right)\right] \psi_j \, dz. \end{aligned} \tag{5.131}$$

The boundary terms vanish for each of the following choices of "homogeneous" boundary conditions (BCs) at $z = a$ *and* b:

$$\psi_i = 0, \qquad \text{Dirichlet BC,} \tag{5.132a}$$

$$\psi_i' = 0, \qquad \text{Neumann BC,} \tag{5.132b}$$

$$\psi_i' + c\psi_i = 0, \qquad \text{mixed BC.} \tag{5.132c}$$

In each case, the two integrals on the right-hand side of Eq. (5.130) become equal:

$$\begin{aligned} \int_a^b \psi_i^* \mathscr{L}\psi_j \, dz &= \int_a^b \psi_j \mathscr{L}\psi_i^* \, dz \\ &= \int_a^b (\mathscr{L}\psi_i)^* \psi_j \, dz \\ &= \int_a^b (\psi_j^* \mathscr{L}\psi_i)^* \, dz, \end{aligned} \tag{5.133}$$

where we have written the right-hand side in three equivalent forms. An operator $\mathscr{L}$ satisfying Eq. (5.133) is said to be *Hermitian* or *self-adjoint*. Thus an operator is Hermitian partly because of the BC (5.132) satisfied by the functions ψ_i^*, ψ_j appearing in Eq. (5.133). For other functions for which the boundary terms in Eq. (5.131) do not vanish, the *same* operator $\mathscr{L}$ is not Hermitian.

There is no reason to believe that one of the BCs (5.132) will always be satisfied for any value of λ. It is possible, however, that it is satisfied for selected values (called a *spectrum*) of $\lambda = \lambda_i$, $i = 0, 1, 2, \ldots, N$. We call these special values the *eigenvalues* of the DEs, and the associated solutions ψ_i their *eigenfunctions*. The eigenvalue spectrum satisfies a number of simple properties (Morse and Feshbach, p. 274):

1. They are infinite in number, that is, $N = \infty$.
2. Their distribution is discrete if $b - a$ is finite. That is, $\lambda_{n+1} - \lambda_n \neq 0$ between two neighboring eigenvalues. The distribution becomes continuous, like points on a line, when $b - a$ becomes infinite.

Now if $\mathscr{L}$ is Hermitian, Eq. (5.130) gives

$$(\lambda_i^* - \lambda_j) \int_a^b w\psi_i^* \psi_j \, dz = 0. \tag{5.134}$$

The integrand is non-negative for $j = i$, since both $w(z)$ and $|\psi_i(z)|^2$ are non-negative. The resulting integral does not vanish, except for the trivial case of $\psi_i(z) = 0$, which we shall exclude. Equation (5.134) then requires that the eigenvalue $\lambda_i^* = \lambda_i$ is real. On the other hand, if $\lambda_j \neq \lambda_i$, Eq. (5.134) can be satisfied only if the integral itself vanishes. Hence

$$\int_a^b w\psi_i^* \psi_j \, dz = \delta_{ij} h_i. \tag{5.135}$$

This shows that the eigenfunctions ψ_i form a system of orthogonal functions in the interval (a,b).

5.8.1. *Eigenfunction expansion*

In exact analogy to the Fourier-series expansion, any piecewise-continuous function $F(z)$ between a and b can be expanded in terms of these eigenfunctions

$$F(z) = \sum_{i=0}^{\infty} c_i \psi_i(z). \tag{5.136}$$

The expression coefficient c_i can be extracted as usual by calculating the inner product or overlap

$$(\psi_i, F) \equiv \int_a^b w\psi_i^* F \, dz = c_i h_i. \tag{5.137}$$

Indeed, the similarity with the Fourier-series expansion can be taken quite literally: If ψ_i are ordered with increasing eigenvalues (i.e., $\lambda_{i+1} > \lambda_i$), the difference between an eigenfunction series and the Fourier series for the same function over the same interval, taken to the same number n of terms, can be shown to be uniformly convergent as $n \to \infty$. (This is demonstrated in Morse and Feshbach, p. 743.)

Eigenfunction expansions have been used in Sections 3.10, 4.8, 4.10–4.12, 5.4, and 5.5.

Problems

5.8.1. Show that the wave function $u(r,\theta,t)$ describing the vibrations of a circular drum has an eigenfunction expansion of the form

$$u(r,\theta,t) = \sum_{m=0}^{\infty} \sum_{i=1}^{\infty} a_{im} J_m(k_{in} r)(\sin m\theta + b_m \cos m\theta) \times (\sin \omega_{im} t + c_{im} \cos \omega_{im} t).$$

How are the eigenvalues m, k_{in}, and ω_{im} determined?

5.8.2. Identify the eigenfunctions of the following Sturm–Liouville equations. Explain why the eigenvalue spectrum is as shown, with $n = 0, 1, 2, \ldots$.

	$p(x)$	$q(x)$	$w(x)$	a	b	λ
(a)	$1-x^2$	0	1	-1	1	$n(n+1)$
(b)	$(1-x^2)^{1/2}$	0	$(1-x^2)^{-1/2}$	-1	1	n^2
(c)	xe^{-x}	0	e^{-x}	0	∞	n
(d)	e^{-x^2}	0	e^{-x^2}	$-\infty$	∞	$2n$

Appendix 5. Tables of mathematical formulas

1. Orthogonal polynomials

Tables 5.1–5.5 in Section 5.4 give Rodrigues' formulas, generating functions, recursion formulas, differential relations, and differential equations for the Legendre, Laguerre, Hermite, and Chebyshev polynomials. Tables 3.1 and 3.2 of Section 3.10 give orthogonality relations and explicit formulas for polynomials of low degrees.

Expansion of $R(x)$ in the orthogonal polynomials $f_n(x)$:

$$R(x) = \sum_{n=0}^{\infty} c_n f_n(x),$$

where

$$c_n = \frac{(f_n, R)}{(f_n, f_n)},$$

$$(f_n, R) = \int_a^b f_n(x) w(x) R(x)\, dx.$$

2. Quantum oscillator

$$\left[-\frac{1}{2}\left(\frac{d^2}{d\xi^2}-\xi^2\right)-(n+\tfrac{1}{2})\right]\psi_n(\xi)=0,$$

$$\psi_n(\xi)=(2^n n!\,\sqrt{\pi})^{-1/2}H_n(\xi)e^{-(1/2)\xi^2},$$

$$a^\dagger\psi_{n+1}(\xi)\equiv-\frac{1}{\sqrt{2}}\left(\frac{d}{d\xi}-\xi\right)\psi_n(\xi)=\sqrt{n+1}\,\psi_{n+1}(\xi),$$

$$a\psi_n(\xi)\equiv\frac{1}{\sqrt{2}}\left(\frac{d}{d\xi}+\xi\right)\psi_n(\xi)=\sqrt{n}\,\psi_{n-1}(\xi),$$

$$[a,a^\dagger]=aa^\dagger-a^\dagger a=1,$$

$$(a^\dagger a-n)\psi_n=0.$$

3. Spherical harmonics

$$Y_l^m(\theta,\phi)=\left(\frac{(l-m)!}{(l+m)!}\frac{2l+1}{4\pi}\right)^{1/2}e^{im\phi}P_l^m(\cos\theta).$$

$$\int Y_{l'}^{m'*}(\theta,\phi)Y_l^m(\theta,\phi)\,d\cos\theta\,d\phi=\delta_{l'l}\delta_{m'm},$$

$$\int_{-1}^{1}P_{l'}^m(x)P_l^m(x)\,dx=\frac{(l+m)!}{(l-m)!}\delta_{l'l}\frac{2}{2l+1}.$$

$$P_l(\cos\alpha)=\frac{4\pi}{2l+1}\sum_{m=-l}^{l}Y_l^{m*}(\Omega')Y_l^m(\Omega).$$

$$\delta(\Omega-\Omega')=\delta(\cos\theta-\cos\theta')\,\delta(\phi-\phi')=\sum_{lm}Y_l^{m*}(\Omega')Y_l^m(\Omega).$$

4. Bessel functions

$$\exp\left[\tfrac{1}{2}x\left(t-\frac{1}{t}\right)\right]=\sum_{n=-\infty}^{\infty}t^nJ_n(x).$$

$$R_nJ_n\equiv\left(\frac{n}{x}-\frac{d}{dx}\right)J_n=J_{n+1},$$

$$L_nJ_n=\left(\frac{n}{x}+\frac{d}{dx}\right)J_n=J_{n-1},$$

$$(L_{n+1}R_n-1)J_n=0.$$

$$J_{-n}(x)=(-1)^nJ_n(x).$$

$$N_\nu(x)=\frac{\cos\nu\pi\,J_\nu(x)-J_{-\nu}(x)}{\sin\nu\pi}$$

$$W=J_\nu N_\nu'-J_\nu'N_\nu=\frac{2}{\pi x}.$$

5. Spherical Bessel functions

$$j_l(z) = \left(\frac{\pi}{2z}\right)^{1/2} J_{l+1/2}(z), \qquad n_l(z) = \left(\frac{\pi}{2z}\right)^{1/2} N_{l+1/2}(z).$$

$$R_l f_l = \left(\frac{l}{z} - \frac{d}{dz}\right) f_l = f_{l+1},$$

$$L_l f_l = \left(\frac{l+1}{z} + \frac{d}{dz}\right) f_l = f_{l-1},$$

$$(L_{l+1} R_l - 1) f_l = 0, \qquad f_l = j_l \qquad \text{or} \qquad n_l.$$

Explicit formulas for some spherical Bessel and Neumann functions are given in Eqs. (5.125)–(5.127).

6. Sturm–Liouville equation

$$\mathscr{L}\psi_i = \left[\frac{d}{dz}\left(p(z)\frac{d}{dz}\right) + q(z)\right]\psi_i = -\lambda_i w(z)\psi_i.$$

The operator $\mathscr{L}$ is Hermitian for each of the following homogeneous boundary conditions:

$$\begin{aligned} &\psi_i = 0, && \text{Dirichlet}, \\ &\psi_i' = 0, && \text{Neumann}, \\ &\psi_i' + c\psi_i = 0, && \text{mixed}, \end{aligned}$$

at the boundaries $z = a$ and b. Then the eigenvalues λ_i are real, and the eigenfunctions are orthogonal with respect to the density function $w(z)$:

$$\int_a^b w(z)\psi_i^*(z)\psi_j(z)\, dz = \delta_{ij}\, h_i.$$

6

FUNCTIONS OF A COMPLEX VARIABLE

6.1. Introduction

The real number system is incomplete. For example, it does not include all the roots of algebraic equations such as

$$x^2 + 1 = 0$$

It was in connection with the study of the roots of cubic equations when the concept of the square root of a negative number became widely used in the early sixteenth century. However, the mathematics of complex numbers was not developed for another two centuries. It was Euler who introduced the symbol $i = \sqrt{-1}$ in 1777. Years later, Gauss first used the notation $a + ib$ to denote a complex number.

In physics, we deal mostly with measurable attributes that can be quantified in terms of real numbers only. Complex numbers are tolerated, and even welcome, for the simplicity and convenience they bring to mathematical manipulations. The real significance of complex numbers in physics is not a matter of convenience, however, but of completeness. The use of complex numbers makes the description of nature mathematically complete. It is this completeness that makes the resulting physical theory elegant and aesthetically satisfying. It also ensures that nothing of importance is accidentally left out, including unexpected relations between distinct physical observables. How this happens is the story of this chapter.

6.2. Function of a complex variable

A real number can be represented conveniently as a point on a straight line. A complex number, containing two real numbers, can be represented by a point in a two-dimensional plane, the *complex plane*. A complex variable $z = x + iy$ can be plotted in the complex plane by using its real and imaginary parts $x = \text{Re}\, z$ and $y = \text{Im}\, z$ as its two rectangular coordinates, as shown in Fig. 6.1. This complex variable can also be represented by the circular polar coordinates (r, θ) with the help of the trigonometric relations

$$x = r\cos\theta, \qquad y = r\sin\theta. \tag{6.1}$$

The dependence of z on the angle θ can be analyzed with the help of

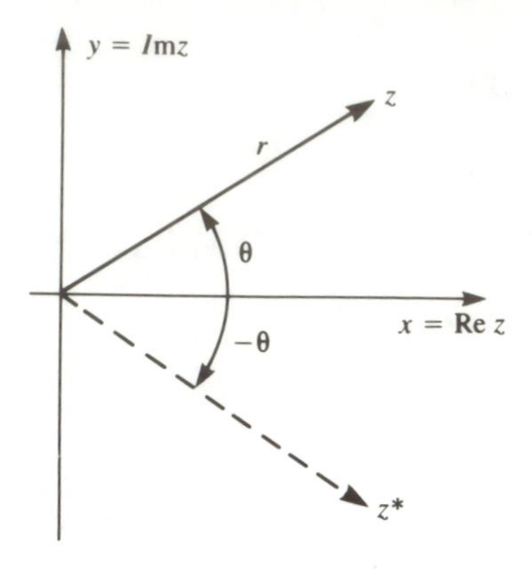

Fig. 6.1. The complex plane.

the Taylor series

$$\cos\theta = 1 - \frac{\theta^2}{2!} + \frac{\theta^4}{4!} - \cdots,$$
$$\sin\theta = \theta - \frac{\theta^3}{3!} + \frac{\theta^5}{5!} - \cdots. \tag{6.2}$$

As a result

$$\cos\theta + i\sin\theta = 1 + i\theta + \frac{(i\theta)^2}{2!} + \frac{(i\theta)^3}{3!} + \cdots$$
$$= e^{i\theta}. \tag{6.3}$$

This *Euler formula* gives rise to the *polar* (or *Argand*) *representation*

$$z = re^{i\theta} \tag{6.4}$$

involving the *modulus* (amplitude or magnitude) r and the *phase* (or argument) θ of z. For example, on the unit circle we find the points

$$e^{i\pi/2} = i, \qquad e^{i\pi} = -1, \qquad e^{i3\pi/2} = -i, \qquad e^{i2\pi} = 1.$$

Other examples are

$$e^i = \cos 1 + i\sin 1 = 0.54 + i0.84;$$
$$i^e = \exp\left(i\frac{\pi}{2}e\right) = \cos\left(\frac{\pi e}{2}\right) + i\sin\left(\frac{\pi e}{2}\right).$$

The multiplication of complex numbers can be done readily by using the rectangular representation, and even more easily by using the polar representation. For example, if

$$z_1 = x_1 + iy_1 = r_1e^{i\theta_1}$$
$$z_2 = x_2 + iy_2 = r_2e^{i\theta_2},$$

then

$$z_1 z_2 = (x_1 + iy_1)(x_2 + iy_2) = x_1 x_2 - y_1 y_2 + i(x_1 y_2 + x_2 y_1), \qquad (6.5)$$

or

$$z_1 z_2 = r_1 r_2 e^{i(\theta_1 + \theta_2)}. \qquad (6.6)$$

By equating the real and imaginary parts of these two equations for the special case of $r_1 = r_2 = 1$, we obtain the trigonometric identities

$$\cos(\theta_1 + \theta_2) = \cos\theta_1 \cos\theta_2 - \sin\theta_1 \sin\theta_2, \qquad (6.7)$$

$$\sin(\theta_1 + \theta_2) = \sin\theta_1 \cos\theta_2 + \cos\theta_1 \sin\theta_2. \qquad (6.8)$$

Trigonometric relations involving multiple angles can be similarly obtained:

$$(\cos\theta + i\sin\theta)^n = (e^{i\theta})^n = \cos n\theta + i\sin n\theta, \qquad (6.9)$$

a result known as *de Moivre's Theorem*.

The replacement $i \to -i$ changes a complex expression into its *complex conjugate*. For example, the complex conjugate of $z = x + iy$ is

$$z^* = x - iy = re^{-i\theta}. \qquad (6.10)$$

This is just the mirror reflection of z about the real axis in the complex plane. The product

$$zz^* = r^2 = z^*z \qquad (6.11)$$

is real and non-negative. The magnitude r is commonly denoted $|z|$.

Functions can be defined for a complex variable in much the same way as for a real variable.

We recall that by a real function of a real variable we mean a relation or mapping whereby one or more real numbers $f(x)$ can be assigned to each value of the real variable x. This functional relationship is shown schematically in Fig. 6.2.

Such a functional relation can be generalized to complex planes by assigning to each point of the complex plane of a complex variable

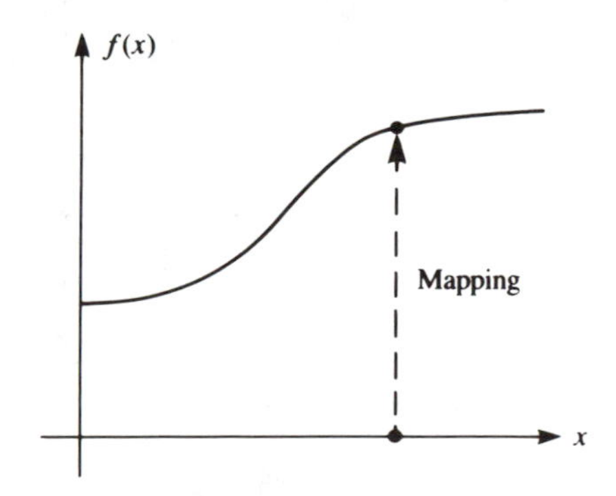

Fig. 6.2. Function of a real variable.

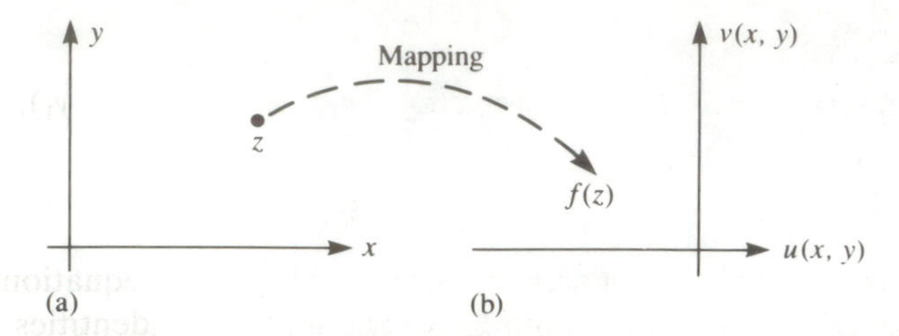

Fig. 6.3. Complex function of a complex variable. (a) Plane of the complex variable $z = x + iy$. (b) Plane of the complex function $f(z) = u(x,y) + iv(x,y)$.

$z = x + iy$ one or more points in a different complex plane, the complex plane of the functional values. This mapping is represented schematically in Fig. 6.3. Each functional value is itself a complex number:

$$f(z) = u(x,y) + iv(x,y). \tag{6.12}$$

This single-valued (or multivalued) function $f(z)$ is called *function of a complex variable* (FCV).

Example 6.2.1.

(a) The function $w = az$ describes a scale change if a is real and a scale change plus a rotation if a is complex.

(b) The function $w = z + b$ describes a translation of the origin of the complex plane.

Example 6.2.2.

$$f(z) = z^2 = r^2e^{2i\theta}. \tag{6.13}$$

This is a single-valued function. The mapping is unique, but not one to one. It is a two-to-one mapping, since there are two variables z_1 and $z_2 = -z_1$ with the same square. This is shown in Fig. 6.4(a), where we have used for simplicity the same complex plane for both z and $f(z)$.

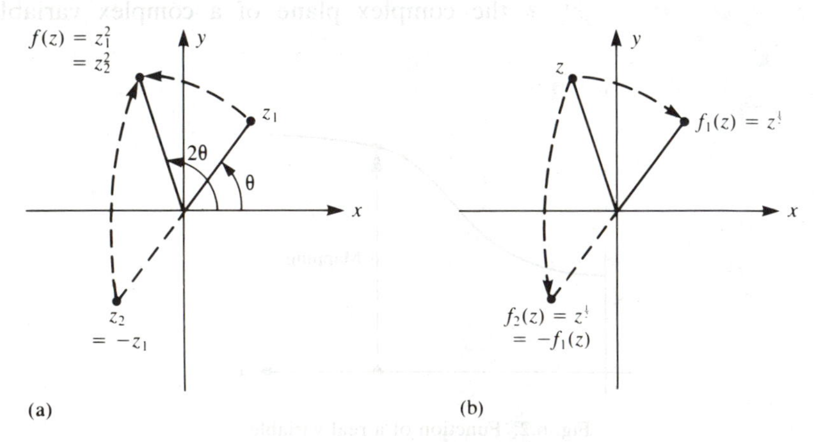

Fig. 6.4. The square and square-root functions. (a) $f(z) = z^2$; (b) $f(z) = z^{1/2}$.

Example 6.2.3.

$$f(z) = z^{1/2}. \tag{6.14}$$

There are two square roots:

$$\begin{aligned} f_1(re^{i\theta}) &= r^{1/2}e^{i\theta/2}, \\ f_2 = -f_1 &= r^{1/2}e^{i(\theta+2\pi)/2}. \end{aligned} \tag{6.15}$$

The function is double valued, and the mapping is one to two, as shown in fig. 6.4(b).

In a similar way, the pth root

$$z^{1/p} = r^{1/p}e^{i(\theta+2\pi n)/p}, \qquad p \text{ a positive integer}, \tag{6.16}$$

gives a one-to-p mapping.

Example 6.2.4.

$$f(z) = e^z = e^{x+iy} = e^x(\cos y + i \sin y). \tag{6.17}$$

The complex exponential function is periodic in y with a period of 2π. The maximum amplitude $|e^z| = e^x$ increases monotonically from zero at $x = -\infty$, through the value of 1 at $x = 0$ to infinity at $x = \infty$. Since its real and imaginary parts do not vanish at the same point, e^z does not vanish for any finite value of z.

Example 6.2.5.

$$f(z) = \sin z = \sin x \cos iy + \cos x \sin iy. \tag{6.18}$$

With the help of the results

$$\cos iy = \tfrac{1}{2}(e^{i(iy)} + e^{-i(iy)}) = \tfrac{1}{2}(e^{-y} + e^{y}) = \cosh y,$$

$$\sin iy = \frac{1}{2i}(e^{-y} - e^{y}) = i \sinh y,$$

the complex sine function can be written as

$$\sin z = \sin x \cosh y + i \cos x \sinh y. \tag{6.19}$$

In a similar way

$$\begin{aligned} \cos z &= \cos x \cos iy - \sin x \sin iy \\ &= \cos x \cosh y - i \sin x \sinh y. \end{aligned} \tag{6.20}$$

Both functions are periodic in x with a period of 2π. Since $\sinh(y = 0) = 0$, these functions are real everywhere on the real axis, as we would expect. Away from the real axis, the real and imaginary parts do not vanish at the same time. As a result, these functions have zeros only along the real axis.

Example 6.2.6.

$$\tan z = \frac{\sin z}{\cos z}, \qquad \cot z = \frac{\cos z}{\sin z}.$$

Both functions are periodic with a period of π. Both functions are finite in the entire complex z plane except on the real axis at the zeros of the function in the denominator.

Problems

6.2.1. Verify that
(a) $(1+i)^4 = -4$;
(b) $i^i = e^{-\pi/2}$;
(c) $i^{e^i} = \exp\left(-\frac{\pi}{2}\sin 1\right)\left[\cos\left(\frac{\pi}{2}\cos 1\right) + i\sin\left(\frac{\pi}{2}\cos 1\right)\right]$;
(d) $e^{i^e} = \exp\left(\cos\frac{\pi e}{2}\right)\left[\cos\left(\sin\frac{\pi e}{2}\right) + i\sin\left(\sin\frac{\pi e}{2}\right)\right]$.

6.2.2. Verify that
(a) $(1+2i) + i(2+i) = 4i$;
(b) $(1+2i)(2+i) = 5i$;
(c) $\dfrac{1+2i}{2-i} = i$;
(d) $[(1+i)(1+2i)(1+3i)]^{-1} = -\frac{1}{10}$.

6.2.3. Show that the roots of the quadratic equation $z^2 + z + 1 = 0$ are $(1 \pm i\sqrt{3})/2$. What are the roots of $z^{*2} + z^* + 1 = 0$?

6.2.4. Find all the zeros of
(a) $\sin z$;
(*b*) $\cosh z$.

6.2.5. If $z = (x,y)$ is considered a vector in the xy plane, show that
(a) $\mathbf{z}_1 \cdot \mathbf{z}_2 = \mathrm{Re}(z_1^* z_2) = \mathrm{Re}(z_1 z_2^*)$;
(b) $\mathbf{z}_1 \times \mathbf{z}_2 = \hat{k}\,\mathrm{Im}(z_1^* z_2) = -\hat{k}\,\mathrm{Im}(z_1 z_2^*)$.

6.2.6. If $(i-1)z = -(i+1)z^*$, show that z must lie on one of the two lines making an angle of 45° with the x axis.

6.2.7. Prove by purely algebraic means that

$$|z_1| - |z_2| \leqslant |z_1 + z_2| \leqslant |z_1| + |z_2|$$

for any two complex numbers z_1 and z_2.

6.2.8. Use de Moivre's theorem to show that
(a) $\cos 3\theta = 4\cos^3\theta - 3\cos\theta$;
(b) $\sin 3\theta = 3\sin\theta - 4\sin^3\theta$.

6.2.9. Use de Moivre's theorem to show that

(a) $$\cos n\theta = \cos^n\theta - \binom{n}{2}\cos^{n-2}\theta\sin^2\theta + \binom{n}{4}\cos^{n-4}\theta\sin^4\theta - \cdots;$$

(b) $\sin n\theta = \binom{n}{1} \cos^{n-1}\theta \sin\theta - \binom{n}{3} \cos^{n-3}\theta \sin^3\theta$

$$+ \binom{n}{5} \cos^{n-5}\theta \sin^5\theta - \cdots,$$

where $\binom{n}{m} = n!/(n-m)!\,m!$ are binomial coefficients.

6.2.10. The geometrical series $\sum_{n=0}^{N} e^{inx}$ for a real variable x can be summed readily. Use the result and Euler's formula to show that

$$\sum_{n=0}^{N} \cos nx = \frac{\sin \frac{1}{2}(N+1)x}{\sin \frac{1}{2}x} \cos \frac{Nx}{2}.$$

Obtain an analogous result for $\sum_{n=0}^{N} \sin nx$.

6.2.11. Show that the equation

$$(z-a)(z^*-a^*) = R^2$$

is the equation of a circle of radius R centered at the point a in the complex plane.

6.2.12. Describe the functions

(a) z^{-1};
(b) $(cz+d)^{-1}$; and
(c) $(az+b)/(cz+d)$,

where a, b, c, and d are complex constants.

6.2.13. Show that

(a) $\sinh(x+iy) = \sinh x \cos y + i \cosh x \sin y$,
$\cosh(x+iy) = \cosh x \cos y + i \sinh x \sin y$.

(b) $|\sinh z|^2 = \sinh^2 x + \sin^2 y$,
$|\cosh z|^2 = \sinh^2 x + \cos^2 y$.

(c) $\tanh z = \dfrac{\sinh 2x + i \sin 2y}{\cosh 2x + \cos 2y}$.

6.2.14. Show that

(a) $(z/z^*)^i = e^{-2\tan^{-1}(y/x)}$;

(b) $z = \dfrac{e^{\tanh^{-1} z} - e^{-\tanh^{-1} z}}{e^{\tanh^{-1} z} + e^{-\tanh^{-1} z}}$, where

$$\tanh^{-1} z = \tfrac{1}{2} \ln\left(\frac{1+z}{1-z}\right);$$

(c) $\sinh^{-1} z = \ln(z + \sqrt{z^2+1})$,
$\cosh^{-1} z = \ln(z + \sqrt{z^2-1})$;

(d) $\sin^{-1} z = -i \ln(iz + \sqrt{1-z^2}) = i \ln(-iz + \sqrt{1-z^2})$,
$\cos^{-1} z = -i \ln(z + \sqrt{z^2-1}) = i \ln(z + \sqrt{z^2-1})$,

$$\tan^{-1} z = \frac{i}{2} \ln\left(\frac{i+z}{i-z}\right).$$

6.3. Multivalued functions and Riemann surfaces

If the angle θ is restricted (arbitrarily) to $-\pi \leqslant \theta \leqslant \pi$ in $z = re^{i\theta}$, its first square root $f_1 = r^{1/2}e^{i\theta/2}$ covers the right half-plane (RHP) $(-\pi/2 \leqslant \theta/2 \leqslant \pi/2)$ of f, while $f_2 = -r^{1/2}e^{i\theta/2}$ covers the left HP. This is shown in Fig. 6.5.

Two points A and B with $z_A = re^{i(-\pi+\delta)}$ and $z_B = re^{i(\pi-\delta)}$, where $\delta > 0$ is small, are actually close together in the complex z plane. But they are widely separated in the complex f plane. This is because A is close to the lower boundary of θ, while B is close to the upper boundary.

To cover the left HP of the second square root f_2, it is necessary to go around the z plane once more, as shown in Fig. 6.6. We see that, although all four points A_1, B_1, A_2, B_2 are close together on the z plane, only A_1 and B_2, and only B_1 and A_2, are close together in the complex f plane, because of our choice of the boundaries for θ.

This is not a problem of FCV alone; it appears also for a real function of a real variable. For example, $x^{1/2}$ has a positive root $|x^{1/2}|$ and a negative root $-|x^{1/2}|$, as shown in Fig. 6.7. Two points C and D that are close together in x can become widely separated in $x^{1/2}$, as shown by the pairs C_1, D_2, and C_2, D_1.

The distinct values of a multivalued function are called its separate *branches*. For example, the first branch of the complex square root, as defined in Fig. 6.6, is on the RHP, while the second branch is on the left HP. In order to differentiate the function at the boundary points such as A_1 or B_1, it is necessary to include another branch of the function, corresponding to the points B_2 and A_2, respectively. Yet in the complex plane of the variable z, the points A_1 and A_2 are indistinguishable if they are at the same location. This causes problems in differential calculus.

6.3.1. The Riemann Surface

To distinguish between identical points in z that give rise to different branches of a function, Riemann (Doctoral Thesis, 1851) suggested the following procedure:

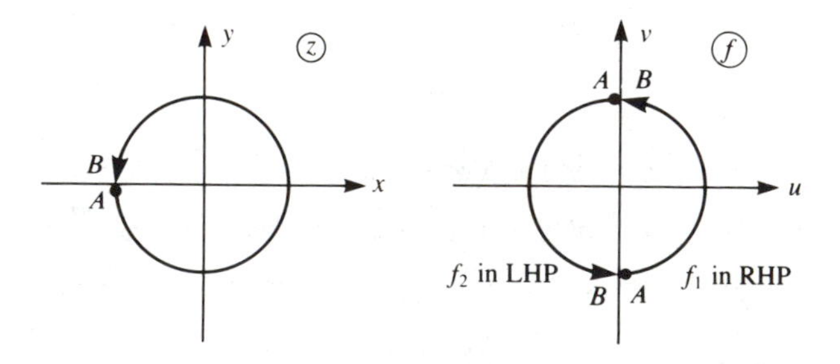

Fig. 6.5. The two roots of $z^{1/2}$.

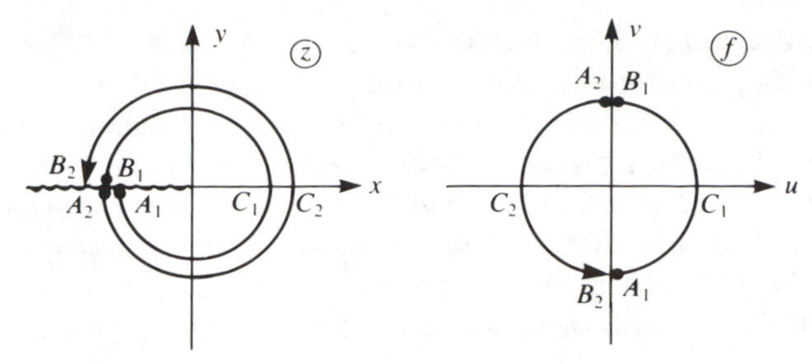

Fig. 6.6. Mapping of points for the function $z^{1/2}$.

For a double-valued function such as $z^{1/2}$, take two sheets or copies of the complex z plane. Make a cut, the *branch cut,* on both sheets from $r=0$ to ∞ along any suitably chosen boundary value of θ. In our example, this is just the negative real axis, as shown by the wavy line on Fig. 6.6. Next, cross-join together the edges of the cut so that B_1 is joined to A_2, and B_2 is joined to A_1. The resulting *Riemann surface* is a continuous two-storied structure with two sets of stairs at the cross-joints. We shall refer to a line of cross joints as a *branch line.* It coincides in location with that of the branch cut.

To explore this structure, we may start at A_1 on the first, say bottom, sheet. Going around the origin once, we reach B_1 on the first sheet. This point is separated from A_1 by an impassable crevice, which is our chosen branch cut. We can, however, take the cross-joining stairs up to A_2 on the second, or top, sheet. A second turn around the origin brings us to B_2 on the second sheet across the abyss from A_2. We see a second set of cross-joining stairs that now takes us down to our starting point A_1 on the first sheet. A complete exploration of this Riemann surface thus requires two complete turns around the origin.

In this way, the area of the complex z plane has been doubled. The mapping is thereby changed from the original one-to-two mapping to two

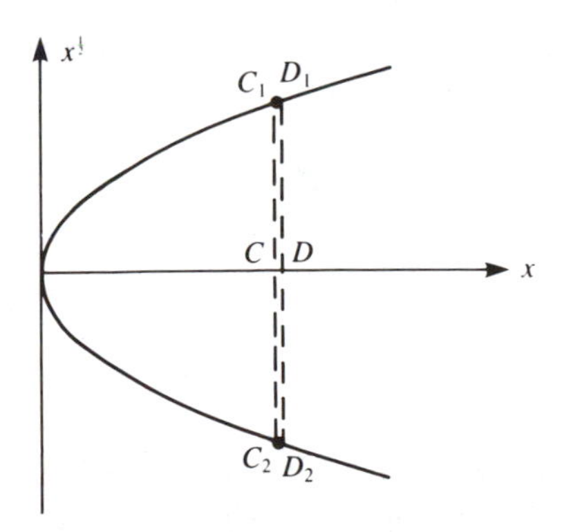

Fig. 6.7. The two branches of $x^{1/2}$.

one-to-one mappings. The double-valued function has been changed to a continuous single-valued function on one two-sheeted Riemann surface.

Thus the antidote for the discontinuities of a multivalued function is a multisheeted Riemann surface for the complex z "plane." A branch cut has to begin or end at a point that of necessity belongs to more than one sheet of the Riemann surface. This point is called a *branch point* of the function. The function is either zero or infinity at a branch point. It is then immaterial whether it is considered to be single, or multiple, valued there.

The position of a branch point is a property of the function, and is independent of the choice of branch cuts. For $z^{1/2}$, the branch points are at $z=0$ and ∞. The point at ∞ is sometimes not considered a branch point, although a branch cut can terminate there. Since one and only one branch of a function can be located on each Riemann sheet, only one branch cut can be drawn through a branch point. Again an exception has to be made for $z=\infty$, where any number of branch cuts can terminate. It can happen that two branch cuts both go to $z=\infty$, but that point is not a branch point. In that case, there is only one branch cut that connects two branch points by way of $z=\infty$.

While a branch cut has to terminate at a branch point, its exact position and shape can be quite arbitrary. The latter depend on the arbitrary choice of the boundary for θ. Figure 6.8 shows other equally acceptable choices of the branch cut for $z^{1/2}$. The feature common to all these choices is that neighboring points on the Riemann surface (such as the points B_1 and A_2, and the points A_1 and B_2 in Fig. 6.6) are always mapped onto neighboring points on the complex $f(z)$ plane.

Concerning the arbitrariness of branch cuts, it has been said that God created branch points, but men made the cuts.

The construction of the Riemann surface can be represented symbolically by a simple notation. We shall denote by S_i the ith Riemann sheet on which the function $f(z)$ is in its ith branch $f_i(z)$. For example, $z^{1/2}$ has

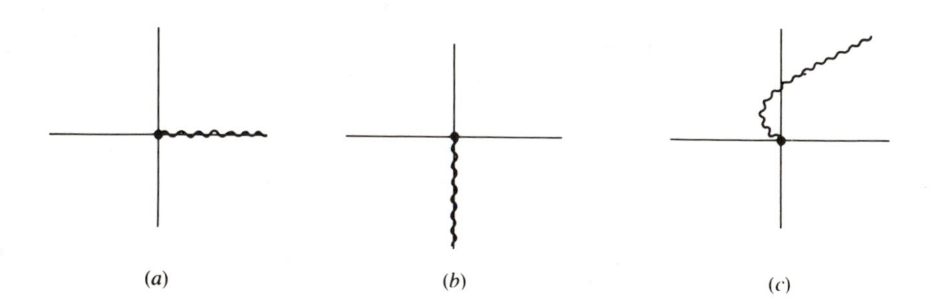

Fig. 6.8. Equally acceptable choices of the branch cut for $f(z)=z^{1/2}$. (a) $0\leq\theta\leq 2\pi$; (b) $-\pi/2\leq\theta\leq 3\pi/2$; (c) θ of branch cut is r dependent.

two branches, and its Riemann surface has two sheets:

$$S_1: \quad f = f_1 = r^{1/2}e^{i\theta/2}$$
$$S_2: \quad f = f_2 = -f_1.$$

For a branch cut we usually use just a straight line L connecting the two branch points P_1 and P_2. Then L may be specified by the symbol (P_1,P_2). For example, a branch cut connecting the origin and infinity is $(0,\infty)$ if it is along the positive real axis, $(-\infty,0)$ if it is along the negative real axis, $(0,i\infty)$ if it is along the positive imaginary axis, and $(-i\infty,0)$ if it is along the negative imaginary axis.

We shall denote by $S_i^{\pm}$ the upper and lower edges of a branch cut on the sheet S_i. The joining of neighboring sheets will be denoted by the pairing (S_i^+,S_{i+1}^-) or (S_{i+1}^+,S_i^-).

Thus a Riemann surface for $f(z)=z^{1/2}$ may be characterized as follows:

Branch points: $z=0,\infty$ (i.e., including $\pm\infty$, $\pm i\infty$)
Branch cut: $L=(-\infty,0)$
Riemann surface: is made by the following joinings of sheets:
at L: (S_1^+,S_2^-), (S_2^+,S_1^-).

These details are shown in Fig. 6.9 where U(L)HP stands for the upper (lower) half-plane.

A number of examples are given below to help the reader become familiar with the concept of Riemann surfaces.

Example 6.3.1.

$$f(z)=(z+1)^{1/2}+(z-1)^{1/2}.$$

This function has four branches:

$$\begin{aligned} f_1(z) &= r_1^{1/2}e^{i\theta_1/2}+r_2^{1/2}e^{i\theta_2/2}, \\ f_2(z) &= r_1^{1/2}e^{i\theta_1/2}-r_2^{1/2}e^{i\theta_2/2}, \\ f_3(z) &= -r_1^{1/2}e^{i\theta_1/2}+r_2^{1/2}e^{i\theta_2/2}, \\ f_4(z) &= -r_1^{1/2}e^{i\theta_1/2}-r_2^{1/2}e^{i\theta_2/2}, \end{aligned} \tag{6.21}$$

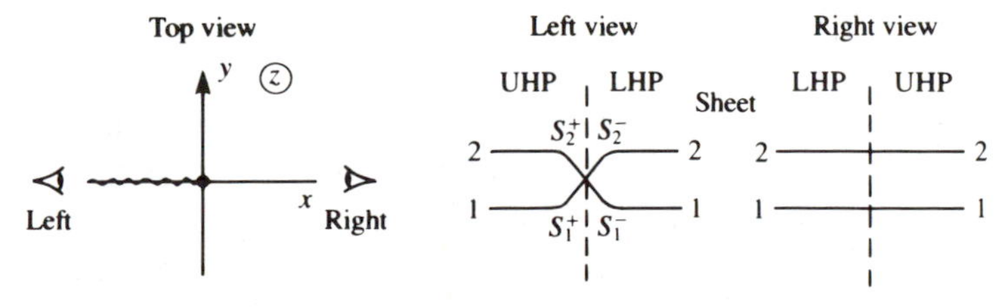

Fig. 6.9. Three views of a Riemann surface of $z^{1/2}$.

where r_i, θ_1, r_2, and θ_2 are defined in Fig. 6.9. The branch points are -1, 1, and ∞. The branch cuts may be chosen to be

$$L_1 = (-\infty, -1) \quad \text{and} \quad L_2 = (1, \infty),$$

these being the branch cuts of the first and second square roots, respectively. The resulting Riemann surface is then obtained by the following joining of sheets:

$$\begin{array}{lllll} \text{at } L_1: & (S_1^+, S_3^-), & (S_3^+, S_1^-), & (S_2^+, S_4^-), & (S_4^+, S_2^-), \\ \text{at } L_2: & (S_1^+, S_2^-), & (S_2^+, S_1^-), & (S_3^+, S_4^-), & (S_4^+, S_3^-). \end{array}$$

Figure 6.10 gives three views of this Riemann surface.

To calculate the actual value of $f(z)$ at each point on its four-sheeted Riemann surface, we must remember that our choices of branch cuts imply that $-\pi \leqslant \theta_1 \leqslant \pi$, $\quad 0 \leqslant \theta_2 \leqslant 2\pi$. For example, at the origin $(z = 0)$ where

$$\mathbf{r}_1 = (r_1, \theta_1) = (1, 0), \qquad \mathbf{r}_2 = (1, \pi),$$

we find

$$\begin{array}{ll} f_1(0) = 1 + e^{i(\pi/2)} = 1 + i, & f_2(0) = 1 - i, \\ f_3(0) = -1 + i, & f_4(0) = -1 - i. \end{array}$$

Example 6.3.2.

$$g(z) = z^{1/3},$$
$$f(z) = z^{2/3} = g^2(z).$$

The function $g(z)$ has three branches

$$g_n(z) = r^{1/3} \exp\left[i\left(\frac{\theta + 2\pi(n-1)}{3}\right)\right]. \tag{6.22}$$

Thus

$$\begin{aligned} g_1(z) &= r^{1/3} e^{i(\theta/3)}, \\ g_2(z) &= k g_1(z), \\ g_3(z) &= k^2 g_1(z), \qquad \text{with} \qquad k = e^{i(2\pi/3)}. \end{aligned}$$

The branch points are at 0 and ∞. As we circle the origin in the z plane in a positive direction, we see these three branches of the function

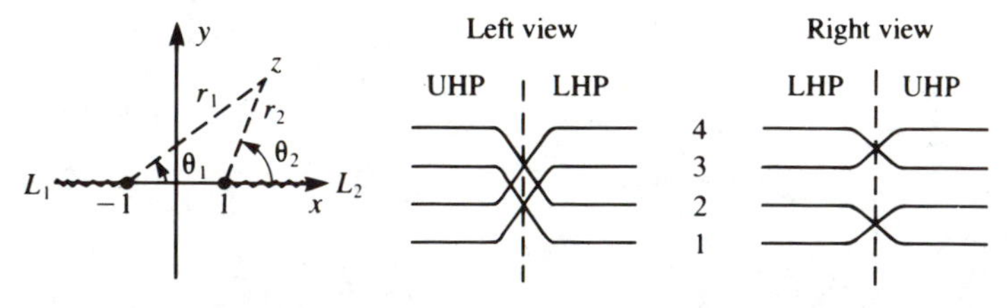

Fig. 6.10. Three views of a Riemann surface for $(z+1)^{1/2} + (z-1)^{1/2}$.

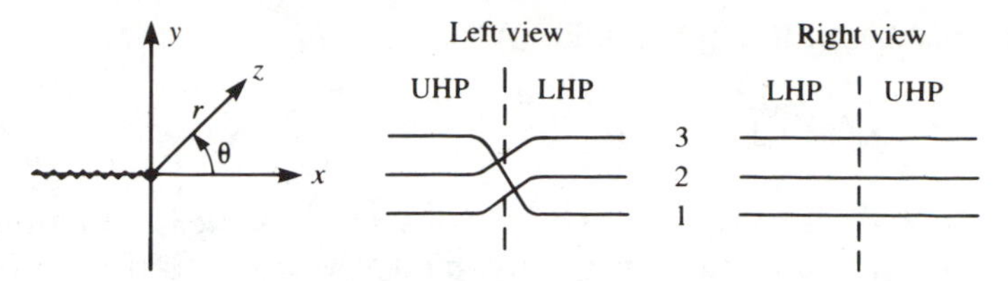

Fig. 6.11. Three views of a Riemann surface for $z^{1/3}$.

repeatedly in the order $g_1 \to g_2 \to g_3 \to g_4 = g_1$, etc. Hence if the branch cut is taken to be $L = (-\infty, 0)$, the Riemann surface should be formed by joining the cut edges as follows

$$(S_1^+, S_2^-), \qquad (S_2^+, S_3^-), \qquad (S_3^+, S_1^-).$$

The result is shown in Fig. 6.11.

The mapping $g = z^{1/3}$ involves three Riemann sheets in z mapping onto a single g plane. The function $f = g^2$ maps the single g plane onto two copies of the f plane. Hence $z^{2/3}$ maps three sheets of the z plane onto two sheets of the f plane, as shown in Fig. 6.12. For the chosen branch cuts, we find that the top half of the first sheet plus the second sheet of z maps onto the first sheet of f, while the bottom half of the first sheet plus the third sheet of z maps onto the second sheet of f.

Example 6.3.3.

$$f(z) = \left(\frac{z+1}{z-1}\right)^{1/3}.$$

This function has two branch points ($z_1 = -1$ and $z_2 = 1$) and three

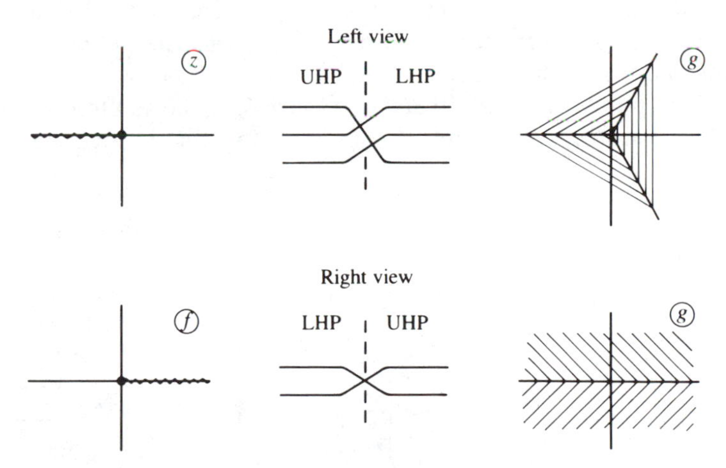

Fig. 6.12. Riemann surfaces for $z^{2/3}$.

distinct branches (for $n = 1$, 2, and 3):

$$f_n(z) = \left(\frac{r_1}{r_2}\right)^{1/3} \exp\left[i\left(\frac{\theta_1 - \theta_2}{3}\right) + \frac{2\pi}{3}(n-1)\right], \tag{6.23}$$

where r_i and θ_i are the polar coördinates of z as measured from the branch point z_i. Note that $n = n_1 - n_2$, mod 3, where n_i is the number of counterclockwise turns around the branch point z_i.

The point $z = \infty$ is not a branch point, since the function is still multivalued there. A branch cut connecting the two branch points can be made by way of infinity, leading to the Riemann surface shown in Fig. 6.13. The connection on the right is dictated by the requirement that, on circling $z_2 = 1$ in a positive (counterclockwise) direction, we see the different branches backwards, that is, in the order $g_1 \to g_3 \to g_2 \to g_1$, etc. This is because $z - 1$ is raised to negative powers, so that increasing n_2 decreases $n = n_1 - n_2$.

The actual functional values are easily calculated with the help of Eq. (6.23). For example:

$$f_1(0) = e^{-i(\pi/3)},$$
$$f_2(0) = kf_1 = e^{i(\pi/3)},$$
$$f_3(0) = k^2 f_1 = e^{+i\pi} = +1.$$

where $k = e^{i(2\pi/3)}$ is a cube root of 1.

Example 6.3.4.

$$f(z) = \ln z.$$

The complex logarithmic function has infinitely many branches:

$$f_n(z) = \ln r + i(\theta + 2\pi n), \tag{6.24}$$

where n is any (positive or negative) integer. The function $f_0(z)$ is called its *principal branch*. Different branches of $\ln z$ occupy different strips of the complex f plane; each strip has a height 2π along the y direction, and an infinite width stretching from $x = -\infty$ to ∞.

The branch points are at $z = 0$ and ∞. Figure 6.14 shows three branches of the function for the choice $L = (-\infty, 0)$ for the branch cut. The

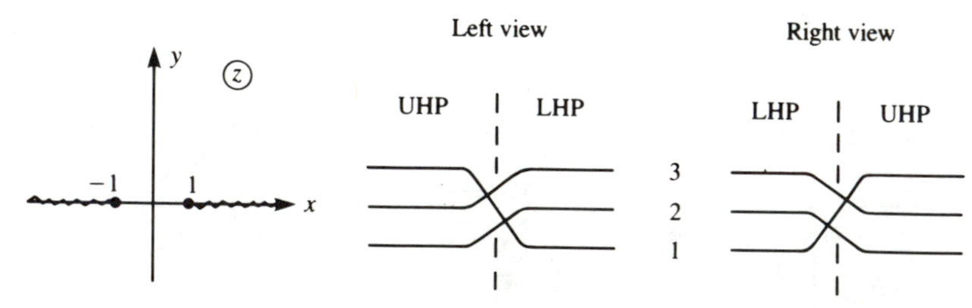

Fig. 6.13. Three views of a Riemann surface for $[(z+1)/(z-1)]^{1/3}$.

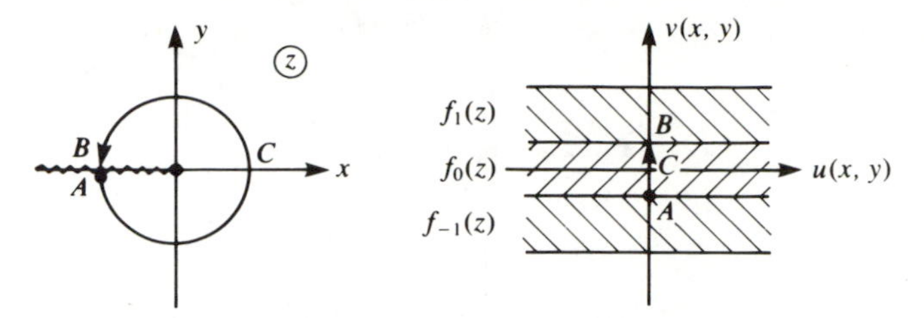

Fig. 6.14. Different branches of the function ln z.

Riemann surface should be joined together as follows

$$\cdots, (S_{-1}^{+}, S_{0}^{-}), (S_{0}^{+}, S_{1}^{-}), (S_{1}^{+}, S_{2}^{-}), \cdots,$$

since in Eq. (6.24) the branch number n increases when θ increases through 2π. The result is shown in Fig. 6.15.

Example 6.3.5.

$$f(z) = \ln[z/(z+1)],$$
$$g(z) = \sqrt{z+1}\ln z.$$

Both functions have branch points at $z = -1$, 0, and ∞. The branch cuts can be chosen to be the same for both functions, as shown in Fig. 6.16(a) and (b). Both functions have an infinity of branches because of the ln.

For $f(z)$, the Riemann surface for Fig. 6.16(a) can be so chosen that the effects of the two ln branch cuts along the positive real axis cancel each other. The function then remains on the same branch on crossing the positive real axis. This is the situation described by Fig. 6.16(c).

The situation is different for $g(z)$, which has a double set of ln branches due to the square root. The branch cut from $z = -1$ to infinity separates two square-root branches of the same ln function, while the branch cut from 0 to ∞ separates two ln branch of the same square-root function. These two branch cuts have different effects. They do not cancel each other out even if they lie together along the positive x axis.

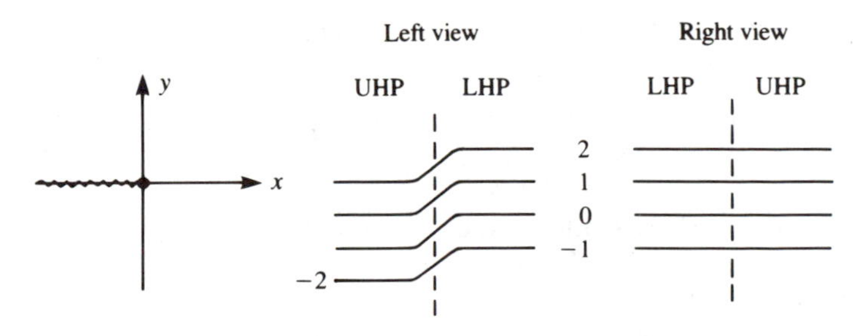

Fig. 6.15. Three views of a Riemann surface for ln z.

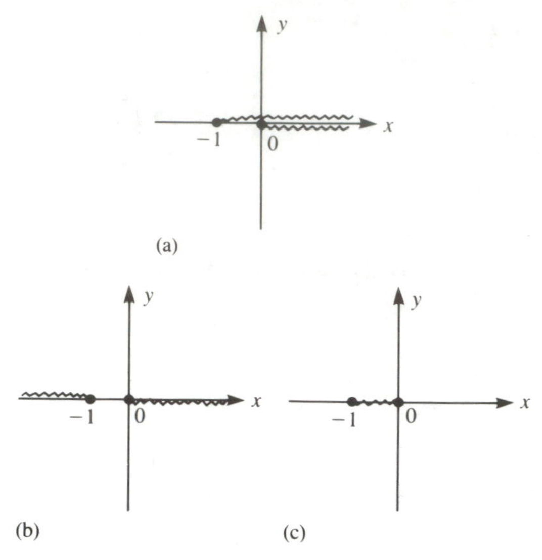

Fig. 6.16. Branch lines for $\ln[z/(z+1)]$; (a) and (b), but not (c), are also valid choices for $\sqrt{z+1}\ln z$.

Example 6.3.6.

$$f(z) = (1 + \sqrt{z})^{1/2}.$$

There are four branches:

$$\begin{aligned} f_1(z) &= (1 + r^{1/2}e^{i\theta/2})^{1/2}, \\ f_2(z) &= -(1 + r^{1/2}e^{i\theta/2})^{1/2}, \\ f_3(z) &= (1 - r^{1/2}e^{i\theta/2})^{1/2}, \\ f_4(z) &= -(1 - r^{1/2}e^{i\theta/2})^{1/2}, \qquad -\pi \leqslant \theta \leqslant \pi. \end{aligned} \tag{6.25}$$

The inside square root gives branch points at $z = 0$ and ∞ on all four sheets. The outside square root has a branch point at $z = 1$, but only on the third and fourth sheets. It is possible to draw the branch line to ∞ in such a way that it appears only on the third and fourth sheets, for example, by using the negative real axis of the complex plane $1 - \sqrt{z}$. This is equivalent to a branch line from $z = 1$ to ∞ on the third and fourth Riemann sheets only. The Riemann surface can then be obtained by cross-joining the cut edges as follows:

$$\begin{aligned} &\text{at } L_1: \quad (S_1^+, S_3^-),\ (S_3^+, S_1^-),\ (S_2^+, S_4^-),\ (S_4^+, S_2^-), \\ &\text{at } L_2: \quad (S_3^+, S_4^-),\ (S_4^+, S_3^-). \end{aligned}$$

This Riemann surface is illustrated in Fig. 6.17.

Example 6.3.7.

$$f(z) = \arcsin z.$$

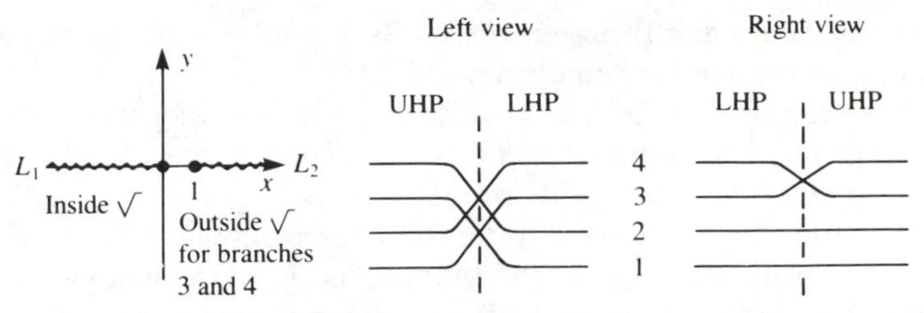

Fig. 6.17. Three views of a Riemann surface for $(1+\sqrt{z})^{1/2}$.

This function is not as simple as it might look. First, it is necessary to define it in terms of more familiar functions:

$$\begin{aligned} f(z) &= i \ln(\sqrt{1-z^2} - iz) \\ &= i \ln[h(z) - iz] \\ &= i \ln g(z). \end{aligned}$$

This shows that the function has infinitely many branches, which can be arranged in the form

$$\begin{aligned} f_{2n}(z) &= i \ln |g_1(z)| - \phi_1(z) - 2\pi n, \\ f_{2n-1}(z) &= i \ln |g_2(z)| - \phi_2(z) - 2\pi n, \end{aligned} \tag{6.26}$$

where $g_i(z) = h_i(z) - iz$ is defined in terms of the ith branch of $h(z) = \sqrt{1-z^2}$, and $\phi_i(z)$ is the polar angle of $g_i(z)$.

The determination of branch points and the choice of branch cuts are also more involved. We first note that the Riemann surface for $g(z) = \sqrt{1-z^2} - iz$ has two sheets joined at the branch lines, which may be taken to be

$$L_1 = (-\infty, -1) \qquad \text{and} \qquad L_2 = (1, \infty).$$

These two sheets are cross-joined at these branch lines in the usual way for the square-root function.

Like the simple square root, the two cross-joints at the same branch line are not equivalent because they map onto different lines in $g(z)$. Consider for example the two copies of the branch line L_2. On that copy lying at the cross joint (S_1^+, S_2^-), the function $g(z)$ has the value of its first branch

$$g_1(z) = g_1(x) = i\sqrt{x^2-1} - ix.$$

Hence this copy of L_2 maps onto that part of the negative imaginary axis of the complex g plane from $-i$ to 0. At the cross-joint (S_2^+, S_1^-), $g(z)$ has the value of its second branch

$$g_2(z) = g_2(x) = -i\sqrt{x^2-1} - ix.$$

This second copy of L_2 therefore maps onto the remaining part of the negative imaginary axis from $-i$ to $-i\infty$. In a similar way, the branch line L_1 maps onto the positive imaginary axis.

We now consider $f(z)$ as a logarithmic function of g. The Riemann surface for $i \ln g$ has infinitely many sheets. That is, the complex g plane is a multisheeted structure with branch points at $g=0$ and ∞. We may choose the branch cut for $i \ln g$ to be the negative imaginary g axis. If so, this coincides with *two* maps of the branch line L_2 in the complex z plane. The complex z plane is now made up of infinitely many copies of the two-sheeted Riemann surface of the last paragraph. They can be joined together into one continuous surface by cutting open both branch lines at L_2 and cross-joining together the edges belonging to different logarithmic branches.

To summarize, a Riemann surface can be constructed according to the scheme:

$$\text{at } L_1: \qquad (S_{2n}^{+}, S_{2n-1}^{-}),\ (S_{2n-1}^{+}, S_{2n}^{-});$$

$$\text{at } L_2: \qquad (S_{2n}^{+}, S_{2n+1}^{-}),\ (S_{2n-1}^{+}, S_{2n+2}^{-}).$$

This Riemann surface is shown in Fig. 6.18, where the two square-root branches belonging to the same ln branch are indicated. Across L_1, the ln part of the function remains on the same branch, while across L_2, both the ln and the square root change branch.

Problems

6.3.1. The branch cuts of the function $f(z)=(z^2-1)^{1/p}$, where p is a positive integer, have been chosen to be

$$L_1=(-\infty,-1),\ L_2=(1,\infty),$$

as shown in the Fig. 6.19. Obtain the value of $f(z)$ for the four points shown, namely

$$z=2\pm i\varepsilon,\ -2\pm i\varepsilon,$$

with $\varepsilon\to 0+$ (i.e., ε approaches zero along the positive real axis)

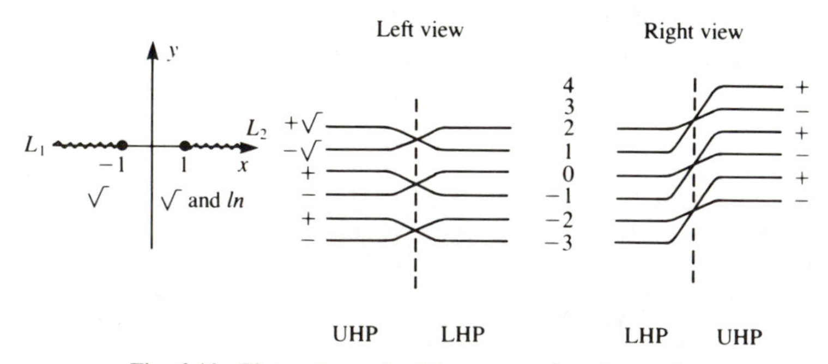

Fig. 6.18. Three views of a Riemann surface for arcsin z.

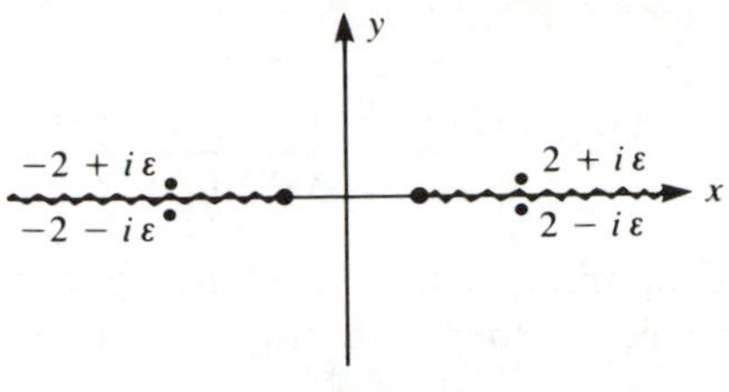

Fig. 6.19.

on the Riemann sheet specified below:

(a) For $p=2$ or $f(z)=(z^2-1)^{1/2}$, on that Riemann sheet on which $f(0)=e^{i(3\pi/2)}$;

(b) For $p=3$ or $f(z)=(z^2-1)^{1/3}$, on that Riemann sheet on which $f(0)=e^{i(5\pi/3)}$;

(c) For $p=4$ or $f(z)=(z^2-1)^{1/4}$, on that Riemann sheet on which $f(0)=e^{i(5\pi/4)}$.

6.3.2. Construct the Riemann surfaces for

(a) $z^{1/2}+z^{1/3}$;
(b) $z^{5/6}$;
(c) $(z-1)^{1/3}+(z+1)^{1/3}$;
(d) $(z^2-1)^{1/3}$;
(e) $\arctan z$;
(f) $\ln(\sqrt{z}+1)$.

The following cases are more complicated:

(g) $(z^{1/3}-1)^{1/2}$;
(h) $(z^{1/3}+1)^{1/2}$;
(i) $\ln(z^{1/2}+1)^{1/3}$.

6.3.3. On which sheet of your chosen Riemann surface does the denominator of

$$f(z)=\frac{1}{(z+1)^{1/2}-2i}$$

have a zero?

6.4. Complex differentiation: Analytic functions and singularities

Given a continuous, single-valued FCV

$$f(z)=u(x,y)+iv(x,y),$$

we define as usual the derivative

$$f'(z)=\frac{d}{dz}f(z)=\lim_{\Delta z\to 0}\frac{f(z+\Delta z)-f(z)}{\Delta z},$$

meaning that

$$f'(z) = \lim_{\Delta x, \Delta y \to 0} \frac{u(x+\Delta x, y+\Delta y) - u(x,y) + i(\text{same for } v)}{\Delta x + i\,\Delta y}. \quad (6.27)$$

There are of course infinitely many ways to approach a point z on a two-dimensional surface, but there are only two *independent* ways—along x or along y:

Along x:

$$f'(z) = \frac{\partial}{\partial x} f(z) = \frac{\partial u}{\partial x} + i\frac{\partial v}{\partial x}$$

Along y:

$$f'(z) = \frac{1}{i}\frac{\partial}{\partial y} f(z) = \frac{1}{i}\left(\frac{\partial u}{\partial y} + i\frac{\partial v}{\partial y}\right).$$

A unique derivative can appear only if these two results are equal. Hence the *Cauchy–Riemann conditions*

$$\frac{\partial u}{\partial x} = \frac{\partial v}{\partial y}, \qquad \frac{\partial v}{\partial x} = -\frac{\partial u}{\partial y} \quad (6.28)$$

must be satisfied. These conditions turn out to be both necessary and sufficient.

If $f(z)$ has a unique and finite derivative, it is said to be *differentiable.* If $f(z)$ is differentiable at z_0, and in a small neighborhood around z_0 in the complex plane, $f(z)$ is said to be an *analytic function* (also called *regular* or *holomorphic*) at $z = z_0$. If $f(z)$ is analytic everywhere in the complex plane within a *finite* distance of the origin, it is an *entire function.*

If $f(z)$ is not differentiable at $z = z_0$, it is said to be singular there. The point z_0 is then called a *singular point* or a *singularity* of $f(z)$.

Example 6.4.1. For $f(z) = z^2 = x^2 - y^2 + 2ixy$:

$$\frac{\partial u}{\partial x} = 2x = \frac{\partial v}{\partial y},$$

$$\frac{\partial u}{\partial y} = -2y = -\frac{\partial v}{\partial x}.$$

Hence the Cauchy–Riemann conditions are satisfied. In addition, the partial derivatives are finite except at $z = \infty$, while the function itself is single valued. Hence the function is analytic for all finite z. That is, it is an entire function.

A much faster check can be made by first assuming the validity of complex differentiations, as illustrated in the following example.

Example 6.4.2. For $f(z)=2z^n$

$$u=\operatorname{Re}f(z)=z^n+z^{*n},$$
$$v=\operatorname{Im}f(z)=(z^n-z^{*n})/i. \tag{6.29}$$

$$\therefore \frac{\partial u}{\partial x}=nz^{n-1}\frac{\partial z}{\partial x}+nz^{*n-1}\frac{\partial z^*}{\partial x}=nz^{n-1}+nz^{*n-1},$$

$$\frac{\partial v}{\partial y}=\frac{1}{i}[nz^{n-1}i-nz^{*n-1}(-i)]=\frac{\partial u}{\partial x},$$

$$\frac{\partial u}{\partial y}=nz^{n-1}i+nz^{*n-1}(-i)=-\frac{\partial v}{\partial x}.$$

The function is single-valued for integral values of n. It is then analytic at every point where the partial derivatives are finite. For positive integers n, this occurs for all finite z. Hence the positive power z^n is an entire function, but there is a singularity at $z=\infty$. For negative integers n, the function z^n is analytic everywhere except at the origin. Finally, for $n=0$ the function is analytic everywhere including $z=\infty$. It is actually just a constant $[f(z)=2]$.

6.4.1. Singularities

Functions of a complex variable have relatively simple structures in regions where they are analytic. Therefore many problems reduce to a study of these functions at singular points where they are not analytic.

If a function is singular at the point $z=a$, but is analytic in all neighborhoods of a, the point $z=a$ is called an *isolated singular point* of the function. If near an isolated singular point $z=a$, the function behaves like $(z-a)^{-n}$, where n is a positive integer, then $z=a$ is a *pole*, or more specifically an *nth-order pole*, of $f(z)$. It is a *simple pole* if $n=1$.

If the order n of the pole goes to infinity, the singularity is said to be an *essential isolated singularity*. A classic example is $\exp(1/z)$ at $z=0$:

$$\text{If } z\to 0 \text{ along the }\begin{cases}\text{positive real axis:} & \exp(1/z)\to\exp(\infty)\to\infty\\ \text{negative real axis:} & \to\exp(-\infty)\to 0\\ \text{imaginary axis:} & \to\exp(\pm i\infty)\\ & \to\text{oscillatory.}\end{cases}$$

Indeed an essential singularity, whether isolated or not, is one at which the function behaves in a wild manner.

If an essential singularity is isolated, it satisfies *Picard's Theorem*:

> A function in the neighborhood of an essential isolated singularity assumes every complex value infinitely many times, except perhaps one particular value.

While the function $\sin(1/z)$ has an essentially isolated singularity at $z=0$, its reciprocal $[\sin(1/z)]^{-1}$ behaves even more wildly at and near

$z = 0$. Indeed it shows an infinity of poles as we approach their limit point $z = 0$. We call this point an *essential singularity* rather than an essential isolated singularity.

A branch point is another singularity that is not isolated. A branch point belongs to two or more branches; it is a point in the neighborhood of which the function is multivalued. A branch of a function is not continuous across a branch *cut* on a Riemann sheet, although the function itself is continuous if we go continuously on the Riemann surface across a branch *line* to another Riemann sheet.

Among the simplest functions is the power z^n. It is an entire function, analytic everywhere within a finite distance of the origin. However, it has an nth-order pole at infinity in the same way that the inverse power z^{-n} has an nth-order pole at the origin.

An entire function that behaves at infinity more weakly than z^n is a polynomial of degree less than n. A function that is everywhere finite and analytic must be a constant. The last result is known as the *Liouville Theorem.*

Thus singularities are the sources of functional behavior. Since physics describes phenomena that are changing in space-time, we must conclude that singularities are also sources of physical phenomena. For this reason, we are more interested in singularities than in analyticity, although conceptually the two must go together.

Problems

6.4.1. Which of the following are analytic functions?

(a) z^{*n};
(b) $\mathrm{Im}\, z = iy$;
(c) $e^{|z|}$;
(d) e^{z^2}.

6.4.2. Show that $z^{1/2}$ is analytic in the complex z plane within a finite distance from the origin except at the branch point $z = 0$ by using

(a) Complex differentiation;
(b) Real differentiation.

For the latter, you might need the relations

$$\cos\frac{\theta}{2} = [\tfrac{1}{2}(\cos\theta + 1)]^{1/2} = \left[\frac{1}{2}\left(\frac{x}{\sqrt{x^2+y^2}} + 1\right)\right]^{1/2}$$

$$\sin\frac{\theta}{2} = \frac{\sin\theta}{\sqrt{2(\cos\theta+1)}} = \frac{y}{\sqrt{x^2+y^2}}\,\frac{1}{[2(x/\sqrt{x^2+y^2}+1)]^{1/2}}.$$

6.4.3. Locate and identify the singularities of the following functions

(a) $(z+i)^{2/3}$;
(b) $\tan(1/z)$;
(c) $\dfrac{1}{\sqrt{z+i}+2i}$;
(d) $\dfrac{z^2(z-1)}{\sin^2 \pi x}$.

6.4.4. Show that in circular polar coordinates the equality of the two independent derivatives of an analytic function $f(z = re^{i\theta})$ requires that

$$\frac{\partial f}{\partial r} = \frac{1}{ir}\frac{\partial f}{\partial \theta}.$$

Hence if $f = Re^{i\Theta}$, the Cauchy–Riemann conditions read

$$\frac{\partial R}{\partial r} = \frac{R}{r}\frac{\partial \Theta}{\partial \theta},$$

$$R\frac{\partial \Theta}{\partial r} = -\frac{1}{r}\frac{\partial R}{\partial \theta}.$$

6.4.5. Show that, if an analytic function is always real, it can only be a constant.

6.4.6. Show that $f(z) = \sinh(x^2 - y^2)\cos(2xy) + i\cosh(x^2 - y^2)\sin(2xy)$ is an entire function.

6.5. Complex integration: Cauchy integral theorem and integral formula

A function $f(z)$ of a complex variable z can be integrated along a path C on the complex z plane

$$\int_C f(z)\,dz = \int_C [u(x,y) + iv(x,y)](dx + i\,dy).$$

The results are particularly simple when $f(z)$ is an analytic function, because of the *Cauchy Integral Theorem*:

If $f(z)$ is analytic in a simply connected domain D, and C is a piecewise smooth, simple closed curve in D, then

$$\oint_C f(z)\,dz = 0, \tag{6.30}$$

where the circle on the integration sign denotes the closure of the integration path C.

To prove this theorem, we note that

$$\oint_C f(z)\,dz = \oint_C (u\,dx - v\,dy) + i\oint_C (v\,dx + u\,dy)$$

$$= \oint_C \mathbf{A}(\mathbf{r})\cdot d\mathbf{r} + i\oint_C \mathbf{B}(\mathbf{r})\cdot d\mathbf{r},$$

if expressed in terms of the two-dimensional vector fields

$$\mathbf{A}(\mathbf{r}) = u\mathbf{i} - v\mathbf{j}, \qquad \mathbf{B}(\mathbf{r}) = v\mathbf{i} + u\mathbf{j}.$$

Stokes's theorem can now be applied to yield

$$\oint_C f(z)\,dz = \int_S\!\!\int d\boldsymbol{\sigma}\cdot(\boldsymbol{\nabla}\times\mathbf{A} + i\,\boldsymbol{\nabla}\times\mathbf{B})$$

$$= \int\!\!\int dx\,dy\left[-\left(\frac{\partial v}{\partial x}+\frac{\partial u}{\partial y}\right)+i\left(\frac{\partial u}{\partial x}-\frac{\partial v}{\partial y}\right)\right].$$

Since $f(x)$ satisfies the Cauchy–Riemann conditions, both the real and the imaginary parts of the integrand vanish, thus proving the Cauchy theorem.

The simply connected domain referred to in the Cauchy theorem is one with no hole in it. A simple curve is one that does not intersect itself. A curve is piecewise smooth if it is made up of a finite number of pieces, each of which is smooth. The closed path C is always taken in a positive, that is, counterclockwise, direction, unless otherwise stated.

Because of the Cauchy theorem, an integration contour can be moved across any region of the complex plane over which the integrand is analytic without changing the value of the integral. It cannot be moved across holes or singularities, but it can be made to collapse around them, as shown in Fig. 6.20. As a result, an integration contour C_0 enclosing n holes or singularities can be replaced by n separately closed contours C_i, each enclosing one hole or singularity:

$$\oint_C f(z)\,dz = \sum_{i=1}^{n}\oint_{C_i} f(z)\,dz.$$

If the integrand is analytic over the whole region enclosed by a closed contour, the contour can be shrunk to a point. Consequently the integral vanishes. This is just the result stated by the Cauchy theorem.

Nontrivial contour integrals involve functions that are not everywhere analytic in the enclosed region. The simplest such functions are of the type $(z-a)^{-n}$, where n is a positive integer (i.e., functions with an nth-order pole at $z=a$). A closed-path integration around such a pole can be evaluated easily by shrinking the contour C to a small circle c of

Fig. 6.20. Collapsing a closed contour around a hole (the shaded area) and a singularity (the dot) without changing the value of an integral.

radius ε around a:

$$\oint_C \frac{1}{(z-a)^n}\,dz = \oint_0^{2\pi} (\varepsilon e^{i\theta})^{1-n} i\,d\theta = \begin{cases} 0, & n \neq 1 \\ 2\pi i, & n = 1, \end{cases} \tag{6.31}$$

since the periodic function $e^{i\theta(1-n)}$ reduces to 1 only for $n = 1$.

More generally, if $f(z)$ is analytic inside and on a closed curve C, and if the point $z = a$ is inside C, then the integral

$$I = \oint_C \frac{f(z)\,dz}{z-a}$$

can again be evaluated by collapsing C into a small circle c around a:

$$\begin{aligned} I &= \oint_c \frac{f(z)-f(a)}{z-a}\,dz + f(a)\oint_c \frac{1}{z-a}\,dz \\ &= \oint_c df(z) + 2\pi i f(a). \end{aligned}$$

Since $df(z)$ is a total differential, the first term vanishes, leaving

$$\oint_C \frac{f(z)\,dz}{z-a} = 2\pi i f(a), \tag{6.32}$$

a result known as the *Cauchy integral formula.*

It is useful to write Eq. (6.32) in the form

$$f(z) = \frac{1}{2\pi i}\oint_C \frac{f(z')\,dz'}{z'-z} \tag{6.33}$$

to emphasize the fact that z can be any point inside the closed path C. Furthermore, it states that an analytic function at z has a value that is a certain average of functional values on any closed path C surrounding z.

Equation (6.33) can be differentiated n times to yield the formula

$$\frac{d^n f(z)}{dz^n} = \frac{n!}{2\pi i}\oint \frac{f(z')\,dz'}{(z'-z)^{n+1}}. \tag{6.34}$$

This result is useful in understanding Taylor expansions of analytic functions in Section 6.7.

Problems

6.5.1. Verify the following contour integrals along the triangular closed contour $(0,0) \to (1,0) \to (1,1) \to (0,0)$:

(a) $\oint \operatorname{Re} z\,dz = -\frac{i}{2}$;

(b) $\oint z\,dz = 0$.

6.5.2. For integration once along the unit circle in the positive direction, show that

(a) $\oint \frac{e^z}{z^5}\,dz = \frac{2\pi i}{4!}$;

(b) $\oint \frac{e^{1/z}}{z^5}\,dz = 0.$

6.5.3. If the Legendre polynomial of degree n is defined by the Rodrigues formula

$$P_n(x) = \frac{1}{2^n n!}\left(\frac{d}{dx}\right)^n (x^2-1)^n,$$

show that

$$P_n(z) = \frac{1}{2^n}\frac{1}{2\pi i}\oint \frac{(t^2-1)^n}{(t-z)^{n+1}}\,dt,$$

called the *Schläfli integral representation.* If the contour is next chosen to be a circle of radius $|(z^2-1)^{1/2}|$ centered at $t=z$ by making a change of variables from t to ϕ with

$$t = z + \sqrt{z^2-1}\,e^{i\phi},$$

show that the integral representation can be written in the form

$$P_n(z) = \frac{1}{\pi}\int_0^{\pi} (z+\sqrt{z^2-1}\cos\phi)^n\,d\phi,$$

first given by Laplace. [Hint: First show that $(t^2-1)/(t-z) = 2(z+\sqrt{z^2-1}\cos\phi)$.]

6.6. Harmonic functions in the plane

In the region of the complex plane z where

$$f(z) = u(x,y) + iv(x,y) \tag{6.12}$$

is analytic, the Cauchy–Riemann equations can be used to write

$$\frac{\partial^2 u}{\partial x^2} = \frac{\partial}{\partial x}\left(\frac{\partial v}{\partial y}\right) = \frac{\partial}{\partial y}\left(\frac{\partial v}{\partial x}\right) = -\frac{\partial^2 u}{\partial y^2}$$

$$\frac{\partial^2 v}{\partial x^2} = \frac{\partial}{\partial x}\left(-\frac{\partial u}{\partial y}\right) = -\frac{\partial}{\partial y}\left(\frac{\partial u}{\partial x}\right) = -\frac{\partial^2 v}{\partial x^2}.$$

Thus both $u(x,y)$ and $v(x,y)$ separately, and hence $f(z)$ as a whole, satisfy the two-dimensional Laplace equation

$$\left(\frac{\partial^2}{\partial x^2} + \frac{\partial^2}{\partial y^2}\right)[u(x,y) + iv(x,y)] = 0,$$

or

$$\nabla^2 f(z) = 0. \tag{6.35}$$

Solutions of Laplace equations are called *harmonic functions*. Hence the real part or the imaginary part of an analytic function is a harmonic function in the plane. More specifically, these two parts of f are said to form a pair of conjugate harmonic functions that are joined together by Eq. (6.12).

Given one harmonic function in the plane, say $u(x,y)$, its conjugate harmonic function $v(x,y)$ can be constructed with the help of one of the Cauchy–Riemann equations if the answer is not obvious on inspection:

$$\begin{aligned} v(x,y) &= \int_{r_0}^{r} dv(x,y) + v(x_0,y_0) \\ &= \int_{r_0}^{r} \left(\frac{\partial v}{\partial x} dx + \frac{\partial v}{\partial y} dy \right) + v(x_0,y_0) \\ &= \int_{r_0}^{r} \left(-\frac{\partial u}{\partial y} dx + \frac{\partial u}{\partial x} dy \right) + v(x_0,y_0). \end{aligned} \tag{6.36}$$

According to the Cauchy integral theorem, the result is path independent provided that no singularity is crossed as the path is deformed.

Example 6.6.1.

(a) Show that $u(x,y) = x^2 - y^2 - y$ is harmonic.

(b) Obtain its conjugate function $v(x,y)$ and the analytic function $f = u + iv$.

(a) $\dfrac{\partial^2 u}{\partial x^2} = 2, \qquad \dfrac{\partial^2 u}{\partial y^2} = -2, \qquad \therefore \nabla^2 u(x,y) = 0$

(b) $v(x,y) = \displaystyle\int_{r_0}^{r} [(2y+1)\,dx + 2x\,dy] + \text{const.}$

Each of these integrals is path-dependent, but their sum is not. For a path parallel first to the x axis and then to the y axis, we find

$$\begin{aligned} v(x,y) &= (2y_0+1)(x-x_0) + 2x(y-y_0) + \text{const} \\ &= 2xy + x + \text{const.} \end{aligned}$$

therefore $f(z) = x^2 - y^2 - y + i(2xy + x) = z^2 + iz$.

Many two-dimensional systems satisfy the Laplace equation in the plane and are described by analytic functions. For example, if $u(x,y)$ is a potential function, the curve along which u is constant is called an *equipotential*. Along an equipotential

$$0 = du = \frac{\partial u}{\partial x} dx + \frac{\partial u}{\partial y} dy = \nabla u \cdot d\mathbf{r}.$$

This shows that the vector field

$$\nabla u = \mathbf{i} \frac{\partial u}{\partial x} + \mathbf{j} \frac{\partial u}{\partial y}$$

gives a vector that is everywhere perpendicular to the equipotential passing through that point. A vector field ∇v can be constructed in a similar way for the harmonic function v conjugate to u.

Now the Cauchy–Riemann conditions can be used to show that the two gradient vectors ∇u and ∇v are everywhere perpendicular to each other:

$$(\nabla u)\cdot(\nabla v) = \frac{\partial u}{\partial x}\frac{\partial v}{\partial x} + \frac{\partial u}{\partial y}\frac{\partial v}{\partial y}$$

$$= \left(\frac{\partial v}{\partial y}\right)\frac{\partial v}{\partial x} + \left(-\frac{\partial v}{\partial x}\right)\frac{\partial v}{\partial y} = 0.$$

Since ∇v is also perpendicular to curves of constant v everywhere, we see that

1. ∇u is everywhere tangent to a curve of constant v, and
2. Curves of constant v are everywhere perpendicular to curves of constant u.

If $u(x,y)$ is a potential function, a curve of constant v whose tangent vector gives the direction of ∇u is called a *line of force* or a *field* (*flow, flux,* or *stream*) *line.* The function $v(x,y)$ itself is called a *stream function.* The function $f(z)$ is sometimes called a *complex potential.*

Some of the results described above are applied in the following example.

Example 6.6.2. Calculate the complex electrostatic potential of an infinite line charge placed at the origin perpendicular to the xy plane.

By symmetry, the potential looks the same on any two-dimensional plane perpendicular to the line charge. If r, θ are the circular coordinates on this plane, the electric field "radiates" from the perpendicular line charge in the form $\hat{E} = E_r\hat{e}_{\mathbf{r}}$. An application of the Gauss law of electrostatics over the surface of a cylinder of unit length surrounding the line charge yields the result

$$\int \mathbf{E}\cdot d\boldsymbol{\sigma} = 2\pi r E_r = \frac{q}{\varepsilon_0},$$

where q is the charge per unit length of the source. Therefore,

$$E_r = \frac{q}{2\pi\varepsilon_0 r} = -\frac{d}{dr}u(\mathbf{r}).$$

Integration of this expression gives the electrostatic potential

$$u(\mathbf{r}) = -\frac{q}{2\pi\varepsilon_0}\ln r. \qquad (6.37)$$

The complex potential that can be constructed from Eq. (6.37) is

$$\begin{aligned} f(z) &= -\frac{q}{2\pi\varepsilon_0}\ln z \\ &= -\frac{q}{2\pi\varepsilon_0}(\ln r + i\theta). \end{aligned} \tag{6.38}$$

It contains an imaginary part

$$v(x,y) = \operatorname{Im} f(z) = -\frac{q}{2\pi\varepsilon_0}\theta, \tag{6.39}$$

which describes the field lines of constant θ radiating from the line charge.

6.6.1. *Conformal mapping*

It is interesting that the real and imaginary parts of an analytic function

$$f(z) = u(x,y) + iv(x,y)$$

are everywhere perpendicular to each other. Thus u and v form a two-dimensional orthogonal curvilinear coordinate system. Indeed the map $z \to f(z)$ is *conformal*, or angle-preserving, in the following sense: If two curves intersect at $z = z_0$ with an angle α, their images in the complex f plane will intersect at $f(z_0)$ with the same angle α. This result arises because the derivative of an analytic function

$$f'(z_0) = |f'(z_0)|\, e^{-i\phi(z_0)} \tag{6.40}$$

is independent of the direction in the complex plane along which it is evaluated. The image of every short segment Δz in the complex f plane:

$$\Delta f \simeq |f'(z_0)|\, e^{i\phi(z_0)}\,\Delta z \tag{6.41}$$

is rotated from Δz by the same angle $\phi(z_0)$ independently of the orientation of Δz itself. This means that the angle between two intersecting segments or curves does not change under the mapping.

Problems

6.6.1. Find the analytic function $f(z)$, given only
(a) Its real part $u(x,y) = x^2 - y^2$;
(b) Its imaginary part $v(x,y) = e^{-y}\sin x$.

6.6.2. Two infinite line charges of equal strength q and opposite signs are placed at $x = \pm a$.
(a) Show that the equipotentials and fields lines are given by the equations $r_1/r_2 = \text{const}$ and $\theta_1 - \theta_2 = \text{const}$.
(b) Show in particular that the equipotentials of value u are two circles of radius R centered at $\pm x_0$:

$$R = |a \operatorname{csch} \alpha|, \qquad x_0 = |a \coth \alpha|,$$

where $\alpha = 2\pi\varepsilon_0 u/q$. These circles do not overlap. The circle centered at x_0 ($-x_0$) lies entirely in the half-plane with $x > 0$ ($x < 0$).

(c) A parallel pair of infinite cylinders of radius r centered at $\pm x_0$ are used as a capacitor. Show that its capacitance is

$$C = \frac{\pi\varepsilon_0}{\text{arc}\cosh(x_0/R)}.$$

(Hint: The complex potential is $f(z) = u + iv = -(q/2\pi\varepsilon_0)[\ln(z - a) - \ln(z + a)]$.)

6.7. Taylor series and analytic continuation

We are familiar with the Taylor expansion of a real function $f(x)$ of a real variable x about the point $x = a$:

$$f(x) = f(a) + (x - a)f' + \cdots + \frac{(x-a)^n}{n!} f^{(n)}(a) + \cdots, \tag{6.42}$$

where

$$f^{(n)}(a) = \frac{d^n}{dx^n} f(x)|_{x=a}.$$

For example, the Taylor expansion of $(1 - x)^{-1}$ about the origin $a = 0$ can be obtained with the help of the derivatives

$$f'(0) = \frac{d}{dx}\frac{1}{1-x}\bigg|_{x=0} = \frac{1}{(1-x)^2}\bigg|_{x=0} = 1,$$

$$f''(0) = \frac{2}{(1-x)^3}\bigg|_{x=0} = 2, \qquad \text{etc.}$$

The result is

$$\frac{1}{1-x} = 1 + x + x^2 + \cdots + x^n + \cdots = \sum_{n=0}^{\infty} x^n. \tag{6.43}$$

Since the ratio of succeeding terms is

$$\left|\frac{x^{n+1}}{x^n}\right| = |x|, \tag{6.44}$$

we see that this Taylor series converges for $|x| < 1$, but fails to converge for $|x| > 1$. For example, at $x = \frac{1}{2}$ the function is 2, while the partial sums of its Taylor series $(1, 1.5, 1.75, 1.875, 1.90625, \ldots.)$ rapidly approach the right answer. On the other hand, at $x = 2$ where the function is -1, the partial sums $(1, 3, 7, 15, 31, \ldots)$ are divergent.

The convergence of the complex series obtained from Eq. (6.42) by replacing x by z is also determined by the ratio test based on Eq. (6.44), namely that $|z| < 1$. Because of this condition, the resulting complex series is said to have a *radius of convergence* of 1.

It is easy to see that the Taylor series for $(1-z)^{-1}$ diverges at $z = 1$, since there is a pole there. The difficulty of the Taylor series elsewhere on the circumference of the circle of radius 1 is not as apparent, but is nevertheless real. For example, at $z = -1$ the partial sums are $1, 0, 1, 0, 1, \ldots$, so that the series is also divergent.

If we now keep clear of this boundary inside the circle of convergence, in the sense that $|z| < R_0 < 1$, the complex Taylor series for $(1-z)^{-1}$ is convergent everywhere in the region. The series is then said to be *uniformly convergent* in this region.

A uniformly convergent infinite series can be differentiated or integrated any number of times. Hence the Taylor series for $(1-z)^{-1}$ is analytic *inside* the circle of convergence. Indeed, it coincides there with the analytic function $(1-z)^{-1}$. Thus convergence implies differentiability.

Conversely, differentiability implies convergence. That is, given a function that is differentiable any number of times at $z = a$, the Taylor series

$$S = \sum_{n=0}^{\infty} (z-a)^n \frac{1}{n!} \frac{d^n f(z)}{dz^n}\bigg|_{z=a} \tag{6.45}$$

should converge in some neighborhood of a. To see this, we first write the Taylor series in the form

$$S = \sum_{n=0}^{\infty} \frac{(z-a)^n}{n} \frac{n!}{2\pi i} \oint_C \frac{f(z')\,dz'}{(z'-a)^{n+1}},$$

where we have used the Cauchy integral formula, Eq. (6.34), for the nth derivative. If the integration contour C encloses the circle of radius $|z-a|$ around a, as shown in Fig. 6.21, then $|z'-a| > |z-a|$ is guaranteed for any point z' on C. Hence the geometrical series

$$\sum_{n=0}^{\infty} \frac{(z-a)^n}{(z'-a)^{n+1}} = \frac{1}{z'-a} \frac{1}{1-(z-a)/(z'-a)} = \frac{1}{z'-z}$$

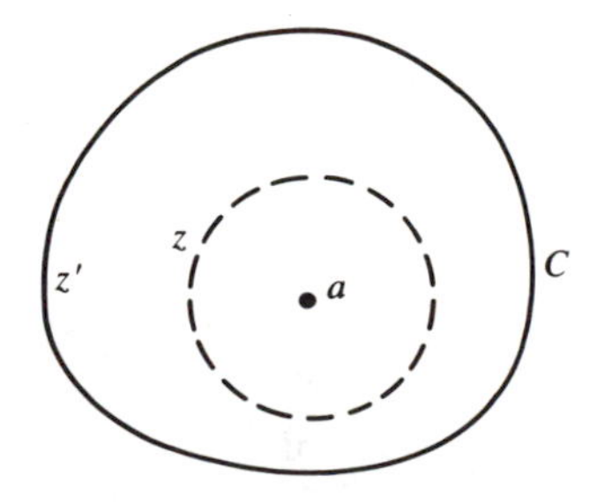

Fig. 6.21. The contour C for the complex integration needed to evaluate the nth derivative.

converges, and can be summed to $(z'-z)^{-1}$. As a result, the Taylor series converges to

$$S = \frac{1}{2\pi i}\oint_C \frac{f(z')\,dz'}{z'-z} = f(z)$$

that is, the function $f(z)$ itself.

We see that every function $f(z)$ analytic at $z=a$ can be expanded in a convergent Taylor series S of Eq. (6.45) in some neighborhood of the point a. This series is unique, because the nth derivative

$$\left.\frac{d^n f(z)}{dz^n}\right|_{z=a} = n!\, c_n$$

is unique for an analytic function. Further, the radius of convergence of this Taylor series is $|z_0 - a|$, where z_0 is the singularity of $f(z)$ nearest a.

Since complex differentiation works in exactly the same way as real differentiation, a complex Taylor series can be obtained in a familiar way. The only difference is that the point a may not lie on the real axis. The following examples illustrate how the knowledge of a Taylor series about the origin (i.e., a *Maclaurin series* in x) can be used.

Example 6.7.1. Expand $(1-z)^{-1}$ about a:

$$\frac{1}{1-z} = \frac{1}{(1-a)-(z-a)} = \frac{1}{1-a}\frac{1}{1-(z-a)/(1-a)}$$
$$= \frac{1}{1-a}\sum_{n=0}^{\infty}\left(\frac{z-a}{1-a}\right)^n.$$

Example 6.7.2. Expand $\ln(a+z)$ about a:

Suppose we know the Maclaurin series

$$\ln(1+x) = x - \frac{x^2}{2} + \frac{x^3}{3} - \cdots,$$

then

$$\ln(1+z) = \ln(1+a+z-a) = \ln(1+a)\left[1+\left(\frac{z-a}{1+a}\right)\right]$$
$$= \ln(1+a) + \ln\left[1+\left(\frac{z-a}{1+a}\right)\right]$$
$$= \ln(1+a) + \left(\frac{z-a}{1+a}\right) - \frac{1}{2}\left(\frac{z-a}{1+a}\right)^2 + \cdots.$$

Example 6.7.3. Expand $\sin z$ about a:

Since

$$\sin x = x - \frac{x^3}{3!} + \cdots,$$
$$\cos x = 1 - \frac{x^2}{2!} + \cdots,$$

we have

$$\begin{aligned}\sin z &= \sin[a + (z - a)] \\ &= \sin a \cos(z - a) + \cos a \sin(z - a) \\ &= \sin a\left(1 - \frac{(z - a)^2}{2!} + \cdots\right) \\ &\quad + \cos a\left((z - a) - \frac{(z - a)^3}{3!} + \cdots\right).\end{aligned}$$

6.7.1. Analytic continuation

Frequently we know an infinite series $S(z)$, but not the function $f(z)$ itself for which it is a Taylor series. The function $f(z)$ could be analytic outside the circle of convergence of $S(z)$. If so, it is of greater interest than $S(z)$. To discover its behavior outside, it is necessary to "extend" $S(z)$ to outside its circle of convergence. The process is called an *analytic continuation* of $S(z)$.

More generally, if we have two functions $g(z)$ and $G(z)$ satisfying the following properties:

1. $g(z)$ is defined on a set (i.e., a part) E of the z plane;
2. $G(z)$ is analytic in the domain D containing E;
3. $G(z)$ coincides with $g(z)$ on E,

we call $G(z)$ the *analytic continuation* of $g(z)$ to the domain D.

Example 6.7.4. The real exponential function

$$g(x) = e^x = \sum_{n=0}^{\infty} \frac{x^n}{n!}$$

converges for all finite real values of x, while the complex exponential function

$$G(z) = e^z = \sum_{n=0}^{\infty} \frac{z^n}{n!}$$

converges for all finite complex values of z. Therefore e^z is the analytic continuation of e^x.

Example 6.7.5. The series

$$S = \sum_{n=0}^{\infty} z^n$$

has a radius of consequence of 1. Hence it cannot be used at a point, say $1 + i$, outside its circle of convergence. We remember, however, that S is the Taylor series of

$$f(z) = \frac{1}{1 - z}$$

about the origin, and that $f(z)$ is analytic everywhere except at $z = 1$, where it has a pole. Hence $f(z)$ is the analytic continuation of S valid everywhere except at $z = 1$. This shows that the analytic continuation of S to $1 + i$ is

$$f(1+i) = \frac{1}{1-(1+i)} = i.$$

The analytic continuation of a Taylor series can be made by changing the point of the Taylor expansion: If the original Taylor series is

$$S_a(z) = \sum_{m=0}^{\infty} c_m (z-a)^m,$$

we look for a Taylor series about another point b,

$$S_b(z) = \sum_{n=0}^{\infty} d_n (z-b)^n.$$

The new Taylor coefficients can be calculated in terms of the old ones by choosing b to be within the circle of convergence of $S_a(z)$:

$$d_n = \frac{1}{n!}\frac{d^n}{dz^n} S_a(z)\bigg|_{z=b} = \frac{1}{n!}\sum_{m=n}^{\infty} c_m \frac{m!}{(m-n)!}(b-a)^{m-n}. \tag{6.46}$$

The new Taylor series $S_b(z)$ has a different circle of convergence, part of which is outside the circle of convergence of $S_a(z)$. In this way, $S_a(z)$ has been analytically continued to this new region. In other words, we can find Taylor expansions all around an isolated singularity z_0, as shown in Fig. 6.22.

Example 6.7.6. Analytically continue the series $S_a = \sum_{n=0}^{\infty} z^n$ to $2i$ with the help of another Taylor series.

The point b about which the second Taylor series is expanded has to be inside the circle of convergence of S_a, which turns out to be the unit circle. Its distance from $2i$ must be less than its distance to the singular point $z = 1$ of S_a. Both conditions are satisfied by the choice $b = 0.9i$.

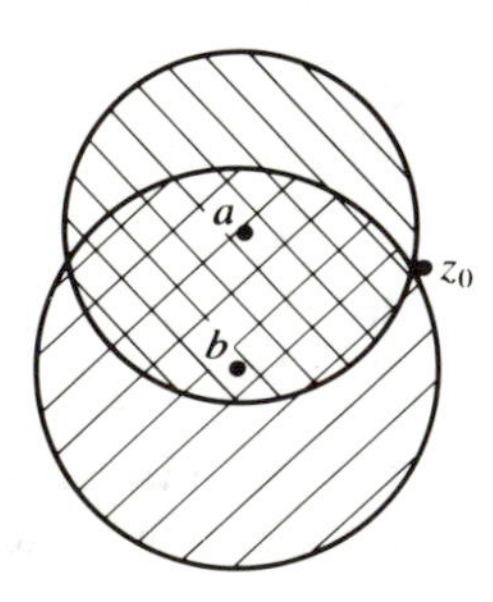

Fig. 6.22. The circles of convergence for two Taylor series $S_a(z)$ and $S_b(z)$ of the same function.

Hence

$$d_n = \frac{1}{n!}\sum_{m=n}^{\infty}\frac{m!}{(m-n)!}(0.9i)^{m-n}$$

$$S_b(2i) = \sum_{n=0}^{\infty} d_n(1.1i)^n.$$

The calculation needed is a little tedious and will be left to the reader. The answer should be the same as

$$\left(\frac{1}{1-z}\right)_{z=2i} = \frac{1}{1-2i} = \tfrac{1}{5}(1+2i).$$

Although in this case the answer can be obtained much more readily by another method, analytic continuation by Taylor series might have to be used if we do not have enough knowledge of S_a to do better.

On the other hand, a wall of singularities can completely obstruct analytic continuation across it. For example, the series

$$\begin{aligned} S(z) &= 1 + z + z^2 + z^4 + z^8 + \cdots \\ &= z + \sum_{n=0}^{\infty} z^{2^n} \end{aligned} \tag{6.47}$$

is convergent for $|z|<1$, but it has a singularity at $z=1$. Actually it has singularities everywhere on the unit circle $|z|=1$. (See Problem 6.7.4.) As a result, there is no analytic continuation across it.

Problems

6.7.1. Expand the following functions in Taylor series about z_0 and determine the radius of convergence in each case.

(a) $\dfrac{1}{z+2}$, $\quad z_0 = 4$;

(b) e^{az}, $\quad z_0 = 2i$;

(c) $\sin(z+i)$, $\quad z_0 = -2$;

(d) $\ln z$, $\quad z_0 = 2$.

6.7.2. Show that $e^z \sin z = \displaystyle\sum_{n=0}^{\infty}\frac{2^{n/2}\sin(n\pi/4)}{n!}z^n$.

6.7.3. Show that the following pairs of series are analytic continuations of each other:

(a) $\displaystyle\sum_{n=0}^{\infty}\frac{z^n}{3^{n+1}}$ and $\displaystyle\sum_{n=0}^{\infty}\frac{(z+a)^n}{(3+a)^{n+1}}$,

where $a \neq -3$ is any complex number;

(b) $\displaystyle\sum_{n=1}^{\infty}(-1)^{n+1}\frac{z^n}{n}$ and $\ln(1+a) + \displaystyle\sum_{n=1}^{\infty}\frac{(-1)^{n+1}}{n}\left(\frac{z-a}{1+a}\right)^n$.

6.7.4. Show that, if the series

$$f(z)=1+z+z^2+z^4+z^8+\cdots=1+\sum_{n=0}^{\infty} z^{2^n}=1+z+\sum_{n=0}^{\infty}(z^2)^{2^n}$$

has a singularity at $z=a$ on the unit circle, then it has singularities at the points $a^{1/2}$, $a^{1/4}$, $a^{1/8}$, etc. Hence there is a wall of singularities on the unit circle.

6.7.5. If the Taylor expansion

$$f(z)=\sum_{n=0}^{\infty} c_n z^n$$

converges for $|z|<R$, show that for any point z_0 inside a circle C of radius $r<R$, centered at the origin, the partial sum

$$s_n(z_0)=\sum_{k=0}^{n} c_k z_0^k$$

is given by the integral

$$s_n(z_0)=\frac{1}{2\pi i}\oint_C \frac{f(z)}{z^{n+1}}\frac{z^{n+1}-z_0^{n+1}}{z-z_0}\,dz.$$

6.8. Laurent series

It is even possible to make a power-series expansion of a function about a point $z=a$ at which the function has a kth-order pole, and even an essential isolated singularity. This is achieved by allowing negative, as well as positive, powers of $(z-a)$, as described by the following theorem:

Every function $f(z)$ analytic in an annulus

$$R_1<|z-a|<R_2$$

can be expanded in a series of positive and negative powers of $(z-a)$

$$\begin{aligned} f(z) &= \sum_{n=-\infty}^{\infty} c_n(z-a)^n \\ &= \sum_{n=0}^{\infty} c_n(z-a)^n+\sum_{n=-\infty}^{-1} c_n(z-a)^n. \end{aligned} \tag{6.48}$$

The series is called a *Laurent series*. It can be separated into two parts: the *regular part* made up of the positive powers of $(z-a)$, and the remaining *principal part* made up of negative powers. If the principle part vanishes, the Laurent series reduces to a Taylor series. Then $f(z)$ is analytic also at $z=a$.

The coefficient c_n of the Laurent series is given by the same formula

$$c_n = \frac{1}{2\pi i} \oint_{\Gamma} \frac{f(z)\,dz}{(z-a)^{n+1}}, \tag{6.49}$$

as that for the Taylor series, with the difference that Γ is a closed curve around $z = a$ lying entirely in the annulus.

It is instructive to derive this integral formula for c_n because of the insight it gives on the origin of the regular and principal parts of the Laurent expansion.

We start by observing that, according to the Cauchy integral formula, $f(z)$ at a point z in the annulus can be expressed as the integral

$$f(z) = \frac{1}{2\pi i} \oint_{\Gamma_1} \frac{f(z')\,dz'}{z'-z} + \frac{1}{2\pi i} \oint_{-\Gamma_2} \frac{f(z')\,dz'}{z'-z},$$

where the second term is needed because the region has a hole in it, as shown in Fig. 6.23. Note the negative sign of the path symbol $-\Gamma_2$ of the second term used to denote the negative (clockwise) direction of integration.

It is convenient to choose for Γ_1 and Γ_2 circles centered at a. then for a point z' on Γ_1

$$\frac{1}{z'-z} = \frac{1}{(z'-a)-(z-a)} = \sum_{n=0}^{\infty} \frac{(z-a)^n}{(z'-a)^{n+1}},$$

while for a point z' on Γ_2

$$\begin{aligned}\frac{1}{z'-z} &= -\frac{1}{z-z'} = -\frac{1}{(z-a)-(z'-a)} \\ &= -\sum_{n=0}^{\infty} \frac{(z'-a)^n}{(z-a)^{n+1}} = -\sum_{n=-\infty}^{-1} \frac{(z-a)^n}{(z'-a)^{n+1}}.\end{aligned}$$

Both geometrical series converge uniformly; they can be integrated term

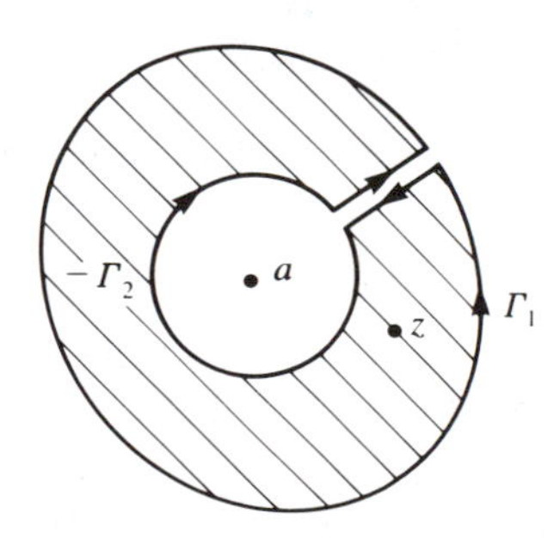

Fig. 6.23. The doubly closed curve specifying Γ_1 and Γ_2 originates as a single closed curve excluding the hole of the annulus.

by term. Hence

$$f(z)=\sum_{n=0}^{\infty}(z-a)^n\left(\frac{1}{2\pi i}\oint_{\Gamma_1}\frac{f(z')\,dz'}{(z'-a)^{n+1}}\right)$$
$$+\sum_{n=-\infty}^{-1}(z-a)^n\left(\frac{1}{2\pi i}\oint_{\Gamma_2}\frac{f(z')\,dz'}{(z'-a)^{n+1}}\right),$$

where we have used the negative sign in front of the second geometrical series to reverse the direction of integration over Γ_2. Since $f(z')$ is analytic in the annulus, the integration contour can be deformed into an arbitrary contour Γ around $z=a$ lying entirely in the annulus. Thus the coefficients of the Laurent series are just those of Eq. (6.49).

The coefficients for negative powers $n=-m$, $m\geqslant 1$

$$c_{-m}=\frac{1}{2\pi i}\oint_{\Gamma}f(z')(z'-a)^{m-1}\,dz'$$

have integrands with non-negative powers of $z'-a$. They vanish whenever $f(z')(z'-a)^{m-1}$ is analytic in the hole of the annulus. For example, if $f(z')$ has a kth-order pole at a, all negative-power coefficients vanish for $m\geqslant k+1$. A function $f(z')$ analytic also in the hole region will involve no negative power in its Laurent series, which is then just a Taylor series. On the other hand, if c_{-m} does not vanish when $m\to\infty$, we say that $f(z')$ has an essential isolated singularity at a.

Finally, we should note that if $f(z)$ has more than one singularity, it will have more than one annular region about any point a. The Laurent series is not the same in different annular regions.

Example 6.8.1. In making power-series expansions of

$$f(z)=\frac{1}{(z-2)(z+1)}\tag{6.50}$$

about the origin, we see first a circular region up to the first pole at $z=-1$, then an annular region up to $z=2$, and finally an annular region outisde $z=2$, as shown in Fig. 6.24.

By expanding $f(z)$ into the partial fractions

$$f(z)=\frac{1}{3}\left(\frac{1}{z-2}-\frac{1}{z+1}\right),$$

we see that convergent power-series expansions can be made separately for $(z-2)^{-1}$ and $(z+1)^{-1}$.

These are

$$\frac{1}{z-2}=-\frac{1}{2(1-z/2)}=-\frac{1}{2}\sum_{n=0}^{\infty}\left(\frac{z}{2}\right)^n,\qquad |z|<2,$$
$$\frac{1}{z-2}=\frac{1}{z(1-2/z)}=\sum_{n=0}^{\infty}\frac{2^n}{z^{n+1}},\qquad |z|>2;$$
$$\frac{1}{z+1}=\sum_{n=0}^{\infty}(-z)^n,\qquad |z|<1,$$

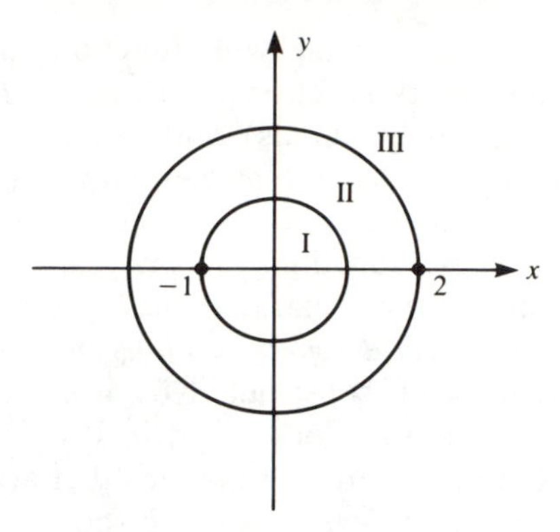

Fig. 6.24. Different regions for power-series expansions of $[(z-2)(z+1)]^{-1}$ about the origin.

and

$$\frac{1}{z+1}=\frac{1}{z(1+1/z)}=\sum_{n=0}^{\infty}\frac{(-1)^n}{z^{n+1}}, \qquad |z|>1.$$

They show that the convergent power-series expansion for region I of Fig. 6.24 is the Taylor series

$$\frac{1}{(z-2)(z+1)}=\frac{1}{3}\sum_{n=0}^{\infty}z^n\left(-\frac{1}{2^{n+1}}-(-1)^n\right), \qquad |z|<1.$$

For region II, it is a Taylor series for $(z-2)^{-1}$ and Laurent series for $(z+1)^{-1}$.

$$\frac{1}{(z-2)(z+1)}=\frac{1}{3}\sum_{n=0}^{\infty}\left(-\frac{z^n}{2^{n+1}}-\frac{(-1)^n}{z^{n+1}}\right), \qquad 1<|z|<2.$$

Finally, in region III, we have the simple Laurent series

$$\frac{1}{(z-2)(z+1)}=\frac{1}{3}\sum_{n=0}^{\infty}\frac{1}{z^{n+1}}[2^n-(-1)^n], \qquad |z|>2.$$

The last series can also be considered as a Taylor series in positive powers of z^{-1}. This is possible because $f(z)$ has no singularity for $|z|>2$.

In a similar way, we can verify that the same function has two different Laurent expansions about the point $z=-1$:

$$\frac{1}{(z-2)(z+1)}=\frac{1}{3}\left[-\frac{1}{3}\sum_{n=0}^{\infty}\left(\frac{z+1}{3}\right)^n-\frac{1}{z+1}\right], \qquad |z+1|<3, \quad (6.51a)$$

and

$$\frac{1}{(z-2)(z+1)}=\frac{1}{(z+1)^2}\sum_{n=0}^{\infty}\left(\frac{3}{z+1}\right)^n, \qquad |z+1|>3. \quad (6.51b)$$

Let us return for a moment to the Cauchy integral formula (6.49) for Laurent coefficients. Besides telling us all about these coefficients, it also has many practical applications. One good example concerns the definition of functions. We are familiar with functions defined as algebraic functions of more elementary functions, with functions defined as power series, both finite (i.e., polynomials) and infinite (Taylor or Laurent series), and with functions defined as derivatives of more elementary functions (e.g., the Rodrigues formula for $P_n(x)$ shown in Problem 6.5.3]. It is also useful to define functions as integrals of more elementary functions, since many of their mathematical properties become more transparent when such *integral representations* are used. Integral representations can sometimes be constructed by using the Cauchy integral formula (6.33) for analytic functions (e.g., in Problem 6.5.3). Equation (6.49) now extends the procedure to functions that are singular inside the path of integration, as the following example shows.

Example 6.8.2. An integral representation of Bessel functions $J_n(t)$ can be calculated from their generating function

$$\exp\left[\frac{1}{2}t\left(z-\frac{1}{z}\right)\right]=\sum_{n=-\infty}^{\infty} J_n(t)z^n=f(z), \qquad z\neq 0. \tag{6.52}$$

This shows that $J_n(t)$ are just the Laurent coefficients of $f(z)$ in a Laurent-series expansion about the origin. Hence from Eq. (6.49)

$$\begin{aligned} J_n(t) &= \frac{1}{2\pi i}\oint_C \frac{f(z)\,dz}{z^{n+1}} \\ &= \frac{1}{2\pi i}\oint_C \frac{\exp[\frac{1}{2}t(z-1/z)]}{z^{n+1}}\,dz. \end{aligned}$$

If we now take the closed path C to be the unit circle, then

$$z=e^{i\theta}, \qquad dz=e^{i\theta}i\,d\theta.$$

Hence

$$\begin{aligned} J_n(t) &= \frac{1}{2\pi i}\int_0^{2\pi} \frac{\exp[\frac{1}{2}t(e^{i\theta}-e^{-i\theta})]e^{i\theta}i\,d\theta}{e^{i\theta(n+1)}} \\ &= \frac{1}{2\pi}\int_0^{2\pi} e^{i(t\sin\theta-n\theta)}\,d\theta. \end{aligned} \tag{6.53}$$

This expression can be simplified further to other expressions. The interested reader is referred to Problem 6.8.2.

Problems

6.8.1. Obtain all possible power-series expansions of the following functions about the point a:

(a) $\dfrac{1}{(z-2)(z+1)}$ about $a=2$;

(b) $\dfrac{1}{(z-2)^2(z+1)^2}$ about $a=2$;

(c) $\dfrac{\sin z}{z^2}$ about $a = 0$;

(d) $\dfrac{\sin(z-1)}{z}$ about $a = 0$ and 1;

(e) $\dfrac{e^z}{(z-1)}$ about $a = 2$.

6.8.2. The Bessel function $J_n(t)$ can be defined in terms of the generating function (valid for $z \neq 0$)

$$\exp\left[\frac{1}{2}t\left(z - \frac{1}{z}\right)\right] = \sum_{n=-\infty}^{\infty} J_n(t)z^n.$$

Show that $J_n(t)$ has the following equivalent integral representations:

$$J_n(t) = \frac{1}{\pi}\int_0^{\pi} \cos(n\theta - t\sin\theta)\, d\theta$$

$$= \frac{i^{-n}}{\pi}\int_0^{\pi} e^{it\cos\phi} \cos n\phi\, d\phi.$$

[Hints: Show that

$$\int_0^{2\pi} \cos(n\theta - t\sin\theta)\, d\theta = 2\int_0^{\pi} \cos(n\theta - t\sin\theta)\, d\theta$$

$$\int_0^{2\pi} \sin(n\theta - t\sin\theta)\, d\theta = 0,$$

and use $\phi = \pi/2 - \theta$.]

6.9. Residues

Many applications of the theory of functions of a complex variable involve complex integrations over closed contours. In the simplest situation, the closed contour encloses *one* isolated singular point z_0 of the integrand $f(z)$. Let the integrand be expanded in a Laurent series about z_0

$$f(z) = \sum_{n=-\infty}^{\infty} c_n(z - z_0)^n.$$

Integrating term by term, we find that the nth power term gives

$$c_n \oint (z - z_0)^n\, dz = c_n \frac{(z - z_0)^{n+1}}{n+1}\bigg|_{z_1}^{z_1} = 0,$$

provided that $n \neq -1$. If $n = -1$, the result is

$$c_{-1} \oint \frac{dz}{z - z_0} = c_{-1} \oint \frac{re^{i\theta} i \, d\theta}{re^{i\theta}} = c_{-1} 2\pi i.$$

Hence

$$\frac{1}{2\pi i} \oint f(z) \, dz = c_{-1} = \mathrm{Res}[f(z_0)]. \tag{6.54}$$

This integral is called the *residue* $\mathrm{Res}[f(z_0)]$ of $f(z)$ at z_0. Its value is just the coefficient of $(z - z_0)^{-1}$ in the Laurent expansion of $f(z)$ about z_0.

If the contour encloses n isolated singularities $\{z_i\}$, it can be deformed into separate contours, each surrounding one isolated singularity, as shown in Fig. 6.25. Hence

$$\oint_C f(z) \, dz = 2\pi i \sum_{i=1}^{n} \mathrm{Res}[f(z_i)], \tag{6.55}$$

a result known as the *residue theorem.*

6.9.1. Calculation of residues

Given a Laurent series about z_0, the coefficient c_{-1}, which gives $\mathrm{Res}[f(z_0)]$, can be obtained by inspection. The resulting residue is for an integral over a closed contour surrounding z_0 in a region where the Laurent series converges. If the contour encloses more than one isolated singularity, c_{-1} gives the sum of the enclosed residues. To avoid possible ambiguities, we shall use the symbol $\mathrm{Res}[f(z_0)]$ to denote the contribution from z_0 alone. In other words, the Laurent series of interest is the one valid in the annular region between z_0 and the next-nearest singularity.

However, it is not necessary to perform a complete Laurent expansion in order to extract c_{-1}. This one coefficient in a series can be picked out with the help of the simple formula

$$\mathrm{Res}[f(z_0)] = \frac{1}{(m-1)!} \lim_{z \to z_0} \left(\frac{d^{m-1}}{dz^{m-1}} [(z - z_0)^m f(z)] \right) \tag{6.56}$$

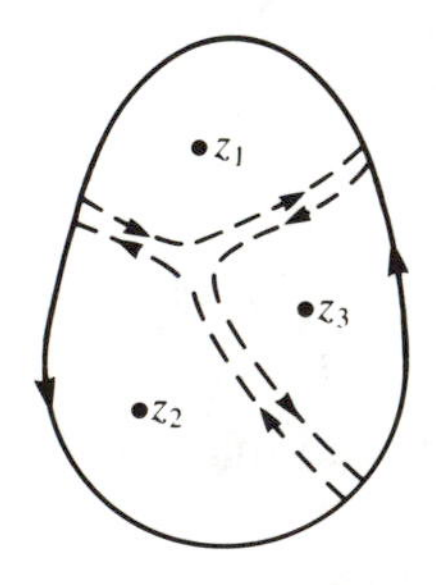

Fig. 6.25. Closing a contour around each isolated singularity.

if $f(z)$ has an mth-order pole at z_0. This formula can easily be derived by noting the $\text{Res}[f(z_0)]$ is also the coefficient d_{m-1} of the term $d_{m-1}(z - z_0)^{m-1}$ in the Taylor expansion about z_0 of the analytic function $(z - z_0)^m f(z)$.

In the case of a simple pole, Eq. (6.56) reads

$$\text{Res}[f(z_0)] = \lim_{z \to z_0} (z - z_0) f(z). \tag{6.57a}$$

A special case of this formula is worth knowing. If $f(z) = P(z)/Q(z)$, where $P(z)$ is analytic at z_0 and $Q(z)$ has a simple zero there, then

$$\text{Res}\left(\frac{P(z)}{Q(z)}\right)_{z_0} = \frac{P(z_0)}{\left(\dfrac{d}{dz} Q(z)\right)_{z_0}}. \tag{6.57b}$$

If z_0 is an essential isolated singularity, Eq. (6.56) cannot be used. Then a direct Laurent expansion is needed, although one does not have to go beyond the c_{-1} term.

Example 6.9.1. $f(z) = e^z/\sin z$ has a simple pole at $z = 0$. Therefore

$$\text{Res}[f(0)] = \lim_{z \to 0} \left(z \frac{e^z}{\sin z}\right) = 1.$$

Alternatively,

$$\text{Res}[f(0)] = \left.\frac{e^z}{\cos z}\right|_{z=0} = 1.$$

Example 6.9.2.

$$f(z) = \frac{1}{(z+1)(z-2)} \quad \text{at} \quad z = -1.$$

$$\text{Res}[f(-1)] = \left.\frac{1}{(z+1)+(z-2)}\right|_{z=-1} = -\tfrac{1}{3}.$$

This is just the coefficient c_{-1} of Eq. (6.51a). The coefficient $c_{-1} = 0$ of Eq. (6.51b) is actually the sum of residues at $z = -1$ and 2 enclosed by a closed contour around both poles of $f(z)$.

Example 6.9.3.

$$f(z) = \frac{e^{tz}}{(z+2)(z-1)^4} \quad \text{at} \quad z = 1.$$

There is a fourth-order pole at $z = 1$. Hence

$$\begin{aligned}
\mathrm{Res}[f(1)] &= \frac{1}{3!} \lim_{z\to 1} \left[\frac{d^3}{dz^3} \left(\frac{e^{tz}}{z+2} \right) \right] \\
&= \frac{1}{3!} \lim_{z\to 1} \left\{ \frac{1}{z+2} \frac{d^3}{dz^3} (e^{tz}) + 3 \left[\frac{d}{dz} \left(\frac{1}{z+2} \right) \frac{d^2}{dz^2} (e^{tz}) \right] \right. \\
&\quad \left. + 3 \left[\frac{d^2}{dz^2} \left(\frac{1}{z+2} \right) \right] \frac{d}{dz} (e^{tz}) + e^{tz} \frac{d^3}{dz^3} \left(\frac{1}{z+2} \right) \right\} \\
&= \frac{1}{3!} e^t \left(\frac{t^3}{3} + 3 \frac{-t^2}{3^2} + 3 \frac{2t}{3^3} - \frac{3!}{3^4} \right),
\end{aligned}$$

where we have used the *Leibnitz formula*

$$\frac{d^m}{dz^m} [f(z)g(z)] = \sum_{n=0}^{m} \binom{m}{n} \left(\frac{d^{m-n}}{dz^{m-n}} f(z) \right) \left(\frac{d^n}{dz^n} g(z) \right), \tag{6.58}$$

where

$$\binom{m}{n} = \frac{m!}{n!\,(m-n)!}$$

is a binomial coefficient.

Example 6.9.4. Locate the singularities of

$$f(z) = \frac{z^2 e^z}{1 + e^{2z}},$$

and evaluate their residues.

The singularities are located at the zeros of the $1 + e^{2z}$, where

$$e^{2z} = -1 = e^{i(2n+1)\pi}.$$

Hence

$$z_n = i(n + \tfrac{1}{2})\pi$$

on the imaginary axis are the singularities. Near z_n

$$1 + e^{2z} \simeq 1 - e^{2(z-z_n)} \simeq -2(z - z_n),$$

so that it is a simple pole. Therefore

$$\begin{aligned}
\mathrm{Res}[f(z_n)] &= \frac{z_n^2 e^{z_n}}{2e^{2z_n}} = \frac{i^2 (n + \frac{1}{2})^2 \pi^2}{2 \exp[i(n + \frac{1}{2})\pi]} \\
&= i(-1)^n (n + \tfrac{1}{2})^2 \frac{\pi^2}{2}.
\end{aligned}$$

Example 6.9.5. $f(z) = e^{1/z}$ at $z = 0$.

The function

$$e^{1/z} = 1 + \frac{1}{z} + \frac{1}{2!\,z^2} + \cdots$$

has an essential isolated singularity at $z = 0$. The Taylor expansion in powers of $1/z$ shows that $c_{-1} = 1$. Hence $\text{Res}(e^{1/z})$ at $z = 0$ is 1.

Problems

6.9.1. Locate the singularities of the following functions and evaluate their residues:

(a) $\sin \dfrac{1}{z}$;

(b) $\cos \dfrac{1}{z}$;

(c) $\dfrac{e^z}{(z-1)^3}$;

(d) $\dfrac{\sin z}{(z+2)^2}$;

(e) $\dfrac{\sin z}{z^2}$;

(f) $\dfrac{e^z}{\sin^2 z}$;

(g) $\dfrac{z+2}{z^2 - z - 1}$;

(h) $\dfrac{\cot z}{z^3}$.

6.10. Complex integration: Calculus of residues

The use of the residue theorem of Eq. (6.55) to evaluate complex integrals is called the *calculus of residues*. One serious restriction of this theorem is that the contour must be closed. Many integrals of practical interest involve integrations over open curves, for example, over parts of the real axis. Their paths of integration must be closed before the residue theorem can be applied. Indeed, our ability to evaluate such an integral depends crucially on how the contour is closed, since it requires knowledge of the additional contributions from the added parts of the closed contour.

A number of techniques are known for closing open contours. A common situation involves the integral along the real axis

$$I = \int_{-\infty}^{\infty} f(x)\, dx.$$

Suppose $f(z)$, the analytic continuation of $f(x)$, is a single-valued function so that there is no branch point or branch line. The following method can often be used.

Method 1. If $|zf(z)| \to 0$ as $|z| \to \infty$, the contour can be closed by a large semicircle either in the upper, or in the lower, half-plane (HP), as shown in Fig. 6.26. In either case

$$I = \pm 2\pi i \sum_{\text{enclosed}} \text{Res}(f) - I_R,$$

where

$$I_R = I(\text{semicircle}) = \lim_{R\to\infty} \int_{\theta=0}^{\theta=\pm\pi} [zf(z)] \frac{dz}{z}$$

$$\leq \lim_{R\to\infty} \pm\pi i \,\text{Max}[zf(z)] = 0.$$

Hence

$$I = \pm 2\pi i \sum_{\text{enclosed}} \text{Res}(f),$$

where the + (−) sign is used for closure in the upper (or lower) HP. The final result is the same in either case. This is because

$$\sum_{\text{all}} \text{Res}(f) = \sum_{\text{UHP}} \text{Res}(f) + \sum_{\text{LHP}} \text{Res}(f) = 0,$$

because there is no contribution from a large circle as $R \to \infty$.

Example 6.10.1.

$$I = \int_{-\infty}^{\infty} \frac{dx}{x^3 + i}.$$

We first verify that

$$\lim_{z\to\infty} \left| \frac{z}{z^3 + i} \right| = 0.$$

We also find that $f(z)$ has three simple poles at $z = i$, $e^{-i\pi/6}$ and $e^{i7\pi/6}$.

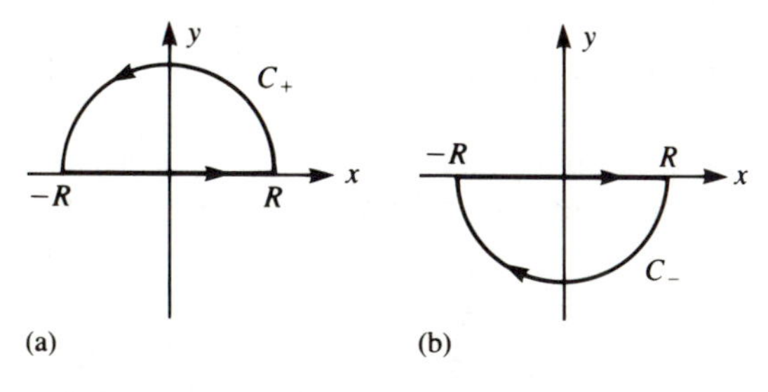

Fig. 6.26. Closure in the (a) upper or (b) lower half-plane if $|zf(z)| \to 0$ as $z \to \infty$.

Hence

$$I = \oint_{C_+} \frac{dz}{z^3 + i} = 2\pi i \,\mathrm{Res}[f(z = i)]$$

$$= 2\pi i \left(\frac{1}{3z^2}\right)_{z=i} = -\frac{2\pi}{3} i,$$

if the contour is closed in the upper HP. For closure in the lower HP,

$$I = -2\pi i\{\mathrm{Res}[f(e^{-i\pi/6})] + \mathrm{Res}[f(e^{i7\pi/6})]\}$$

$$= -\frac{2\pi i}{3}(e^{i\pi/3} + e^{-i\pi/3}) = -\frac{2\pi i}{3},$$

as expected.

Method 2. Complex Fourier integrals have the form

$$I = \int_{-\infty}^{\infty} g(x) e^{i\lambda x}\, dx, \tag{6.59}$$

where λ is a real constant. The integration contour of these integrals can often be closed with the help of *Jordan's Lemma*:

> If $\lim_{|z| \to \infty} g(z) = 0$, the integration contour in Eq. (6.59) can be closed by a large semicircle in the upper HP if $\lambda > 0$, and in the lower HP if $\lambda < 0$. The contribution on this semicircle will vanish, giving
>
> $$I = \pm 2\pi i \sum_{\text{enclosed}} \mathrm{Res}[g(z) e^{i\lambda z}]. \tag{6.60}$$

Since Fourier integrals are rather important in physics, it is instructive to prove this useful result. It is sufficient to consider the case of $\lambda > 0$. The idea is to show that the contribution I_R on the large semicircle of radius R in the upper HP does not exceed an upper bound which itself vanishes as $g(z) \to 0$.

On this large semicircle

$$z = Re^{i\theta} = R\cos\theta + iR\sin\theta, \qquad dz = Re^{i\theta} i\, d\theta.$$

Therefore

$$I_R = \int_0^{\pi} g(z) e^{\lambda R(i\cos\theta - \sin\theta)} Re^{i\theta} i\, d\theta.$$

If $\varepsilon = \mathrm{Max}\, g(z = Re^{i\theta})$ is the upper bound of g on this semicircle, then

$$I_R \leq \varepsilon R \int_0^{\pi} e^{-\lambda R \sin\theta}\, d\theta$$

$$= 2\varepsilon R \int_0^{\pi/2} e^{-\lambda R\sin\theta}\, d\theta,$$

since $\sin\theta$ is symmetric about $\pi/2$.

Now between 0 and $\pi/2$, $\sin\theta > 2\theta/\pi$, as shown in Fig. 6.27. Because of this inequality, the upper bound of the integral can be simplified to

$$\begin{aligned} |I_R| &\leqslant 2\varepsilon R \int_0^{\pi/2} e^{-\lambda R(2\theta/\pi)}\, d\theta \\ &= 2\varepsilon R \frac{\pi}{2\lambda R}(1 - e^{-\lambda R}) \\ &\leqslant \frac{\pi\varepsilon}{\lambda}. \end{aligned} \tag{6.61}$$

Hence

$$\lim_{R\to\infty} |I_R| \leqslant \frac{\pi}{\lambda} \lim_{|z|\to\infty} |g(z)| = 0.$$

With $\lambda > 0$, the contour cannot be closed in the lower HP, because the upper bound in Eq. (6.61) would then become an exponentially increasing function of R.

Example 6.10.2.

(a) $I = \int_{-\infty}^{\infty} [e^{i\lambda z}/(x + ia)]\, dx$, where λ and a are both real and positive. The contour can be closed in the upper HP, giving

$$I = \oint_{\text{UHP}} \frac{e^{i\lambda z}}{z + ia}\, dz = 2\pi i \sum_{\text{UHP}} \text{Res}(f) = 0, \tag{6.62a}$$

since the pole is in the LHP.

(b) $I = \int_{-\infty}^{\infty} [e^{-i\lambda z}/(z + ia)]\, dx$, where λ and a are both real and positive. The contour can be closed in the LHP, giving

$$I = \oint_{\text{LHP}} \frac{e^{-i\lambda z}}{z + ia}\, dz = -2\pi i e^{-a\lambda}. \tag{6.62b}$$

(c) Since the sine function can be written in terms of exponential functions, we can write

$$I = \int_{-\infty}^{\infty} \frac{\sin \lambda x}{x + ia}\, dx = \frac{1}{2i} \int_{-\infty}^{\infty} \frac{e^{i\lambda x} - e^{-i\lambda z}}{x + ia}\, dx.$$

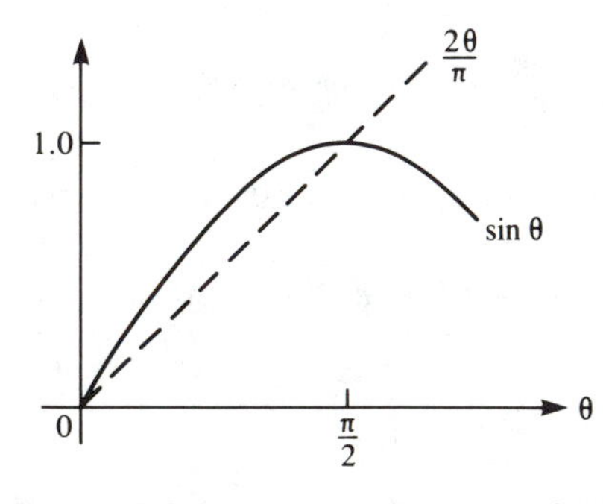

Fig. 6.27. Graphical demonstration of $\sin\theta > 2\theta/\pi$ for $0 < \theta < \pi/2$.

The two Eqs. (6.62) can now be used to give

$$I = -\frac{1}{2i}(-2\pi i e^{-a\lambda})$$
$$= \pi e^{-a\lambda}. \qquad (6.63)$$

The integral

$$\int_0^\infty \frac{\sin kx}{x}\,dx = \lim_{a\to 0}\frac{1}{2}\int_{-\infty}^{\infty}\frac{\sin kx}{x+ia}\,dx = \frac{\pi}{2}$$

is a special case of Eq. (6.63).

Method 3. If the integrand $f(z)$ along another path is proportional to its value along the original path, these two paths can often be joined to form a useful closed contour. This technique is best described with the help of examples.

Example 6.10.3.

$$I = \int_0^\infty \frac{dx}{x^2+a^2}.$$

$$I = \frac{1}{2}\int_{-\infty}^{\infty}\frac{dx}{x^2+a^2} = \frac{1}{2}\oint\frac{dz}{z^2+a^2} = \frac{\pi}{2a},$$

where closure with a large semicircle can be made in either the upper or lower HP. This is a rather trivial case, but it does illustrate the idea, which in this case consists of adding the contribution from the negative real axis.

Example 6.10.4.

$$I = \int_0^\infty \frac{dx}{1+x^3}.$$

The identity $z^3 = r^3$ holds not only along the positive real axis, but also along the lines $re^{i2\pi/3}$ and $re^{-i2\pi/3}$. Either of these can be added to the original path. The contour can then be closed with a large arc on which the contribution vanishes, as shown in Fig. 6.28. We then find that

$$\oint\frac{dz}{1+z^3} = I + I'$$

where I', the contribution along the radius $re^{-i2\pi/3}$, is

$$I' = \int_\infty^0 \frac{e^{-i2\pi/3}\,dr}{1+r^3} = -e^{-i2\pi/3}I.$$

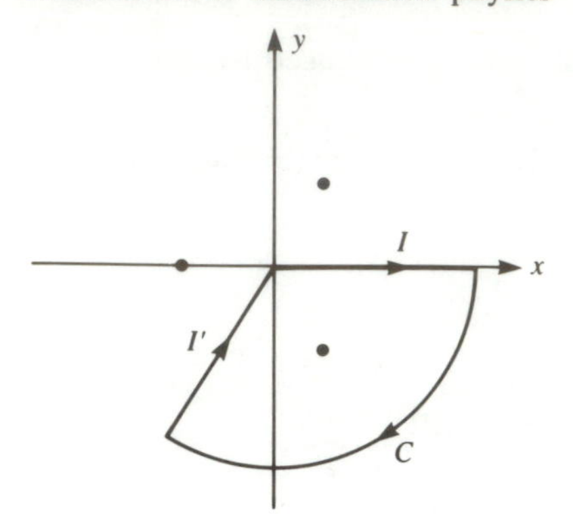

Fig. 6.28. Closing a contour along $re^{-i2\pi/3}$.

Hence

$$(1 - e^{-i2\pi/3})I = \oint_C \frac{dz}{1+z^3} = -\frac{2\pi i}{3(e^{-i\pi/3})^2},$$

$$\therefore I = \frac{2\pi i}{3} \frac{1}{e^{i2\pi 3} - e^{-i2\pi/3}} = \frac{\pi}{3\sin(2\pi/3)} = \frac{2\pi}{3\sqrt{3}}. \tag{6.64}$$

This gives a rather nice illustration of the technique.

Example 6.10.5.

$$I = \int_{-\infty}^{\infty} \frac{e^{ax}}{1+e^x}\,dx, \qquad 0 < a < 1.$$

The complex exponential function e^z is periodic in y with a period of 2π. Hence the analytic continuation of the integrand satisfies the periodicity property

$$f(x + i2\pi n) = e^{i2\pi n a} f(x). \tag{6.65}$$

This means that if C_n is the closed path shown in Fig. 6.29, then

$$\oint_{C_n} f(z)\,dz = I + I' + I_+ + I_-,$$

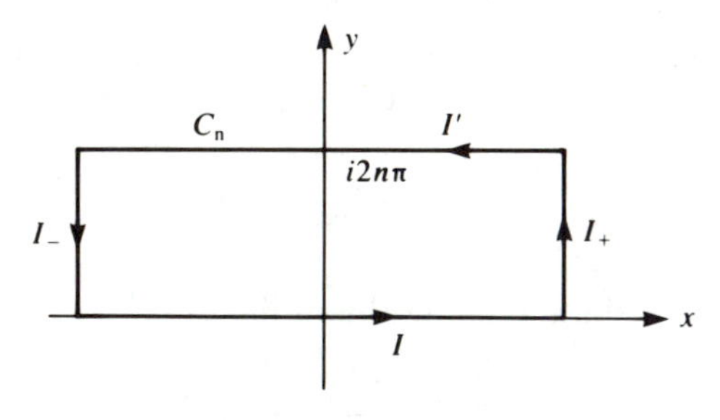

Fig. 6.29. A possible closed contour that makes use of the periodicity property Eq. (6.65).

where

$$I' = \int_{\infty+i2\pi n}^{-\infty+i2\pi n} f(x + i2\pi n)\, dx = -e^{i2\pi na} I.$$

On the two short sides of the rectangular contour

$$f(z) = \frac{e^{a(x+iy)}}{1+e^{x+iy}} = \begin{cases} e^{(a-1)(x+iy)}, & x \to \infty \\ e^{a(x+iy)}, & x \to -\infty. \end{cases}$$

Hence $f(z) \to 0$ as $|x| \to \infty$, and both I_+ and I_- vanish. This leaves

$$I = \frac{1}{1 - e^{i2\pi na}} \oint_{C_n} f(z)\, dz.$$

The residue theorem can now be applied. The integrand has poles at the solutions of the equation

$$e^{z_n} = -1 = e^{i(2n+1)\pi},$$

where n is any integer; that is,

$$z_n = i(2n+1)\pi.$$

Hence

$$I = \frac{2\pi i}{1 - e^{i2\pi na}} \sum_{m=0}^{n-1} (-1) e^{i(2m+1)\pi a}.$$

The result is actually independent of n. The simplest expression is obtained with $n = 1$, for which

$$I = \frac{2\pi i}{1 - e^{i2\pi a}} (-1) e^{i\pi a} = \frac{\pi}{\sin \pi a}.$$

The same result is obtained for other values of n, even in the limit $n \to \infty$. In this limit, the term $\exp(i2\pi na)$ in the denominator oscillates wildly, and may be replaced by its average value of 0. Hence

$$\begin{aligned} I &= -2\pi i \sum_{m=0}^{\infty} e^{i(2m+1)\pi a} \\ &= -2\pi i \frac{e^{i\pi a}}{1 - e^{i2\pi a}} = \frac{\pi}{\sin \pi a}. \end{aligned}$$

This result can also be obtained by simply closing the contour by a large semicircle at infinity in the upper (or lower) HP.

Example 6.10.6.

$$I = \int_0^\infty \frac{\sqrt{x}\, dx}{1 + x^2}.$$

On the negative real axis $f(-x) = if(x)$ is in one of its two branches. Hence

$$I = \frac{1}{1+i} \int_{-\infty}^{\infty} \frac{\sqrt{x}\, dx}{1+x^2} = \frac{1}{1+i} 2\pi i \operatorname{Res}[f(z = i)]$$

if we close in the upper HP. Therefore

$$I = \frac{1}{1+i} 2\pi i \left(\frac{\sqrt{z}}{2z}\right)_{z=i} = \frac{\pi}{\sqrt{2}}.$$

The presence of branch cuts may not be a hindrance to complex integration. They are often of help when the function just below the cut differs from, but is simply proportional to, the function just above the cut, as the following example shows.

Example 6.10.7. The integral in the previous example can be evaluated with the help of the closed contour shown in Fig. 6.30. The branch cut for $\sqrt{x}$ has been chosen to be $(0,\infty)$. Below the cut, but on the same branch, the square root changes sign

$$\sqrt{z(\text{below cut})} = -\sqrt{x}.$$

Hence the contribution from the path just below the cut is

$$I_- = \int_\infty^0 \frac{(-\sqrt{x})\,dx}{1+x^2} = I = I_+.$$

There is no contribution from either the small circle (I_0) or the large circle (I_∞) because

$$|zf(z)| \to \begin{cases} \left|\dfrac{z^{3/2}}{z^2}\right| \to 0, & |z| \to \infty \\ |z^{3/2}| \to 0, & |z| \to 0. \end{cases}$$

Hence

$$I = \frac{1}{2} \oint_C \frac{\sqrt{z}\,dz}{1+z^2} = \pi i \{\text{Res}[f(i)] + \text{Res}[f(-i)]\}$$

$$= \pi i \left(\frac{e^{i\pi/4}}{2i} + \frac{e^{i3\pi/4}}{-2i}\right) = \frac{\pi}{\sqrt{2}}.$$

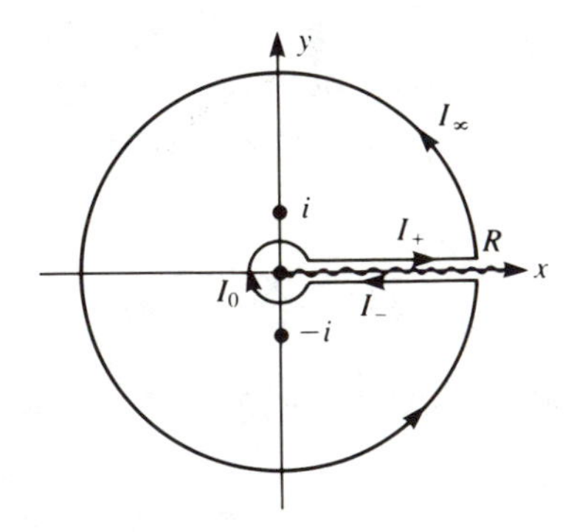

Fig. 6.30. A closed contour around a branch cut.

Method 4. A branch cut is so useful that it is sometimes manufactured for the purpose. For example, the integral

$$I=\int_0^\infty f(x)\,dx$$

can sometimes be evaluated by first considering another integral

$$J=\oint f(z)\ln z\,dz=J_+ + J_- + J_0 + J_\infty$$

along the same closed contour shown in Fig. 6.30. The contributions J_0 and J_∞ on the small and large circles vanish *if*

$$|zf(z)\ln z|\rightarrow 0$$

as $|z|\rightarrow 0$ and ∞, respectively. We are then left with the contributions above and below the branch cut:

$$J_+=\int_0^\infty f(x)\ln x\,dx \qquad \text{and} \qquad J_-=\int_\infty^0 f(x)(\ln x+2\pi i)\,dx.$$

The first term in J_- exactly cancels J_+, leaving

$$J=2\pi i\int_\infty^0 f(x)\,dx=-2\pi i I.$$

Hence

$$I=-\sum_{\text{enclosed}}\text{Res}[f(z)\ln(z)]. \tag{6.66}$$

Example 6.10.8.

$$I=\int_0^\infty \frac{dx}{1+x^3}=-\sum_{n=1}^{3}\text{Res}\left(\frac{\ln z}{1+z^3}\right)_{z=z_n},$$

where z_n are the three roots of $1+z^3=0$:

$$z_n=e^{i\pi/3}, \qquad e^{i\pi}=-1, \qquad e^{i5\pi/3}.$$

The residues turn out to be $(i\pi/9)\exp(-i2\pi/3)$, $i\pi/3$, and $(i5\pi/9)\exp(-i10\pi/3)$ for $n=1$, 2, and 3, respectively. Equation (6.66) then gives $2\pi/\sqrt{27}$, in agreement with Eq. (6.64). Exactly the same result is obtained if the cube roots $\exp(i7\pi/3)$, $\exp(i9\pi/3)$, and $\exp(i11\pi/3)$ are used. This choice actually means that $\ln z$ is evaluated on the next-higher sheet of the Riemann surface. The result is actually the same on every sheet of the Riemann surface. (See Problem 6.10.2.)

Method 5. Another common situation involves angular integrations of the type

$$I=\int_0^{2\pi} G(\sin\theta,\cos\theta)\,d\theta.$$

This can be converted into a contour integration over a unit circle with the help of the change of variables

$$z = e^{i\theta}, \qquad dz = e^{i\theta} i \, d\theta,$$

$$G\left(\frac{1}{2i}\left(z - \frac{1}{z}\right), \frac{1}{2}\left(z + \frac{1}{z}\right)\right) = f(z)$$

into

$$I = \oint f(z) \frac{dz}{iz}. \tag{6.67}$$

Example 6.10.9.

$$I = \int_0^{2\pi} \frac{d\theta}{a + b \sin \theta}, \qquad a > |b| > 0.$$

$$I = \oint \frac{dz}{iz} \frac{1}{a + b(1/2i)(z - 1/z)}$$

$$= \frac{2}{b} \oint \frac{dz}{(z^2 - 1) + 2i(a/b)z}.$$

The poles of the integrand are at

$$z_\pm = \left[-\frac{a}{b} \pm \left(\frac{a^2}{b^2} - 1\right)^{1/2}\right] i.$$

Since $z_+ z_- = 1$ and $a > |b|$, one of these poles is inside the unit circle, while the other is outside. For example, if $b > 0$, a/b is greater than 1. Hence

$$-iz_- = -\frac{a}{b} - \left(\frac{a^2}{b^2} - 1\right)^{1/2} < -\frac{a}{b} < -1;$$

that is, z_- is outside the unit circle, and z_+ is inside. In a similar way, if $b < 0$, z_- is inside. In either case, the result is

$$I = \frac{2}{b} 2\pi i \operatorname{Res}[f(z_{\text{inside}})]$$

$$= \frac{2\pi}{(a^2 - b^2)^{1/2}}.$$

Problems

6.10.1. Use contour integration to verify the following results:

(a) $\displaystyle\int_0^\infty \frac{dx}{x^4 + 1} = \frac{\pi}{2\sqrt{2}};$

(b) $\displaystyle\int_0^\infty \frac{dx}{(x^2+1)^2} = \frac{\pi}{4};$

(c) $\displaystyle\int_0^\infty \frac{dx}{(x^2+1)^2(x^2+4)} = -\frac{\pi}{18};$

(d) $\displaystyle\int_0^\infty \frac{dx}{(x^2+1)(x^2+4)^2} = \frac{5\pi}{288};$

(e) $\displaystyle\int_0^\infty \frac{x^6}{(x^4+1)^2}\,dx = \frac{3\pi\sqrt{2}}{16};$

(f) $\displaystyle\int_0^\infty \frac{\sin^2 x}{x^2}\,dx = \frac{\pi}{2};$

(g) $\displaystyle\int_0^\infty \frac{\cos \pi x}{x^2+1}\,dx = \frac{\pi}{2}e^{-\pi};$

(h) $\displaystyle\int_0^\infty \frac{x^{a-1}}{1+x}\,dx = \frac{\pi}{\sin a\pi},$ for $0<a<1.$

(i) $\displaystyle\int_0^\infty \frac{x^{a-1}\ln x}{1+x}\,dx = -\frac{\pi^2 \cos a\pi}{\sin^2 a\pi},$ for $0<a<1.$

(j) $\displaystyle\int_0^1 x^a(1-x)^{1-a}\,dx = \frac{\pi a(1-a)}{2\sin \pi a},$ for $-1<a<2.$

(k) $\displaystyle\int_0^{2\pi} \frac{\sin^2\theta\,d\theta}{a+b\cos\theta} = \frac{2\pi(a-\sqrt{a^2-b^2})}{b^2},$ for $0<b<a.$

(l) $\displaystyle\int_0^\infty \frac{x^{2m}}{1+x^{2n}}\,dx = \frac{\pi/2n}{\sin\{[(2m+1)/2n]\pi\}},$

for positive integers m and $n>m$.

6.10.2. Show that the integral of Example 6.10.8 has the same value on every sheet of the Riemann surface of $\ln z$.

6.11. Poles on the contour and Green functions

6.11.1. Simple pole on the contour

If there is a simple pole located right on a closed contour of integration, the integral is not uniquely defined. Two different results are possible dependent on whether the small semicircle around the pole z_0 on the contour is completed in the positive (counterclockwise) or negative direction:

$$I_\pm = \oint_\pm f(z)\,dz = ⨍ f(z)\,dz + I_\pm(z_0). \tag{6.68}$$

Here the symbol $⨍$ denotes the common contribution exclusive of the small semicircle around z_0, while $I_\pm(z_0)$ denotes the contribution from the semicircle:

$$I_\pm(z_0) = \lim_{z\to z_0} \int_\pm [(z-z_0)f(z)] \frac{dz}{z-z_0}$$
$$= \text{Res}[f(z_0)] \int i\,d\theta = \pm\pi i \,\text{Res}[f(z_0)].$$

The final result is obtained by noting that the polar angle θ of $z - z_0$ goes through a range of $\pm\pi$ on the semicircle. The situation is illustrated in Fig. 6.31.

The unique quantity is not $I_\pm$, but the *Cauchy principal value*

$$⨍ f(z)\,dz = P\int f(z)\,dz = I_\pm - I_\pm(z_0)$$
$$= 2\pi i\left(\sum_{\text{enclosed}} \text{Res}(f) + \frac{1}{2}\sum_{\substack{\text{on}\\ \text{contour}}} \text{Res}(f)\right). \tag{6.69}$$

We should note that the on-contour contribution is either $(2\pi i - \pi i)$ Res or $(0 + \pi i)$ Res for either the positive or negative semicircle at z_0. Thus in a sense the on-contour pole is counted as half inside and half outside the contour for the principal-value integral.

Example 6.11.1. Show that

$$⨍_{-\infty}^{\infty} \frac{e^{ikx}}{x-x_0}\,dx = \text{sgn}\,(k)\pi i e^{ikx_0},$$

where sgn(k) is the sign of k.

The contour can be closed with the help of an infinite semicircle into

$$I = ⨍ \frac{e^{ikz}}{z-x_0}\,dx.$$

If $k > 0$, Jordan's lemma requires closure in the UHP. Hence

$$I = \pi i \,\text{Res}\left(\frac{e^{ikz}}{z-x_0}\right) = \pi i e^{ikx_0}.$$

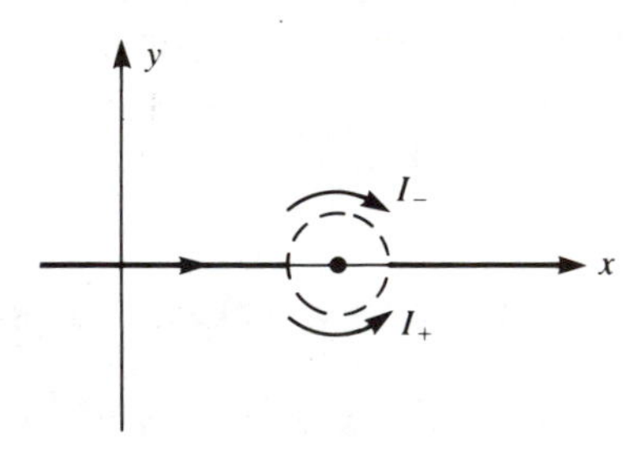

Fig. 6.31. Two possible paths around a pole on the contour of integration.

If $k<0$, we must close in the LHP with the result

$$I=-\pi i\ \mathrm{Res}\left(\frac{e^{ikz}}{z-x_0}\right)=-\pi i e^{ikx_0}.$$

The special case

$$⨍_{-\infty}^{\infty}\frac{e^{ix}}{x}\,dx=\pi i$$

yields the results

$$\int_{-\infty}^{\infty}\frac{\cos x}{x}\,dx=0$$

and

$$\int_{-\infty}^{\infty}\frac{\sin x}{x}\,dx=\pi.$$

The former integral vanishes because the integrand is odd. The latter agrees with Eq. (6.63).

6.11.2. Moving a pole off the contour

The integral I_+ (or I_-) also gives the result when the simple pole on the contour is moved to inside (or outside) the enclosed region. If the simple pole is originally on the real axis, the result can be written explicitly as

$$I_{\pm}=\lim_{\varepsilon\to 0}\int\frac{g(x)}{x-(x_0\pm i\varepsilon)}\,dx=⨍\frac{g(x)}{x-x_0}\,dx\pm i\pi g(x_0) \qquad (6.70)$$

if we substitute $f(x)=g(x)/[x-(x_0\pm i\varepsilon)]$ in Eq. (6.68).

Equation (6.70) appears frequently in physics, where all observables are real quantities. For this reason, it is instructive to derive it by another procedure that illuminates the origin of the on-contour contribution. We start with the identity

$$\frac{1}{x-(x_0\pm i\varepsilon)}=\frac{x-x_0}{(x-x_0)^2+\varepsilon^2}\pm i\frac{\varepsilon}{(x-x_0)^2+\varepsilon^2}.$$

In the limit $\varepsilon\to 0$

$$\lim_{\varepsilon\to 0}\frac{1}{x-(x_0\pm i\varepsilon)}=\frac{P}{x-x_0}\pm i\pi D(x-x_0),$$

where the quantity

$$D(x-x_0)=\lim_{\varepsilon\to 0}\frac{\varepsilon/\pi}{(x-x_0)^2+\varepsilon^2}$$

has the properties that

$$D(x-x_0)\propto\begin{cases}\varepsilon\to 0, & x\neq x_0\\ 1/\varepsilon\to\infty, & x=x_0,\end{cases}$$

and

$$\int_{-\infty}^{\infty} D(x-x_0)\,dx = \frac{\varepsilon}{\pi}\oint \frac{dz}{z^2+\varepsilon^2} = 1,$$

a result obtained by closing the contour in either the upper or the lower HP. This is just the properties of a Dirac δ function first studied in Chapter 3. Hence

$$D(x-x_0) = \delta(x-x_0),$$

and

$$\lim_{\varepsilon\to 0} \frac{1}{x-(x_0 \pm i\varepsilon)} = \frac{P}{x-x_0} \pm i\pi\,\delta(x-x_0). \tag{6.71}$$

The fact that the value of a contour integral is changed on moving a pole off the contour has some interesting consequences in the theory of certain inhomogeneous linear differential equations.

6.11.3. Inhomogeneous linear differential equations

Consider, for example, the one-dimensional wave equation

$$\left(\frac{\partial^2}{\partial x^2} - \frac{1}{c^2}\frac{\partial^2}{\partial t^2}\right)u(x,t) = 0. \tag{6.72}$$

By separating variables, we can write the solution in the form

$$u(x,t) = f(x)e^{-i\omega_0 t}, \tag{6.73}$$

provided that the spatial factor $f(x)$ (called a time-independent wave function) satisfies the one-dimensional Helmholtz equation

$$\left(\frac{d^2}{dx^2} + k_0^2\right)f(x) = 0, \qquad \text{where} \qquad k_0 = \omega_0/c. \tag{6.74}$$

Equation (6.72) or (6.74) describes a wave propagating in free space. If we would like to include the source of the wave disturbance in our description, we must solve an inhomogeneous differential equation such as

$$\left(\frac{d^2}{dx^2} + k_0^2\right)f(x) = \rho(x), \tag{6.75}$$

with a nonzero *source function* $\rho(x)$. A rather similar situation occurs in the classical mechanics of a point mass m moving under the influence of a linear restoring force and an additional driving force

$$\left(\frac{d^2}{dt^2} + \omega_0^2\right)x(t) = \frac{F_d(t)}{m}. \tag{6.76}$$

In this case, we may refer to the right-hand-side term as the *driving term*.

6.11.4. *Green functions*

In solving an inhomogeneous linear DE, it is useful to write it compactly in the form

$$\mathscr{L}(x)f(x) = \rho(x), \tag{6.77}$$

where $\mathscr{L}(x)$ is a linear differential operator such as $d^2/dx^2 + k_0^2$ in Eq. (6.75). We now show that the solution $f(x)$ has the simple integral representation

$$f(x) = \int_{-\infty}^{\infty} G(x - x_0)\rho(x_0)\, dx_0, \tag{6.78}$$

where $G(x - x_0)$ satisfies the inhomogeneous linear DE

$$\mathscr{L}(x)G(x - x_0) = \delta(x - x_0). \tag{6.79}$$

This is easily done by direct substitution into Eq. (6.77):

$$\begin{aligned}\mathscr{L}(x)\int_{-\infty}^{\infty} G(x - x_0)\rho(x_0)\, dx_0 &= \int_{-\infty}^{\infty} \mathscr{L}(x)G(x - x_0)\rho(x_0)\, dx_0 \\ &= \int_{-\infty}^{\infty} \delta(x - x_0)\, \rho(x_0)\, dx_0 \\ &= \rho(x).\end{aligned}$$

The function $G(x - x_0)$ is called a *Green function*. It is a solution arising from a point source at x_0, or an impulsive driving force if we are dealing with an equation of motion in classical mechanics such as Eq. (6.76). According to Eq. (6.78), it is that part of the solution $f(x)$ that is completely independent of the source function $\rho(x)$. As a result, the same Green function can be used with different sources to generate different solutions by superposing point-source solutions in different ways, as shown in Eq. (6.78). This nice feature is a consequence of the linearity of the differential equation.

A unique solution of a differential equation, whether homogeneous or inhomogeneous, requires the specification of a number of boundary (or initial) conditions. For example, a second-order DE requires two boundary conditions such as the value and derivative of the solution at one point. Each choice of these two boundary conditions gives rise to a unique solution, and hence also to a unique Green function. However, all these solutions can be expressed in terms of a particular solution and a linear combination of the two linearly independent solutions of the homogeneous equation, as shown explicitly in Section 4.2. As a result, there are only two linearly independent Green functions for the inhomogeneous equation.

To illustrate these remarks, let us calculate the Green functions for the inhomogeneous Helmholtz equation

$$\left(\frac{d^2}{dx^2} + k_0^2\right)G(x - x_0) = \delta(x - x_0).$$

We can first solve this equation with the point source at the origin (i.e., $x_0 = 0$):

$$\left(\frac{d^2}{dx^2} + k_0^2\right)G(x) = \delta(x). \tag{6.80}$$

The general solution is then obtained afterwards by simply substituting $x - x_0$ for x in $G(x)$. This method works because the differential operator involved is invariant under a translation in x; that is,

$$\mathscr{L}(x - x_0) = \frac{d^2}{d(x - x_0)^2} + k_0^2 = \mathscr{L}(x).$$

Equation (6.80) can be handled conveniently by the method of Fourier transformation, which transforms the differential equation into an algebraic equation. In our case, it works in the following way:

We start by recalling the "derivative" property of a Fourier transform

$$\mathscr{F}\left\{\frac{d^2}{dx^2}G(x - x_0)\right\} = (ik)^2\mathscr{F}\{G(x - x_0)\}$$

tabulated in Table 3.1. Hence the left-hand side of Eq. (6.80) transforms as

$$\mathscr{F}\left\{\left(\frac{d^2}{dx^2} + k_0^2\right)G(x)\right\} = (-k^2 + k_0^2)\tilde{G}(k),$$

while the right-hand side becomes

$$\mathscr{F}\{\delta(x)\} = \frac{1}{\sqrt{2\pi}}.$$

Therefore

$$\tilde{G}(k) = -\frac{1}{\sqrt{2\pi}}\frac{1}{k^2 - k_0^2}.$$

The Green function $G(x)$ is the inverse Fourier transform of this:

$$\begin{aligned} G(x) &= \frac{1}{\sqrt{2\pi}}\int_{-\infty}^{\infty} \tilde{G}(k)e^{ikx}\,dk \\ &= -\frac{1}{2\pi}\int_{-\infty}^{\infty} \frac{e^{ikx}}{k^2 - k_0^2}\,dk. \end{aligned} \tag{6.81}$$

The integrand in Eq. (6.81) has poles at $k = \pm k_0$ on the real axis. The integral is therefore of a type discussed in this section. The contour can be closed in the upper (or lower) HP if $x > 0$ (or $x < 0$) with the help of Jordan's lemma, since for complex k

$$\lim_{|k|\to\infty} \left|\frac{1}{k^2 - k_0^2}\right| = 0.$$

If we evaluate the integral in Eq. (6.81) as a Cauchy principal value, we get the *principal-value Green function*:

$$G_P(x) = -\frac{1}{2\pi} ⨍ \frac{e^{ikx}}{k^2 - k_0^2}\, dk$$
$$= -\frac{i}{2}\,\mathrm{sgn}(x)\,\{\mathrm{Res}[\tilde{G}(k_0)] + \mathrm{Res}[\tilde{G}(-k_0)]\}.$$

Here sgn(x) appears because we must close the contour in the upper HP if $x > 0$, and in the lower HP if $x < 0$. The result is

$$G_P(x) = -\frac{i}{2}\,\mathrm{sgn}(x)\left(\frac{e^{ik_0x}}{2k_0} - \frac{e^{-ik_0x}}{2k_0}\right) = \frac{1}{2k_0}\sin k_0\,|x|. \tag{6.82}$$

The function

$$G_P(x - x_0) = \frac{1}{2k_0}\sin k_0\,|x - x_0| \tag{6.83}$$

is sketched in Fig. 6.32. We see that the function is continuous but not smooth at the position $x = x_0$ of its point source. Rather its slope is discontinuous at $x = x_0$, increasing by exactly 1 unit as x increases through x_0. Without this discontinuity the point-source term in the DE cannot be reproduced.

This important feature of the slope of the Green function of a second-order linear DE can be seen directly from Eq. (6.80) even before $G(x)$ itself is calculated:

$$1 = \int_{-\varepsilon}^{\varepsilon} \delta(x)\, dx = \int_{-\varepsilon}^{\varepsilon} \left(\frac{d^2}{dx^2} + k_0^2\right) G(x)\, dx$$
$$= \frac{d}{dx} G(x)\bigg|_{-\varepsilon}^{\varepsilon}. \tag{6.84}$$

Before discussing the boundary conditions under which $G_P(x)$ appears, let us first calculate two other Green functions obtained by giving k_0 a

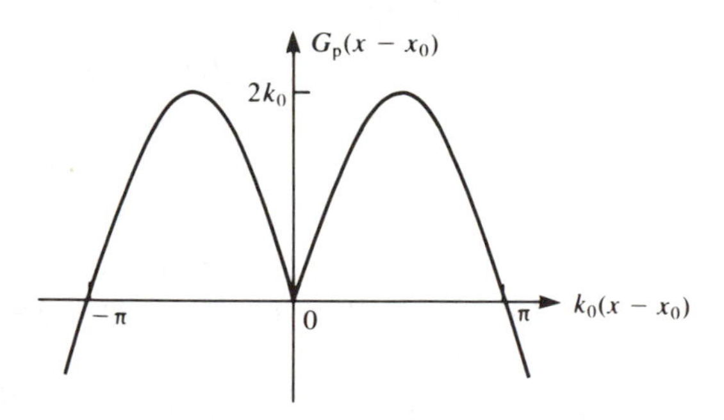

Fig. 6.32. The principal-value Green function for the one-dimensional Helmholtz equation.

small positive or negative imaginary part:

$$k_\pm = k_0 \pm i\varepsilon.$$

The resulting Green functions

$$G_\pm(x) = -\frac{1}{2\pi}\int_{-\infty}^{\infty} \frac{e^{ikx}}{k^2 - k_\pm^2}\,dk$$

can be evaluated by noting that G_+ involves a pole k_+ in the UHP and a pole $-k_+$ in the LHP, while G_- involves a pole k_- in the LHP and a pole $-k_-$ in the UHP. As a result,

$$G_\pm(x) = \pm\frac{1}{2ik_\pm}\,e^{\pm ik_\pm |x|}.$$

In the limit $\varepsilon \to 0$,

$$G_\pm(x) = \pm\frac{1}{2ik_0}\,e^{\pm ik_0 |x|} \tag{6.85}$$

will thus satisfy Eq. (6.79), including the requirement that its slope has a unit discontinuity at $x = 0$. Both these Green functions differ from $G_P(x)$.

The physical situation described by these Green functions can be understood readily by first constructing the actual solution for a given source function $\rho(x)$:

$$\begin{aligned} f_+(x) &= \int_{-\infty}^{\infty} G_+(x - x')\rho(x')\,dx \\ &= \frac{1}{2ik_0}\left(\int_{-\infty}^{x} e^{ik_0(x-x')}\rho(x')\,dx'\right. \\ &\quad \left. + \int_{x}^{\infty} e^{ik_0(x'-x)}\rho(x')\,dx'\right). \end{aligned} \tag{6.86}$$

For example, if $\rho(x')$ is localized near the origin in the sense that

$$\rho(x') = 0 \qquad \text{if} \qquad |x'| > X,$$

then

$$f_+(x) = \frac{1}{2ik_0} \times \begin{cases} e^{ik_0 x}\displaystyle\int_{-X}^{X} e^{-ik_0 x'}\rho(x)\,dx' & x > X \\ e^{-ik_0 x}\displaystyle\int_{-X}^{X} e^{ik_0 x'}\rho(x)\,dx', & x < -X. \end{cases} \tag{6.87}$$

To get a physical picture, we go all the way back to the time-dependent wave function $u(x,t)$ of Eq. (6.73) by reinserting the time dependence $\exp(-ik_0 ct)$ to obtain

$$u_+(x,t) = \text{const} \times \begin{cases} e^{ik_0(x-ct)}, & x > X \\ e^{-ik_0(x+ct)}, & x < -X. \end{cases} \tag{6.88}$$

This result shows that a point of constant phase of the wave satisfies the equation $x - ct = \text{const}$ for $x > X$, and $x + ct = \text{const}$ for $x < -X$. The wave is therefore traveling with the velocity

$$v = \frac{dx}{dt} = \begin{cases} c, & x > X \\ -c, & x < -X. \end{cases}$$

That is, it is moving away from the source (i.e., the inhomogeneity term) near the origin. We call this solution the *outgoing-wave solution*, and the Green function $G_+(x - x')$ the *outgoing-wave Green function*.

In a similar way, the solution $f_-(x)$ obtainable from $G_-(x - x')$ is proportional to $\exp(-ik_0x)$ for $x > X$, and to $\exp(ik_0x)$ for $x < -X$. When used with the time factor $\exp(-ik_0ct)$, it describes a wave traveling toward the source. It therefore gives an *ingoing-wave solution*, and $G_-(x - x')$ is the *ingoing-wave Green function*.

Why do we use a time factor of $\exp(-i\omega t)$ and not $\exp(i\omega t)$? It is just a convention, but one that is useful in special relativity, where the combination $kx - \omega t$ appears naturally. If we had used $\exp(i\omega t)$ instead, the solution $f_-(x)$ would have been the outgoing-wave solution.

What about the principal-wave Green function? According to Eqs. (6.82) and (6.85), it is half ingoing wave and half outgoing wave. It is a *standing-wave Green function*, and it gives rise to a *standing-wave solution*.

The two-traveling-wave Green functions $G_\pm$ are linearly independent of each other in the sense that one cannot be expressed in terms of the other. On the other hand, the standing-wave Green function G_P can be expressed in terms of $G_\pm$, and is therefore linearly dependent on them. This is not unexpected, since a second-order linear DE has only two linearly independent solutions. As a result, any solution or Green function can be expressed as a linear combination of two linearly independent solutions or Green functions. For wave motion, it is convenient to express an arbitrary Green function in terms of the traveling-wave Green functions

$$G(x - x') = aG_+(x - x') + (1 - a)G_-(x - x'), \tag{6.89}$$

because of the simplicity of the physical situations they describe. Note that this $G(x - x')$ also has one unit step discontinuity in its slope at $x = x'$.

Other choices of the two linearly independent Green functions might be more useful for other differential equations. These correspond to other ways of moving the poles off the contour. For example, the DE

$$\left(\frac{d^2}{dt^2} + \omega_0^2\right)G(t - t') = \delta(t - t') \tag{6.90}$$

satisfied by a mechanical system under an impulsive driving force is mathematically identical to Eq. (6.80). As a result, it can be written in a form analogous to Eq. (6.81):

$$G(t-t') = -\frac{1}{2\pi}\int_{-\infty}^{\infty} \frac{e^{-i\omega(t-t')}}{\omega^2 - \omega_0^2}\,d\omega. \tag{6.91}$$

Equation (6.91) is actually identical to Eq. (6.81) except for the use of $-i$ instead of i in the Fourier transform. This is the result of the convention mentioned earlier that is designed to lead naturally to the combination $kx - \omega t$ of significance in special relativity. If Eq. (6.91) is evaluated after moving *both* poles to just below the real axis, we get

$$G_r(t-t') = \begin{cases} \dfrac{1}{\omega_0}\sin \omega_0(t-t'), & t > t' \\ 0 & t < t'. \end{cases} \tag{6.92}$$

This is called a *retarded Green function,* because it describes a response of the system occurring *after* the disturbance at time t'.

In physics, we insist that the results of a disturbance cannot be *seen* before the disturbance itself takes place. This common-sense requirement that effects always follow causes is referred to as the *principle of causality*. Causality is guaranteed if the transformed Green function, in our case

$$\tilde{G}_r(\omega) = -\frac{1}{\sqrt{2\pi}}\frac{1}{(\omega-\omega_0+i\varepsilon)(\omega+\omega_0+i\varepsilon)} \tag{6.93}$$

is analytic in the upper half ω plane. Of course, $G(\omega)$ should *not* be analytic in the lower half ω plane also, otherwise it would simply be a constant, which gives rise to no response at all for $t > t'$. In other words, measurable responses of a physical system to disturbances come from singularities in the lower half ω plane.

There is more than meets the eye, however. The physical or retarded Green function is only one of the two linearly independent Green functions needed to provide a *complete* solution of the second-order linear DE (6.90). The second Green function may be taken to be the *advanced Green function* obtained from Eq. (6.91) by moving both poles to above the real axis. The result is:

$$G_a(t-t') = \begin{cases} 0, & t > t' \\ -\dfrac{1}{\omega_0}\sin \omega_0(t-t'), & t < t'. \end{cases} \tag{6.94}$$

It describes a science-fictional situation in which the response takes place entirely *before* the disturbance. We may call this an unphysical situation that gives rise to unphysical solutions. Like a movie run backwards, the scenario described by G_a is unreal, in the sense that we can never see it in real life. It is nevertheless important in helping to provide a mathematically complete description of the system. For this reason, the properties of a system in unphysical regions are also of interest in the mathematical description of a physical theory.

Problems

6.11.1. Use contour integration to verify the following principal-value integrals:

(a) $\displaystyle ⨍_{-\infty}^{\infty} \frac{\cos x}{x-a}\,dx = -\pi \sin a;$

(b) $\displaystyle ⨍_{0}^{\infty} \frac{\cos x}{x^2-a^2}\,dx = -\frac{\pi}{2a} \sin a.$

6.11.2. Show that the Green function for the first-order DE

$$\frac{d}{dx}G(x) + aG(x) = \delta(x)$$

is

$$G(x) = \begin{cases} 0, & x<0 \\ e^{-ax}, & x>0. \end{cases}$$

6.11.3. Show that the retarded Green function for the driven oscillator equation with dissipation

$$\frac{d^2}{dt^2}G(t) + 2\beta\frac{d}{dt}G(t) + \omega_0^2 G(t) = \delta(t)$$

is

$$G_r(t) = \begin{cases} 0, & t<0 \\ \omega_1^{-1}e^{-\beta t}\sin\omega_1 t, & t>0, \end{cases}$$

where

$$\omega_1 = (\omega_0^2 - \beta^2)^{1/2}.$$

6.12. Laplace transforms

A Fourier transform

$$\mathscr{F}\{f(x)\} = \frac{1}{\sqrt{2\pi}}\int_{-\infty}^{\infty} f(x)e^{-ikx}\,dx$$

may not converge because $f(x)$ does not vanish sufficiently rapidly at infinity. This means that often a problem in which $f(x)$ appears cannot be treated by the method of Fourier transforms.

If $f(x)$ is such that it does not diverge more rapidly than $\exp(\sigma_0 x)$, with $\sigma_0 \geqslant 0$, as $x\to\infty$, then it can be rendered well behaved by adding a *convergence factor* $\exp(-\sigma x)$, $\sigma > \sigma_0$. Of course, this factor will diverge for negative values of x, so we must restrict ourselves to $x>0$. This restriction can be achieved by the multiplication with a *Heaviside step function*

$$\Theta(x)=\begin{cases}0, & x<0\\ 1, & x>0\\ \frac{1}{2}, & x=0.\end{cases} \tag{6.95}$$

The Fourier transform of the modified function

$$\begin{aligned}\tilde{f}(k)&=\int_{-\infty}^{\infty}[f(x)e^{-\sigma x}\Theta(x)]e^{-ikx}\,dx\\ &=\int_{0}^{\infty}f(x)e^{-(\sigma+ik)x}\,dx\\ &=g(s=\sigma+ik)\end{aligned} \tag{6.96}$$

is then well defined.

The transform (6.86) can be inverted with the help of the Fourier inversion formula

$$f(x)e^{-\sigma x}\Theta(x)=\frac{1}{2\pi}\int_{-\infty}^{\infty}\tilde{f}(k)e^{ikx}\,dk.$$

Moving the factor $e^{-\sigma x}$ to the right, we find

$$f(x)\Theta(x)=\frac{1}{2\pi i}\int_{\sigma-i\infty}^{\sigma+i\infty}g(s)e^{sx}\,ds, \qquad s=\sigma+ik. \tag{6.97}$$

We call $g(s)$ the *Laplace transform* (LT) of $f(x)$, and Eq. (6.97) the *Laplace inversion formula.*

We have assumed that the function $f(x)$ does not have other divergence problems elsewhere. Otherwise its Laplace transform may not exist. For example, the integral

$$\int_{0}^{\infty}t^{-n}\,dt$$

diverges at $t=0$ if $n\geqslant 1$. Hence t^{-n}, $n\geqslant 1$, does not have an LT.

We shall use the notation

$$\mathscr{L}\{f(t)\}=\int_{0}^{\infty}e^{-st}f(t)\,dt \tag{6.98}$$

for an LT.

Example 6.12.1. The calculation of Laplace transforms is illustrated below with the help of a number of simple examples.

(a) $\mathscr{L}\{1\} = \int_0^\infty e^{-st}\,dt = \frac{1}{s}.$

(b) $\mathscr{L}\{t\} = \int_0^\infty te^{-st}\,dt = -\frac{d}{ds}\int_0^\infty e^{-st}\,dt$

$$= \frac{1}{s^2}.$$

(c) $\mathscr{L}\{e^{i\lambda t}\} = \int_0^\infty e^{i\lambda t - st}\,dt = \frac{1}{s - i\lambda}.$ (6.99)

6.12.1. *Properties of Laplace transforms*

The basic properties of Laplace transforms are summarized in Table 6.1. Both translation and attenuation properties can be derived by a change of variables. In translating the argument of f from t to $t-a$, we must require that $a>0$. This is because if $a=-|a|<0$, the beginning of the time integration at $t=0$ would correspond to the point $t-a=-a=|a|$, as shown in Fig. 6.33. As a result, that part of $f(t-a)$ between 0 and $|a|$ would be excluded from the Laplace transform. The remaining part then differs from the original function $f(t)$.

The derivative property arises from an integration by parts:

$$\int_0^\infty e^{-st}\frac{d}{dt}f(t)\,dt = e^{-st}f(t)|_0^\infty - \int_0^\infty \left(\frac{d}{dt}e^{-st}\right)f(t)\,dt$$

$$= -f(0) + sg(s). \tag{6.100}$$

Repeated applications of Eq. (6.100) lead to the formulas for the higher derivatives.

Table 6.1. Basic properties of Laplace transforms

Property	$F(t)$	$G(s)$	Conditions
Definition	$f(t)\Theta(t)$	$g(s)$	$\mathrm{Re}\,s > \sigma_0$, if $\lvert e^{-\sigma_0 t}f(t)\rvert < M$
Translation	$f(t-a)\Theta(t-a)$	$e^{-as}g(s)$	$a>0$, $\mathrm{Re}\,s > \sigma_o$
Attenuation	$e^{-at}f(t)$	$g(s+a)$	$\mathrm{Re}\,s > \sigma_0 - a$
Derivatives	$\frac{d}{dt}f(t)$	$sg(s) - f(0)$	
	$\frac{d^2}{dt^2}f(t)$	$s\mathscr{L}\left\{\frac{d}{dt}f(t)\right\} - f'(0)$ $= s^2g(s) - sf(0) - f'(0)$	
	$\frac{d^n}{dt^n}f(t)$	$s^ng(s) - \sum_{k=1}^{n} s^{k-1}\left(\frac{d^{n-k}}{dt^{n-k}}f(t)\right)_{t=0}$	

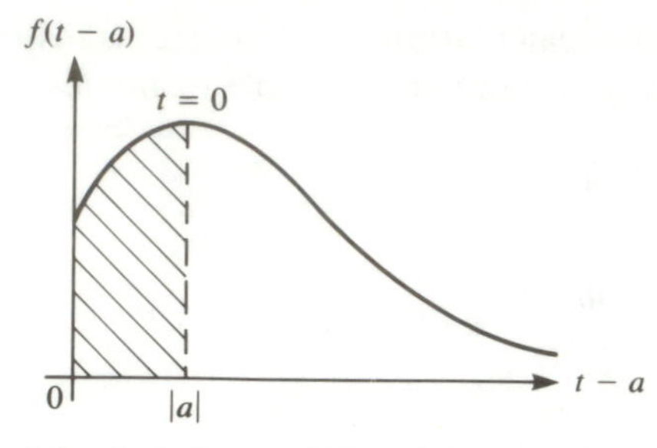

Fig. 6.33. Exclusion of the shaded part of $f(t-a)$ from the Laplace transform if $a<0$.

A number of related Laplace transforms can be deduced from Eq. (6.99):

$$\begin{aligned}
\mathscr{L}\{\cos \lambda t\} &= \mathscr{L}\{\tfrac{1}{2}(e^{i\lambda t}+e^{-i\lambda t})\} = \frac{1}{2}\left(\frac{1}{s-i\lambda}+\frac{1}{s+i\lambda}\right) = \frac{s}{s^2+\lambda^2},\\
\mathscr{L}\{\sin \lambda t\} &= \mathscr{L}\left\{\frac{1}{2i}(e^{i\lambda t}-e^{-i\lambda t})\right\} = \frac{1}{2i}\left(\frac{1}{s-i\lambda}-\frac{1}{s+i\lambda}\right) = \frac{\lambda}{s^2+\lambda^2},\\
\mathscr{L}\{\cosh \kappa t\} &= \mathscr{L}\{\cos i\kappa t\} = \frac{s}{s^2-\kappa^2},\\
\mathscr{L}\{\sinh \kappa t\} &= \mathscr{L}\left\{\frac{1}{i}\sin i\kappa t\right\} = \frac{\kappa}{s^2-\kappa^2},\\
\mathscr{L}\{t\cos \lambda t\} &= \mathscr{L}\left\{\frac{d}{d\lambda}\sin \lambda t\right\} = \frac{d}{d\lambda}\left(\frac{\lambda}{s^2+\lambda^2}\right) = \frac{s^2-\lambda^2}{(s^2+\lambda^2)^2}.
\end{aligned} \tag{6.101}$$

6.12.2. Solution of differential equations

An important application of LTs is in the solution of DEs. We illustrate this by working out a simple example.

Example 6.12.2. Solve the differential equation

$$\left(\frac{d^2}{dt^2}-3\frac{d}{dt}+2\right)y(t) = at^2 \tag{6.102}$$

using an LT.

Let us denote

$$\mathscr{L}\{y(t)\} = \bar{y}(s).$$

The LT changes the differential equation into the algebraic equation

$$[s^2\bar{y}(s)-sy(0)-y'(0)]-3[s\bar{y}(0)-y(0)]+2\bar{y}(s) = \frac{2}{s^3}a,$$

or

$$(s^2-3s+2)\bar{y}(s)+(3-s)y(0)-y'(0) = \frac{2a}{s^3}.$$

This can be solved to get

$$\tilde{y}(s) = \frac{2a/s^3 + y'(0) - (3-s)y(0)}{s^2 - 3s + 2}. \tag{6.103}$$

This result contains a number of interesting features:

1. The first term on the right is a particular solution of the original inhomogeneous DE, Eq. (6.102). In particular, it satisfies the boundary conditions $y'(0) = 0$ and $y(0) = 0$.
2. The second term gives the homogeneous solution for the boundary conditions $y'(0) \neq 0$ and $y(0) = 0$.
3. The third term gives the homogeneous solution for the boundary conditions $y'(0) = 0$ and $y(0) \neq 0$.
4. $\tilde{y}(s)$ is the LT of the general solution satisfying the boundary conditions $y'(0) \neq 0$ and $y(0) \neq 0$.

This example shows that the Laplace transform has an advantage over the Fourier transform in treating differential equations by making explicit the dependence on the boundary conditions.

The solution $\tilde{y}(s)$ must be inverted to obtain the actual solution

$$y(t) = \mathscr{L}^{-1}\{\tilde{y}(s)\}.$$

This will be done in the next section.

Problems

6.12.1. Verify the following Laplace transforms:

(a) $\mathscr{L}\left\{\frac{1}{a^3}(at - \sin at)\right\} = \frac{1}{s^2(s^2+a^2)}$;

(b) $\mathscr{L}\left\{\frac{1}{2a^3}(\sin at - at\cos at)\right\} = \frac{1}{(s^2+a^2)^2}$;

(c) $\mathscr{L}\left\{\frac{e^{bt} - e^{at}}{t}\right\} = \ln\left(\frac{s-a}{s-b}\right)$;

(d) $\mathscr{L}\{t^{-1/2}\} = \left(\frac{\pi}{s}\right)^{1/2}$;

(e) $\mathscr{L}\left\{\frac{e^{bt} - e^{at}}{t^{3/2}}\right\} = \sqrt{4\pi}\,(\sqrt{s-a} - \sqrt{s-b})$;

(f) $\mathscr{L}\{(1-\cos at)/t\} = \frac{1}{2}\ln\left(\frac{s^2+a^2}{s^2}\right)$;

(g) $\mathscr{L}\{\cos 2\sqrt{kt}/\sqrt{t}\} = \sqrt{\pi/s}\exp\left(-\frac{k}{s}\right)$;

(h) $\mathscr{L}\left\{\frac{1}{\sqrt{t}}\exp\left(-\frac{k^2}{4t}\right)\right\} = \sqrt{\pi/s}\exp(-k\sqrt{s}), \qquad k \geqslant 0.$

6.12.2. If $f(x)$ is periodic with a period of T, $f(x+T)=f(x)$, show that

$$\mathscr{L}\{f(x)\} = \left(\int_0^T e^{-sx} f(x)\, dx\right) \Big/ (1-e^{-sT}).$$

6.12.3. With the help of the integral

$$\int_0^\infty e^{-x^2}\, dx = \frac{\sqrt{\pi}}{2},$$

show that

(a) $\mathscr{L}\{t^{-1/2}\} = \sqrt{\pi/s}$,

(b) $\mathscr{L}\{t^{n-1/2}\} = \dfrac{(2n-1)!!}{2^n} \dfrac{1}{s^n} \sqrt{\pi/s}$.

6.13. Inverse Laplace transform

The inverse Laplace transform

$$f(t) = \mathscr{L}^{-1}\{g(s)\} = \frac{1}{2\pi i} \int_{\sigma - i\infty}^{\sigma + i\infty} g(s) e^{st}\, ds \tag{6.104}$$

can be obtained by consulting a table of LTs or by evaluating the complex integral (called a Bromwich integral) with the help of the calculus of residues.

6.13.1. Using a table of Laplace transforms

This is not as trivial as it appears at first sight. The reason is that a table such as Table 6.2 gives the transforms for only a limited number of basic

Table 6.2. A table of basic Laplace transforms

$f(t)$	$g(s) = \mathscr{L}\{f(t)\Theta(t)\}$	Restrictions
t^n	$\dfrac{n!}{s^{n+1}}$	$s>0,\ n>-1$
e^{at}	$\dfrac{1}{s-a}$	$s>a$
$t^{n-1/2}$	$\dfrac{(2n-1)!!\sqrt{\pi}}{2^n} \dfrac{1}{s^{n+1/2}}$	$n=1, 2, 3, \ldots$
$\dfrac{1}{t}(e^{bt}-e^{at})$	$\ln\left(\dfrac{s-a}{s-b}\right)$	$s>b$
$\dfrac{1}{\sqrt{t}} \cos 2\sqrt{kt}$	$\sqrt{\pi/s}\, \exp\left(-\dfrac{k}{s}\right)$	
$\dfrac{1}{\sqrt{t}} \exp\left(-\dfrac{k^2}{4t}\right)$	$\sqrt{\pi/s}\, \exp(-k\sqrt{s})$	$k \geqslant 0$

functions. A certain amount of manipulation is needed before the table entries can be used.

Example 6.13.1.

(a)
$$\mathscr{L}^{-1}\left\{\frac{1}{s-a}\right\} = e^{at}\mathscr{L}^{-1}\left\{\frac{1}{s}\right\}$$
$$= e^{at}\Theta(t) = \begin{cases} e^{at}, & t>0 \\ 0, & t<0, \end{cases}$$

with the help of the attenuation property shown in Table 6.1.

(b)
$$\mathscr{L}^{-1}\left\{\frac{1}{(s-a)^n}\right\} = e^{at}\mathscr{L}^{-1}\left\{\frac{1}{s^n}\right\}$$
$$= e^{at}\frac{t^{n-1}}{(n-1)!}\Theta(t),$$

where use has been made of the first entry of Table 6.2.

(c) $\mathscr{L}^{-1}\{\arctan(k/s)\}$:
We first note that

$$\arctan(k/s) = \frac{1}{2i}\ln e^{2i\arctan(k/s)}$$
$$= \frac{1}{2i}\ln\left(\frac{s+ik}{s-ik}\right).$$

The fourth entry of Table 6.2 can now be used with $a=-ik$, $b=ik$, to give

$$\mathscr{L}^{-1}\{\arctan(k/s)\} = \frac{1}{2it}(e^{ikt}-e^{-ikt})\Theta(t)$$
$$= \frac{1}{t}\sin kt\,\Theta(t).$$

Many LT $g(s)$ involve the inverse of a polynomial in s. These can readily be handled by decomposition into partial fractions:

$$g(s) = \frac{1}{p_n(s)} = \frac{1}{\prod_{i=1}^{m}(s-a_i)} = \sum_{i=0}^{n}\frac{c_i}{s-a_i}, \qquad (6.105)$$

if $p_n(s)$ has no multiple root. [Note that the coefficients c_i are just the residues of $g(s)$ at $s=a_i$.] If one of the roots, say a_1, occurs m times, its contribution in the final partial fraction decomposition should be replaced by the expression

$$\sum_{k=1}^{m}\frac{c_{1,k}}{(s-a_1)^k},$$

which is just the principal part of the Laurent expansion of $g(s)$ near a_1. Similar changes should be made for other multiple roots.

We now have to find the inverse LT of a linear combination of basic functions. This can be done easily because both the LT and its inverse are linear operators satisfying the linearity property

$$\mathscr{L}^{-1}\left\{\sum_i c_i g_i(s)\right\} = \sum_i c_i \mathscr{L}^{-1}\{g_i(s)\}$$
$$= \sum_i c_i f_i(t). \qquad (6.106)$$

This procedure is illustrated in the following example.

Example 6.13.2. Find the inverse Laplace transforms of the following functions of s:

(a) $$g_1(s) = \frac{1}{s^2 - 3s + 2} = \frac{1}{(s-2)(s-1)}$$
$$= \frac{1}{s-2} - \frac{1}{s-1},$$

where the partial-fraction decomposition has been made by comparing the coefficients appearing in the numerator functions on both sides of the equation. Therefore,

$$f_1(t) = \mathscr{L}^{-1}\{g_1(s)\} = \mathscr{L}^{-1}\left(\frac{1}{s-2}\right) - \mathscr{L}^{-1}\left\{\frac{1}{s-1}\right\}$$
$$= (e^{2t} - e^t)\Theta(t).$$

(b) $$g_2(s) = \frac{s-3}{s^2 - 3s + 2} = -\frac{1}{s-2} + \frac{2}{s-1}$$
$$f_2(t) = \mathscr{L}^{-1}\{g_2(s)\} = (-e^{2t} + 2e^t)\Theta(t).$$

(c) $$g_p(s) = \frac{2a}{s^3(s^2 - 3s + 2)} = 2a\left(\frac{1}{2}\frac{1}{s^3} + \frac{3}{4}\frac{1}{s^2} + \frac{7}{8}\frac{1}{s} + \frac{1}{8}\frac{1}{s-2} - \frac{1}{s-1}\right)$$
$$f_p(t) = \mathscr{L}^{-1}\{g_p(s)\} = 2a\left[\frac{1}{2}\left(\frac{t^2}{2}\right) + \tfrac{3}{4}t + \tfrac{7}{8} + \tfrac{1}{8}e^{2t} - e^t\right]\Theta(t).$$

The three functions considered above are just those appearing in the three terms of Eq. (6.103). Hence the general solution to the DE (6.102) is just

$$y(t) = \mathscr{L}^{-1}\{\bar{y}(s)\} = \mathscr{L}^{-1}\{g_p(s) + y'(0)g_1(s) + y(0)g_2(s)\}$$
$$= f_p(t) + y'(0)f_1(t) + y(0)f_2(t).$$

This completes the solution of Eq. (6.102) by the LT. It can further be verified that the solution obtained is also valid for negative t. Hence the step function can be dropped.

6.13.2. Bromwich integrals

When an available table of LTs turns out to be inadequate, the complex integral

$$f(t) = \mathscr{L}^{-1}\{g(a)\} = \frac{1}{2\pi i}\int_{\sigma-i\infty}^{\sigma+i\infty} g(s)e^{st}\,ds$$

might have to be evaluated exactly or approximately.

The path of integration is a line in the complex s plane parallel to the imaginary axis. It should be chosen to the right of the rightmost singularity s_0 of $g(s)$; that is,

$$\sigma > \operatorname{Re} s_0 = \sigma_0, \tag{6.107}$$

as shown in Fig. 6.34. This will ensure that $f(t)$ vanishes for $t<0$. To see this result, let us close the contour of integration and evaluate the integral by residual calculus. If $|g(s)| \to 0$ as $|s| \to \infty$, Jordan's lemma can be used for this purpose. Writing

$$e^{st} = e^{i(-is)t},$$

we see that if $t<0$, we should close in the lower HP of $(-is)$. This is the half-plane to the right of the vertical line of integration. Since there is no singularity in this right HP, the integral vanishes for all $t<0$.

For $t>0$, the contour may be closed in a large semicircle to the left, thus enclosing all the singularities responsible for the functional behavior of $f(t)$. This feature is illustrated in the examples below.

Example 6.13.3.

(a) $\mathscr{L}^{-1}\left\{\dfrac{1}{s}\right\} = \dfrac{1}{2\pi i}\displaystyle\int_{\sigma-i\infty}^{\sigma+i\infty} e^{st}\frac{1}{s}\,ds.$

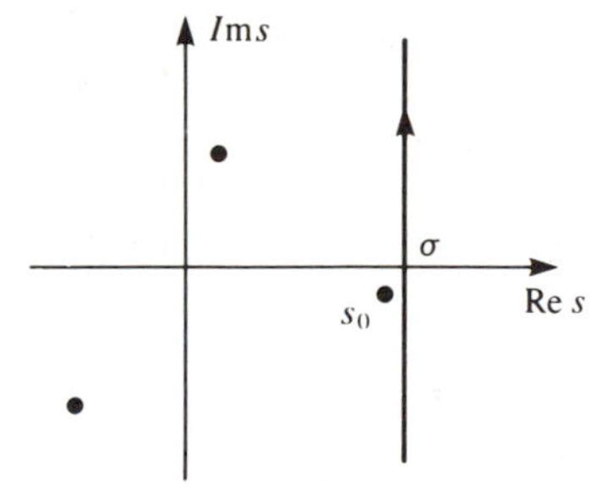

Fig. 6.34. The integration path used in the inverse Laplace transform should be to the right of the rightmost singularity at $s = s_0$ of $g(s)$ in the complex s plane.

Since s^{-1} has a simple pole at $s=0$ with residue 1, we choose $\sigma>0$ to get

$$\mathscr{L}^{-1}\left\{\frac{1}{s}\right\}=\begin{cases}1, & t>0\\ 0, & t<0.\end{cases}$$

(b) $$f(t)=\mathscr{L}^{-1}\left\{\frac{k^2}{s(s^2+k^2)}\right\}=\Theta(t)\left\{\sum \operatorname{Res}\left(\frac{k^2e^{st}}{s(s^2+k^2)}\right)\quad \text{at}\quad s=0,\pm ik\right\},$$

where $\sigma>0$ has been used. Hence

$$\begin{aligned}f(t)&=\Theta(t)\left(1+\frac{e^{ikt}k^2}{ik(2ik)}+\frac{e^{-ikt}k^2}{-ik(-2ik)}\right)\\&=\Theta(t)(1-\cos kt).\end{aligned}$$

Example 6.13.4.

$$f(t)=\mathscr{L}^{-1}\left\{\frac{1}{\sqrt{s+1}}\right\}=\frac{1}{2\pi i}\int_{\sigma-i\infty}^{\sigma+i\infty}\frac{e^{st}}{\sqrt{s+1}}\,dx. \tag{6.108}$$

We shall first look up the result for the third entry of Table 6.2:

$$\mathscr{L}^{-1}\{s^{-1/2}\}=(\pi t)^{-1/2}\Theta(t).$$

With the help of the "attenuation" property shown in Table 6.1, this yields (with $a=1$)

$$\mathscr{L}^{-1}\{(s+1)^{-1/2}\}=(\pi t)^{-1/2}e^{-t}\Theta(t).$$

Let us now evaluate the Bromwich integral directly. The branch point $s=-1$ is placed to the left of the integration path by choosing $\sigma>-1$. If the contour is closed in the way shown in Fig. 6.35, there is no contribution from the large semicircle or the small circle around $s=-1$. The closed contour also does not enclose any singularity. As a result,

$$\begin{aligned}f(t)&=-(I_++I_-)\\&=-\frac{1}{2\pi i}\left(\int_{-\infty}^{-1}\frac{e^{st}}{\sqrt{s+1}}\,ds+\int_{-1}^{-\infty}\frac{e^{st}}{\sqrt{s+1}}\,ds\right).\end{aligned}$$

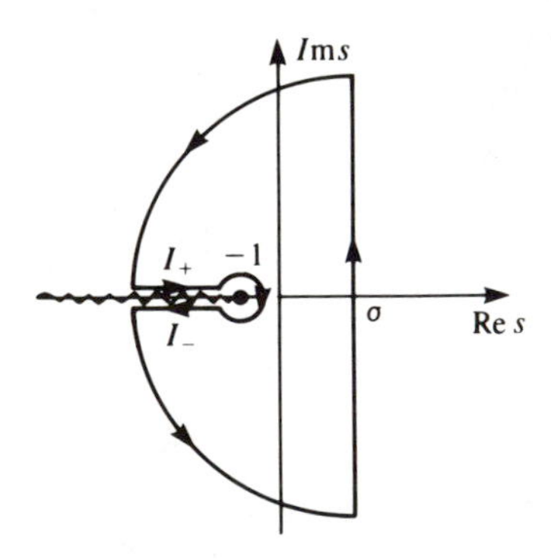

Fig. 6.35. Closing the contour for the Bromwich integral of Eq. (6.108).

These integrals can be evaluated by the following changes of variables: For I_+ above the branch cut on the negative real axis:

$$s+1=re^{i\pi}, \qquad ds=-dr,$$

$$I_+ = \frac{1}{2\pi i}\int_\infty^0 \frac{e^{-(r+1)t}}{r^{1/2}e^{i\pi/2}}(-dr),$$

$$= -\frac{1}{2\pi}\int_0^\infty \frac{e^{-(r+1)t}}{r^{1/2}}\,dr.$$

For I_- below the branch cut:

$$s+1=re^{-i\pi}, \qquad ds=-dr,$$

$$I_{-1} = \frac{1}{2\pi i}\int_0^\infty \frac{e^{-(r+1)t}}{r^{1/2}e^{-i\pi/2}}(-dr),$$

$$= -\frac{1}{2\pi}\int_0^\infty \frac{e^{-(r+1)t}}{r^{1/2}}\,dr = I_+.$$

Hence

$$f(t) = -2I_+$$

$$= \frac{e^{-t}}{\pi}\int_0^\infty \frac{e^{-rt}}{r^{1/2}}\,dr$$

$$= \frac{e^{-t}}{\pi}2\int_0^\infty e^{-u^2 t}\,du, \tag{6.109}$$

where $u=r^{1/2}$ is a real variable. We now need to evaluate a real integral of the type

$$\int_0^\infty e^{-x^2}\,dx = \frac{1}{2}\left[\left(\int_{-\infty}^\infty e^{-x^2}\,dx\right)\left(\int_{-\infty}^\infty e^{-y^2}\,dy\right)\right]^{1/2}$$

$$= \frac{1}{2}\left(\pi\int_0^\infty e^{-r^2}\,dr^2\right)^{1/2}$$

$$= \frac{\sqrt{\pi}}{2}. \tag{6.110}$$

Hence

$$f(t) = \frac{e^{-t}}{\pi}\sqrt{\pi/t} = \frac{e^{-t}}{\sqrt{\pi t}} \qquad \text{for} \qquad t>0.$$

This result agrees with that extracted from the third entry of Table 6.2.

Problems

6.13.1. Verify the second, fifth, and sixth entries of Table 6.2 by evaluating Bromwich integrals.

6.13.2. Show that $\mathscr{L}^{-1}\{(s^2+1)^{-1/2}\} = J_0(t)$ with the help of the integral representation of the Bessel functions:

$$J_0(t) = \frac{1}{2\pi i}\oint \exp\left[\frac{t}{2}\left(u - \frac{1}{u}\right)\right]u^{-1}\,du.$$

6.13.3. Evaluate the Bromwich integrals for

(a) $(s+a)^{-1}(s+b)^{-1}$, $\quad b \neq a$;

(b) $\ln s/s$.

6.13.4. The one-dimensional diffusion equation for thermal conduction is

$$\left(\frac{\partial^2}{\partial x^2} - \frac{1}{\kappa}\frac{\partial}{\partial t}\right)T(x,t) = 0,$$

where κ is the thermal conductivity.

(a) Show that a general solution of this equation can be written in the form

$$T(x,t) = \mathscr{L}^{-1}\{g_1(s)e^{-\sqrt{s/\kappa}\,x} + g_2(s)e^{\sqrt{s/\kappa}\,x}\}.$$

(b) Show by evaluating a Bromwich integral that

$$\mathscr{L}^{-1}\left\{\frac{1}{s}e^{-\sqrt{s/\kappa}\,x}\right\} = \operatorname{erfc}\left(\frac{x}{2\sqrt{\kappa t}}\right),$$

where

$$\operatorname{erfc}(y) = \frac{2}{\sqrt{\pi}}\int_y^\infty e^{-u^2}\,du$$

is the complementary error function. (Hint: There is a branch point and a pole, both at $s=0$. The contour is similar to Fig. 6.35. There are contributions from the small circle around the pole at the origin, and on the negative real axis both above and below the branch cut.)

(c) One end (at $x=0$) of a one-dimensional infinitely long conductor is suddenly brought into contact with a heat reservoir at a constant temperature T_0 at $t=0$. Show that the subsequent ($t>0$) temperature distribution of the conductor ($x>0$) is

$$T(x,t) = T_0\mathscr{L}^{-1}\left\{\frac{1}{s}e^{-\sqrt{s/\kappa}\,x}\right\}$$

$$= T_0 \operatorname{erfc}\left(\frac{x}{2\sqrt{\kappa t}}\right).$$

(d) Use a table of error functions such as that found in Abramowitz and Stegun, *Handbook of Mathematical Functions*, to sketch $\operatorname{erfc}(y)$ for $y = 0$–10. Describe the temperature distribution obtained in (c) when $x^2 \gg \kappa t$ and when $x^2 \ll \kappa t$.

6.14. Construction of functions and dispersion relations

Research in physics has some similarity to detective work. In both professions, one tries to reconstruct a situation from clues. In physics, the situations of interest are functions describing physical attributes. Since singularities are the sources of functional behavior, we must examine properties of functions near their singularities. In physics, this is done by performing meaningful experiments!

The mathematical construction of functions from their singularities can be achieved with the help of the Cauchy integral formula

$$f(z) = \frac{1}{2\pi i} \oint_\Gamma \frac{f(z')}{z' - z} dz'$$

over a contour surrounding z and enclosing a region in which $f(z')$ is analytic. Although all singularities are thus excluded, they are nevertheless important, as the following example shows.

Example 6.14.1. Construct a function $f(z)$ satisfying the following properties:

(a) $f(z)$ is analytic except for a simple pole of residue R at $z = a$ and a branch cut $(0,\infty)$ at which the function has a discontinuity

$$f(x + i\varepsilon) - f(x - i\varepsilon) = 2\pi i g(x), \qquad x \geqslant 0.$$

(b) $|f(z)| \to 0$ as $|z| \to \infty$, and $|zf(z)| \to 0$ as $|z| \to 0$.

The closed contour appearing in the Cauchy formula comes in two pieces, Γ_1 and Γ_2, as shown in Fig. 6.36. There is no contribution from the small circle in Γ_2. The contribution from the large circle also vanishes by virtue of property (b). Hence

$$\begin{aligned} f(z) &= \frac{1}{2\pi i} \oint_{\Gamma_1} \frac{f(z')}{z' - z} dz' + \frac{1}{2\pi i} \int_0^\infty \frac{f_+(x') - f_-(x')}{x' - x} \\ &= \frac{R}{z - a} + \int_0^\infty \frac{g(x')}{x' - z} dx'. \end{aligned} \qquad (6.111)$$

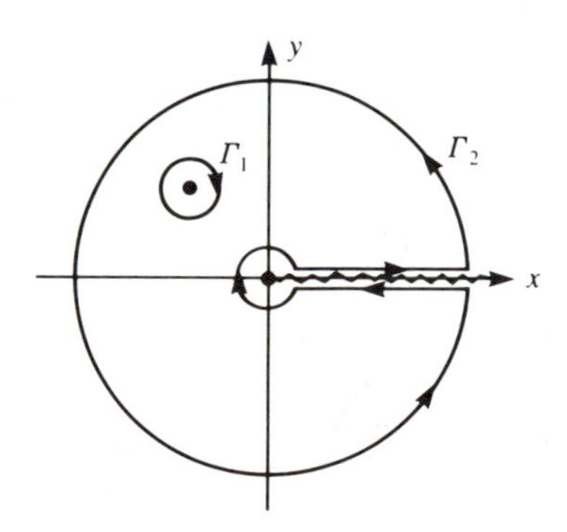

Fig. 6.36. A closed contour around a pole and a branch cut.

Thus the sources of functional behavior for $f(z)$ are the residue at the pole and the discontinuity across the branch cut. The corresponding terms in Eq. (6.111) are sometimes called the *pole contribution* and the *continuum contribution* to $f(z)$. If R and $g(x')$ are known, we can perform the integration and complete the task of finding $f(z)$.

Example 6.14.2. Construct a function $f(z)$ satisfying the following properties:

(a) $f(z)$ is analytic except for a simple pole at $z=-1$ with residue $\frac{1}{2}$, and a branch cut $(0,\infty)$ where the function has a discontinuity:

$$f(x+i\varepsilon)-f(x-i\varepsilon)=\frac{2\pi i}{1+x^2}.$$

(b) $|f(z)|\to 0$ as $|z|\to\infty$, and $|zf(z)|\to 0$ as $|z|\to 0$.

According to Eq. (6.111), the answer is

$$f(z)=\frac{1}{2(z+1)}+\int_0^\infty \frac{1}{x'-z}\frac{1}{1+x'^2}\,dx'.$$

The integral can be evaluated with the help of Eq. (6.66):

$$\begin{aligned}\int_0^\infty \frac{1}{x'-z}\frac{1}{1+x'^2}\,dx' &= -\sum_{\text{enclosed}} \operatorname{Res}\left(\frac{\ln z'}{z'-z}\frac{1}{1+z'^2}\right)\\ &= -\frac{1}{1+z^2}\left(\ln_0 z+\frac{\pi}{2}(z-2i)\right),\end{aligned}$$

where $\ln_0 z$ is the principal branch of the ln function. The result turns out to be the same on every sheet of the Riemann surface of $\ln z'$ or $\ln z$ (Problem 6.14.1). The reader should also verify explicitly that

$$f(z)=\frac{1}{2(z+1)}-\frac{1}{1+z^2}\left(\ln z+\frac{\pi}{2}(z-2i)\right) \tag{6.112}$$

is analytic at $z=\pm i$.

6.14.1. Dispersion relations

In physics, information on the R and $g(x')$ of a function $f(z)$ describing a physical attribute of interest is to be deduced from meaningful measurements. This is not easy to do, since our experimental knowledge of a physical situation is usually quite incomplete. In many cases where detailed physical theories have not yet been formulated, we may not even know the possible singularity structures of $f(z)$. Under the circumstances, a much more indirect procedure might have to be used.

On the other hand, the functions of physical interest are usually relatively simple. In many cases they are complex functions of a real variable $f(x)$, $-\infty\leqslant x\leqslant\infty$. Suppose its analytic continuation $f(z)$ vanishes at infinity, say in the UHP. The Cauchy integral formula can then be applied over the contour shown in Fig. 6.37 to give

$$f(x)=\frac{1}{2\pi i}\int_{-\infty}^{\infty}\frac{f(x')}{x'-x}\,dx'+\tfrac{1}{2}f(x), \tag{6.113}$$

where $\frac{1}{2}f(x)$ on the right is the contribution of the semicircular contour below the point x on the real axis.

Equation (6.113) gives two relations between two real functions $\operatorname{Re} f(x)$ and $\operatorname{Im} f(x)$ of the real variable x:

$$\begin{aligned}\operatorname{Re} f(x)&=\frac{1}{\pi}\int_{-\infty}^{\infty}\frac{\operatorname{Im} f(x')}{x'-x}\,dx',\\ \operatorname{Im} f(x)&=-\frac{1}{\pi}\int_{-\infty}^{\infty}\frac{\operatorname{Re} f(x')}{x'-x}\,dx'.\end{aligned} \tag{6.114}$$

Thus, knowledge of $\operatorname{Im} f(x)$ for all real x will give use $\operatorname{Re} f(x)$, or vice versa. Two functions $\operatorname{Re} f(x)$ and $\operatorname{Im} f(x)$ that are related to each other in this way are said to be *Hilbert transforms* of each other. In a sense, they are two versions of the same mathematical structure.

Although $\operatorname{Re} f(x)$ and $\operatorname{Im} f(x)$ involve the same mathematical structure, they may have distinct physical manifestations, each of which can be measured independently of the other. In this way, knowledge of one property permits information to be deduced on the second. This possibility was first recognized by Kronig and by Kramers in connection with the study of the dispersion of light by a medium. For this reason, they are known in physics as *dispersion relations*.

It is useful to see how $\operatorname{Re} f(x)$ and $\operatorname{Im} f(x)$ could affect two separate physical properties. In the case of optical dispersion, we are interested in the refractive index $n(\omega)$ for a plane wave of frequency ω. This function appears in the optical wave function in the form

$$\exp\left[i\omega\left(n(\omega)\frac{z}{c}-t\right)\right]=\exp\left[i\omega\left(\operatorname{Re} n(\omega)\frac{z}{c}-t\right)-\frac{\omega}{c}\operatorname{Im} n(\omega)z\right], \tag{6.115}$$

if the wave propagates in the $+z$ direction. Here c is the speed of light in vacuum. The actual speed in the medium is not c, but the phase velocity,

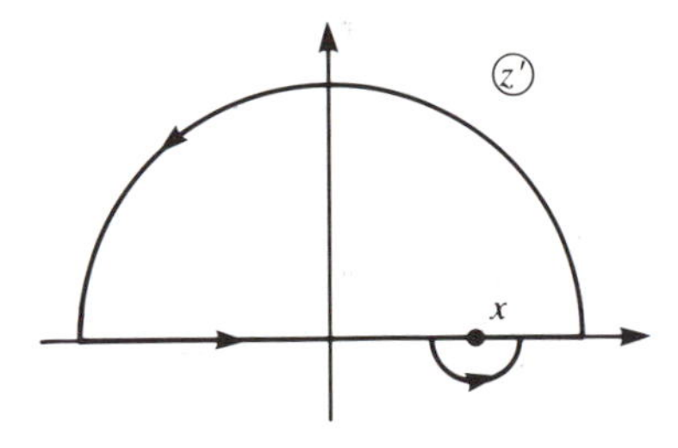

Fig. 6.37. The closed contour for Eq. (6.113) in the complex z' plane.

that is, the velocity of the wave front of constant phase

$$\frac{\operatorname{Re} n(\omega)}{c} z - t = \text{const.} \tag{6.116}$$

Differentiation of this expression with respect to t gives

$$v(\omega) = \frac{dz}{dt} = \frac{c}{\operatorname{Re} n(\omega)}. \tag{6.117}$$

It is a physical fact that $v(\omega)$ decreases below c in a medium. This decrease causes a light ray to bend as it enters the medium from the vacuum at an angle. The phase velocity differs for different frequencies or colors. This causes colors to disperse when lights of different frequencies are refracted to different angles by a prism.

Equation (6.115) also shows that $\operatorname{Im} n(\omega)$ is concerned not with wave propagation or refraction, but with its attenuation in the medium. This attenuation of the wave *intensity* (i.e., the square of the wave function) is described by an absorption coefficient

$$\alpha(\omega) = \frac{2\omega}{c} \operatorname{Im} n(\omega). \tag{6.118}$$

Refraction and absorption can be measured independently. Thus we have independent probes into the same mathematical structure.

To construct dispersion relations for $n(\omega)$, we also need to known how $n(\omega)$ behaves as $\omega \to \infty$. Now light is known to be just an electromagnetic (EM) radiation in a certain range of frequencies. In the limit $\omega \to \infty$, the EM radiation becomes penetrating, and its speed then approaches the vacuum value. That, $n(\omega) \to 1$ as $\omega \to \infty$. This means that dispersion relations of the form of Eq. (6.114) cannot be constructed for $\operatorname{Re} n(\omega)$, but only for $\operatorname{Re} n(\omega) - 1$, which does vanish as $\omega \to \infty$. [The atomic theory of the index of refraction shows that $\operatorname{Re} n(\omega) - 1$ is proportional to ω^{-2} for large ω.] Hence the dispersion relations should be written as

$$\begin{aligned} \operatorname{Re} n(\omega) &= 1 + \frac{1}{\pi} \int_{-\infty}^{\infty} \frac{\operatorname{Im} n(\omega')}{\omega' - \omega} d\omega', \\ \operatorname{Im} n(\omega) &= -\frac{1}{\pi} \int_{-\infty}^{\infty} \frac{\operatorname{Re} n(\omega') - 1}{\omega' - \omega} d\omega'. \end{aligned} \tag{6.119}$$

One final complication is that Eqs. (6.119) involve negative frequencies. All physically measurable quantities such as $\operatorname{Re} n(\omega)$ and $\alpha(\omega)$ are independent of the sign of ω. This means that $\operatorname{Re} n(\omega)$ must be even in ω, while $\operatorname{Im} n(\omega)$ is odd in ω. As a result

$$\int_{-\infty}^{0} \frac{\operatorname{Im} n(\omega')}{\omega' - \omega} d\omega' = \int_{0}^{\infty} \frac{\operatorname{Im} n(|\omega'|)}{|\omega'| + \omega} d\,|\omega'|,$$

and

$$\operatorname{Re} n(\omega) = 1 + \frac{1}{\pi} \int_0^\infty \operatorname{Im} n(\omega') \left(\frac{1}{\omega' - \omega} + \frac{1}{\omega' + \omega} \right) d\omega'$$

$$= 1 + \frac{2}{\pi} \int_0^\infty \frac{\omega' \operatorname{Im} n(\omega')}{\omega'^2 - \omega^2} d\omega'$$

$$= 1 + \frac{c}{\pi} \int_0^\infty \frac{\alpha(\omega')\, d\omega'}{\omega'^2 - \omega^2}. \tag{6.120}$$

In a similar way, we obtain

$$\alpha(\omega) = -\frac{(2\omega)^2}{\pi c} \int_0^\infty \frac{\operatorname{Re} n(\omega') - 1}{\omega'^2 - \omega^2} d\omega'. \tag{6.121}$$

The first of these relations is the original Kramers–Kronig dispersion relation connecting refraction to absorption.

Dispersion relations are not substitutes for physical theories. They only make manifest the connection between related observables. Of course, a great deal of physical insight is needed just to recognize the relation (if any) between two distinct observables. Once recognized, these relations can be checked experimentally by measuring both related observables. Alternatively, if a theoretical relation is assumed, experimental measurements of theoretically related quantities can be checked against each other for consistency in the theoretical description. Finally, quantities that have not yet been measured can sometimes be constructed from known quantities with the help of dispersion relations.

The important observation made in this section is that these dispersion relations between distinct physical properties arise solely from the analyticity of $f(z)$ on and inside the chosen contour of integration. The function $f(z)$ describing these physical properties must be nontrivial, and therefore has singularities. But these singularities must lie entirely outside the chosen integration contour.

Problems

6.14.1. Show that Eq. (6.112) is the function specified in Example 6.14.2 on every sheet of the Riemann surface of $\ln z'$ or $\ln z$.

6.14.2. Construct a function $f(z)$ having the following properties:

(a) $f(z)$ is analytic except for

(i) a simple pole of residue 5 at $z = -3$,

(ii) a branch cut from 0 to ∞ along the real axis where

$$f(x + i\varepsilon) - f(x - i\varepsilon) = 2\pi i[(x+1)(x+2)]^{-1}.$$

(b) $|f(z)| \to 0$ as $|z| \to \infty$, and $|zf(z)| \to 0$ as $|z| \to 0$.

Answer: $f(z) = \dfrac{5}{z+3} - \left(\dfrac{\ln_0 z}{(z+1)(z+2)} - \dfrac{i\pi}{1+z} - \dfrac{i\pi + \ln 2}{2+z} \right).$

6.14.3. From an appropriate dispersion relation, obtain the sum rule

$$\operatorname{Re} f(0) = \frac{1}{\pi} \int_{-\infty}^{\infty} \frac{\operatorname{Im} f(x')\, dx'}{x'}$$

if $|f(x)| \to 0$ as $|x| \to \infty$. Evaluate the sum rules

$$\frac{1}{\pi} \int_{-\infty}^{\infty} \frac{\operatorname{Im} f(x')}{x'^n}\, dx', \qquad n = 2, 3, \ldots .$$

Answer: $\dfrac{1}{(n-1)!} \dfrac{d^{n-1}}{dx^{n-1}} \operatorname{Re} f(x)\big|_{x=0}$.

6.14.4. Construct dispersion relations for $(1/x)f(x)$. What are the least restrictions that can be placed on $f(z)$ to ensure that there is no contribution on the large and small circles in the complex z plane? Use your results to derive the "subtracted" dispersion relations

$$\frac{\operatorname{Re}[f(x) - f(x_0)]}{x - x_0} = \frac{1}{\pi} \int_{-\infty}^{\infty} \frac{\operatorname{Im} f(x')}{(x' - x_0)(x' - x)}\, dx'$$

$$\frac{\operatorname{Im}[f(x) - f(x_0)]}{x - x_0} = -\frac{1}{\pi} \int_{-\infty}^{\infty} \frac{\operatorname{Re} f(x')}{(x' - x_0)(x' - x)}\, dx'.$$

Answer: $f(z) \to$ const as $|z| \to \infty$, $f(z) \to 0$ as $z \to 0$.

6.14.5. Construct a function $f(z)$ having the following properties:

(a) $f(z)$ is analytic except for

(i) a single pole of residue R at $z = a$;

(ii) branch cuts from $-\infty$ to -1 and from 1 to ∞ on the real axis.

(b) $f(z) \to 0$ as $|z| \to \infty$ and $(z - b)f(z) \to 0$ as $|z| \to b = \pm 1$.

(c) $f(x + i\varepsilon) - f(x - i\varepsilon) = 2\pi i (x^2 + 4)^{-1}$ across both branch cuts.

Answer: $\dfrac{R}{z - a} - \dfrac{1}{z^2 + 4} [\ln(z - 1) - \ln(-z - 1) + z \tan^{-1} 2]$

Appendix 6. Tables of mathematical formulas

1. Function of a complex variable

$$f(z) = f(x + iy) = u(x,y) + iv(x,y).$$

The FCV $f(z) = (z + 1)^{1/2} + (z - 1)^{1/2}$ has four branches:

$$f_1 = r_1^{1/2} e^{i\theta_1/2} + r_2^{1/2} e^{i\theta_2/2}$$
$$f_2 = r_1^{1/2} e^{i\theta_1/2} - r_2^{1/2} e^{i\theta_2/2}$$
$$f_3 = -r_1^{1/2} e^{i\theta_1/2} + r_2^{1/2} e^{i\theta_2/2}$$
$$f_4 = -r_1^{1/2} e^{i\theta_1/2} - r_2^{1/2} e^{i\theta_2/2}$$

residing on a single, continuous Riemann surface z made up of four Riemann sheets. The branch points are at -1, 1, and ∞, where all four branches have the same value. The four Riemann sheets can be cut from or to the branch points at ± 1. They are then cross-joined together at these cuts to form a continuous Riemann surface. The actual value of $f(z)$ in each of its four branches depends on the angular ranges of θ_1 and θ_2, that is, on the way the branch cuts are made.

2. Complex differentiation and analytic functions

Cauchy–Riemann conditions: $\dfrac{\partial u}{\partial x} = \dfrac{\partial v}{\partial y}, \quad \dfrac{\partial v}{\partial x} = -\dfrac{\partial u}{\partial y}.$

Isolated singularities: mth-order pole: $R/(z-a)^m$,
essential singularity: $\sin(1/z)$.
Essential singularity: $[\sin(1/z)]^{-1}$.
Entire function: z^n.

Liouville theorem: a function everywhere finite and analytic is a constant.

Singularities are sources of functional behavior.

3. Complex integration

Cauchy integral theorem: $\oint_C f(z)\, dz = 0.$

Cauchy integral formulas: $f(z) = \dfrac{1}{2\pi i} \oint_C \dfrac{f(z')\, dz'}{z'-z},$

$$\frac{d^n f(z)}{dz^n} = \frac{n!}{2\pi i} \oint_C \frac{f(z')\, dz'}{(z'-z)^{n+1}}.$$

4. Harmonic functions in the plane

$$\left(\frac{\partial^2}{\partial x^2} + \frac{\partial^2}{\partial y^2}\right) f(x+iy) = 0,$$

$$f(z) = u(x,y) + iv(x,y).$$

$u = \text{const(equipotentials)} \perp \nabla u\text{(lines of force)},$
$v = \text{const} \perp \nabla v,$
$\nabla v \perp \nabla u.$

5. Taylor series and analytic continuation

$$S_a(z) = f(a) + (z-a)f'(a) + \cdots + \frac{(x-a)^n}{n!} f^{(n)}(a) + \cdots$$

$$= \sum_{n=0}^{\infty} c_n (z-a)^n$$

is convergent in a circular region from a to the nearest singularity of $f(z)$ if a is not a singularity of $f(z)$. Similarly,

$$S_b(z) = \sum_{n=0}^{\infty} d_n(z-b)^n$$

has a circle of convergence around b. If these circles overlap and $S_b(z) = S_a(z)$ in the region of intersection of the two circles, the two Taylor series are analytic continuations of each other.

6. Laurent series and residues

A Laurent series is a power series

$$f(z) = \sum_{n=-\infty}^{\infty} c_n(z-a)^n$$

with negative powers. It describes an analytic function in an annular region about a (where the function is singular).

The coefficient c_{-1} of a Lautent series is

$$c_{-1} = \frac{1}{2\pi i}\oint f(z)\,dz = \sum_{\text{enclosed}} \text{Res}[f(z_i)].$$

The residue at z_i

$$\text{Res}[f(z_i)] = \frac{1}{m-1}\lim_{z\to z_i}\left(\frac{d^{m-1}}{dz^{m-1}}[(z-z_i)^m f(z)]\right)$$

is the coefficient c_{-1} of a Laurent series in the annular region immediately surrounding the mth-order pole at z_i. For simple poles

$$\begin{aligned}\text{Res}[f(z_i)] &= \lim_{z\to z_i}[(z-z_i)f(z)]\\ &= \lim_{z\to z_i}\left[P(z_i)\Big/\left(\frac{d}{dz}Q(z)\Big|_{z_i}\right)\right].\end{aligned}$$

7. Complex integration: Calculus of residues

Integrals can be evaluated by calculating residues if the integration path can be closed in the complex plane.

Method 1.

$$I = \int_{-\infty}^{\infty} f(x)\,dx = \oint f(z)\,dz$$

in the UHP or LHP if $|zf(z)| \to 0$ as $z \to \infty$.

Method 2 (Jordan's lemma).

$$I = \int_{-\infty}^{\infty} g(x)e^{i\lambda x}\,dx = \oint_C g(z)e^{i\lambda z}\,dx$$

if $|g(z)| \to 0$ as $z \to \infty$. Use C in the UHP (LHP) if $\lambda > 0$ (<0).

Method 3. Clever return paths.

Method 4. Adding a branch cut

$$I=\int_0^\infty f(x)\,dx, \qquad J=\oint \ln z\, f(z)\,dz,$$

$$I=-\sum_{\text{enclosed}} \text{Res}[f(z)\ln z].$$

Method 5. Use a unit circle

$$I=\int_0^{2\pi} G(\sin\theta,\cos\theta)\,d\theta$$

$$=\oint G\left[\frac{1}{2i}\left(z-\frac{1}{z}\right),\frac{1}{2}\left(z+\frac{1}{z}\right)\right]\frac{dz}{iz}.$$

8. Poles on contour and Green functions

$$I_\pm = ⨍ f(z)\,dz + I_\pm(z_0)$$

$$\lim_{\varepsilon\to 0}\frac{1}{x-(x_0\pm i\varepsilon)}=\frac{1}{x-x_0}\pm i\pi\,\delta(x-x_0).$$

If $\mathscr{L}(x)f(x)=\rho(x)$, then

$$f(x)=\int G(x-x')\rho(x')\,dx',$$

where $\mathscr{L}(x)G(x-x')=\delta(x-x')$.

If $\mathscr{L}(x)=d^2/dx^2+k_0^2$,

$$G(x)=-\frac{1}{2\pi}\int_{-\infty}^{\infty}\frac{e^{ikx}}{k^2-k_0^2}\,dk:$$

$$G_\pm(x)=\pm\frac{1}{2ik_0}e^{\pm ik_0|x|}, \qquad \text{ingoing-(outgoing-)wave GF;}$$

$$G_P(x)=\frac{1}{2k_0}\sin k_0|x|, \qquad \text{principal-value (standing-wave) GF;}$$

If $\mathscr{L}(t)=d^2/dt^2+\omega_0^2$,

$$G(t)=-\frac{1}{2\pi}\int_{-\infty}^{\infty}\frac{e^{-i\omega t}}{\omega^2-\omega_0^2}\,d\omega:$$

$$G_{r,a}(t)=\frac{1}{\omega_0}\sin\omega_0 t\times\begin{cases}\Theta(t), & \text{retarded GF}\\ -\Theta(-t), & \text{advanced GF.}\end{cases}$$

9. Laplace transform

$$\mathscr{L}\{f(x)\} = \int_0^\infty f(x)e^{-sx}\,dx = g(s)$$

$$\mathscr{L}^{-1}\{g(s)\} = f(x)\Theta(x)$$
$$= \frac{1}{2\pi i}\int_{\sigma-i\infty}^{\sigma+i\infty} g(s)e^{sx}\,ds \qquad \text{(Bromwich integrals).}$$

The integration path is to the right of all singularities.

10. Construction of functions and dispersion relations

$$f(z) = \frac{1}{2\pi i}\oint \frac{f(z')}{z'-z}\,dz'.$$

Examples:

(a) $f(z) = \dfrac{R}{z-a} + \displaystyle\int_0^\infty \frac{g(x')}{x'-z}\,dx',$

where R is the residue at the pole at a and $2\pi i g(x)$ is the discontinuity across the branch cut on the positive real axis.

(b) $f(x) = (1/\pi i)⨍_{-\infty}^{\infty} [f(x')/(x'-x)]\,dx'$ gives the Hilbert transforms

$$\operatorname{Re} f = \frac{1}{\pi}⨍ \frac{\operatorname{Im} f(x')}{x'-x}\,dx',$$

$$\operatorname{Im} f = -\frac{1}{\pi}⨍ \frac{\operatorname{Re} f(x')}{x'-x}\,dx'.$$

(c) Kramers–Kronig dispersion relations for the complex index of refraction $n(\omega)$ in $\exp[i\omega n(\omega)z/c]$ with $n(\omega)\to 1$ as $\omega\to\infty$:

$$\operatorname{Re} n(\omega) = 1 + \frac{c}{\pi}⨍_0^\infty \frac{\alpha(\omega')\,d\omega'}{\omega'^2-\omega^2},$$

where $\alpha(\omega) \equiv (2\omega/c)\operatorname{Im} n(\omega)$;

$$\alpha(\omega) = -\frac{(2\omega)^2}{\pi c}⨍_0^\infty \frac{\operatorname{Re} n(\omega')-1}{\omega'^2-\omega^2}\,d\omega'.$$

REFERENCES

The following list of books has been selected with the needs of students in mind. Each reference is described in full in the Bibliography.

General
Arfken, Boas, Bradbury, Harper, Jones, Kraut, Margenau/Murphy, Mathews/Walker, Morse/Feshbach, Potter/Goldberg

Chapter 1 Vectors and fields in space
Marion, Spiegel (*Vector Analysis*)

Chapter 2 Transformation, matrices and operators
Ayres, Pettofrezzo

Chapter 3 Fourier series and Fourier transform
Carslaw, Spiegel (*Fourier Analysis*), Tolstov

Chapter 4 Differential equations in physics
Hochstadt (*Differential Equations*)

Chapter 5 Special functions
Hochstadt (*The Functions of Mathematical Physics*), Spiegel (*Fourier Analysis*)

Chapter 6 Functions of a complex variable
Churchill/Brown/Verhey, Dettman, Spiegel (*Complex Variables*)

Tables
Abramowitz/Stegun

Dictionaries and handbooks
Abramowitz/Stegun, Gellert et al., *Handbook of Chemistry and Physics*

Biographies
Ball, Debus, Encyclopaedia Britannica (1985), Gillispie, Kline, *Physics Today*, Struik

BIBLIOGRAPHY

Abramowitz, Milton, and Irene A. Stegun (eds.), *Handbook of Mathematical Functions,* National Bureau of Standards, Applied Mathematics Series 55, U.S. Government Printing Office, Washington, D.C., 1964.

Arfken, George, *Mathematical Methods for Physicists,* Academic Press, New York, 2nd ed., 1970; 3rd ed., 1985.

Ayres, Frank, Jr., *Matrices,* Schaum's Outline Series, McGraw-Hill, New York, 1962.

Ball, W. W. Rouse, *A Short Account of the History of Mathematics,* 6th ed., Dover, New York, 1960.

Boas, Mary, L., *Mathematical Methods in the Physical Sciences,* Wiley, New York, 2nd ed., 1983.

Bradbury, Ted Clay, *Mathematical Methods with Applications to Problems in the Physical Sciences,* Wiley, New York, 1984.

Carslaw, H. S., *An Introduction to the Theory of Fourier Series and Integrals,* 3rd ed., Dover, New York, 1950.

Churchill, Ruel V., James W. Brown, and Roger F. Verhey, *Complex Variables and Applications,* McGraw-Hill, New York, 4th ed., 1984.

Debus, Allen G. (ed.), *World Who's Who in Science,* Marquis Who's Who, Chicago, 1968.

Dettman, John W., *Applied Complex Variables,* Dover, New York, 1984.

Dirac, P. A. M., *The Principles of Quantum Mechanics,* Clarendon Press, Oxford, 4th ed., 1958.

Encyclopaedia Britannica, Encyclopaedia Britannica, Inc., Chicago, 15th ed., 1985.

Erdélyi, A., *et al.* (eds.), *Table of Integral Transforms* (Bateman Manuscript Project), McGraw-Hill, New York, 1954.

Gellert, W., *et al.* (eds.), *The VNR Concise Encyclopedia of Mathematics,* Van Nostrand Reinhold, New York, 1975.

Gillispie, Charles Coulston (ed.), *Dictionary of Scientific Biography,* Charles Schribner's Sons, New York, 1980.

Handbook of Chemistry and Physics, CRC Press, Boca Raton, 70th ed., 1989–90.

Harper, Charlie, *Introduction to Mathematical Physics,* Prentice-Hall, Englewood Cliffs, 1976.

Hochstadt, Harry, *Differential Equations,* Dover, New York, 1975.

Hochstadt, Harry, *The Functions of Mathematical Physics,* Dover, New York, 1986.

Jones, Lorella M., *An Introduction to Mathematical Methods of Physics,* Benjamin/Cummings, Menlo Park, 1979.

Kline, Morris, *Mathematical Thought from Ancient to Modern Times,* Oxford University Press, New York, 1972.

Kraut, Edgar A., *Fundamentals of Mathematical Physics,* McGraw-Hill, New York, 1967.

Margenau, Henry, and George Moseley Murphy, *The Mathematics of Physics and Chemistry,* Van Nostrand, New York, 2nd ed., 1976.

Marion, Jerry B., *Principles of Vector Analysis,* Academic Press, New York, 1965.

Martens, P. C. H., *Physics Reports* **115,** 315 (1985).
Mathews, Jon, and Robert L. Walker, *Mathematical Methods of Physics,* Benjamin, New York, 2nd ed., 1970.
May, Robert M., *Nature* **261,** 459 (1976).
Morse, Philip M., and Herman Feshbach, *Methods of Theoretical Physics,* McGraw-Hill, New York, 1953.
Pettofrezzo, Anthony J., *Matrices and Transformations,* Dover, New York, 1978.
Physics Today, American Institute of Physics, Woodbury, N.Y.
Potter, Merle C., and Jack Goldberg, *Mathematical Methods,* Prentice-Hall, Englewood Cliffs, 2nd ed., 1987.
Spiegel, Murray, R., *Complex Variables,* Schaum's Outline Series, McGraw-Hill, New York, 1964.
Spiegel, Murray R., *Fourier Analysis,* Schaum's Outline Series, McGraw-Hill, New York, 1974.
Spiegel, Murray R., *Vector Analysis,* Schaum's Outline Series, McGraw-Hill, New York, 1959.
Struik, Dirk J., *A Concise History of Mathematics,* Dover, New York, 1967.
Tolstov, Georgi P., *Fourier Series,* Dover, New York, 1976.

NAME INDEX

SUBJECT INDEX